DISCARD
GEORGE WASHINGTON UNIVERSITY
PAUL HIMMELFARB HEALTH SCIENCES LIBRARY
2300 EYE STREET, N. W.
WASHINGTON, D. C.
AF784535

DISCARD
GEORGE WASHINGTON UNIVERSITY
PAUL HIMMELFARB HEALTH SCIENCES LIBRARY
2300 EYE STREET, N. W.
WASHINGTON, D. C. 20037

Drugs of Abuse and Neurobiology

Ronald R. Watson, Ph.D.
Research Professor
Department of Family and Community Medicine
University of Arizona School of Medicine
Tucson, Arizona

CRC Press
Boca Raton Ann Arbor London Tokyo

Library of Congress Cataloging-in-Publication Data

Drugs of abuse and neurobiology / editor, Ronald R. Watson.
p. cm. — (Physiology of drug abuse)
Includes bibliographical references and index.
ISBN 0-8493-7934-2
1. Neurotoxicology. 2. Drug abuse—Physiological aspects.
I. Watson, Ronald R. (Ronald Ross) II. Series.
[DNLM: 1. Nervous System—drug effects. 2. Nicotine-
-pharmacology. 3. Psychotropic Drugs—pharmacology.
QV 77 D7946]
RC347.5.D78 1992
615′.78—dc20
DNLM/DLC
for Library of Congress 91-26904
CIP

This book represents information obtained from authentic and highly regarded sources. Reprinted material is quoted with permission, and sources are indicated. A wide variety of references are listed. Every reasonable effort has been made to give reliable data and information, but the author and the publisher cannot assume responsibility for the validity of all materials or for the consequences of their use.

All rights reserved. This book or any parts thereof, may not be reproduced in any form without written consent from the publisher.

Direct all inquiries to CRC Press, Inc., 2000 Corporate Blvd., N.W., Boca Raton, Florida 33431.

© 1992 by CRC Press, Inc.

International Standard Book Number 0-8493-7934-2

Library of Congress Card Number 91-26904
Printed in the United States of America 1 2 3 4 5 6 7 8 9 10

C
47.5
D78
992

PREFACE

Public attention and research efforts are being driven by an ever-increasing understanding of the problems and magnitude of substance abuse. Drugs of abuse — including alcohol, cocaine, amphetamines, marijuana, tobacco, among others — cause pathologic changes in different organs and systems in the body. Two out of three high school students report that they used alcohol during the previous month. Smaller but significant numbers of youth regularly use a variety of drugs of abuse. This continuing source of potential abusers, plus the 10 million alcohol-abusing adults, with the several million abusers of other drugs, costs the people of the U.S. more than 200 billion dollars in lost wages, health care, theft, and shortened lifespan.

This volume is the first in a series dealing with the effects of drugs of abuse on various systems. As the neurologic systems are the primary sites of action of these psychotropic compounds, the effects of drugs of abuse on specific brain sites and neural pathways will be explored initially. The neurobiology and neurochemistry of these high-profile drugs is examined, with chapters focusing on such aspects as drug-induced euphoria, sensorimotor integration, pharmacodynamics of addiction, and other neurologic sequelae. Collectively, the volume offers a definitive and authoritative review of current research and thought in the field.

Future volumes will explore other aspects of physiology modified by drugs of abuse. Clearly, commonly used drugs of abuse cause a wide variety of biochemical changes including: immunomodulation, altered disease resistance, membrane fluidity and functional modifications, teratogenic effects, modifications of hormonal regulation and production with major functional consequences. The progress of research over the past decade is encouraging as we begin to summarize and evaluate, in detail, advances in understanding changes in the biochemistry and physiology caused by these drugs. This information will assist the researcher, clinician, and student in comprehending the complex changes caused by direct and indirect effect of single drugs at the cellular level.

PHYSIOLOGY OF DRUG ABUSE SERIES

Edited by: Ronald R. Watson

Forthcoming Titles:

Alcohol and Cancer

Alcohol and Neurobiology: Brain Development and Hormone Regulation

Alcohol and Neurobiology: Receptors, Membranes, and Channels

Marijuana/Cannabinoids: Neurobiology and Neurophysiology

THE EDITOR

Ronald R. Watson, Ph.D., initiated and directs the Specialized Alcohol Research Center at the University of Arizona School of Medicine. The main theme of this NIAAA center grant is to understand the role of ethanol-induced immunosuppression on disease and disease resistance in animals. He has edited 14 books on drug and alcohol abuse. He has worked for several years on research for the U.S. Navy Alcohol and Substance Abuse Program.

Dr. Watson attended the University of Idaho, but graduated from Brigham Young University in Provo, Utah with a degree in Chemistry in 1966. He completed his Ph.D. in 1971 in Biochemistry from Michigan State University. His postdoctoral schooling was completed at the Harvard School of Public Health in Nutrition and Microbiology. He was an Assistant Professor of Immunology and did research at the University of Mississippi Medical Center in Jackson from 1973 to 1974. He was an Assistant Professor of Microbiology and Immunology at the Indiana University Medical School from 1974 to 1978. He was an Associate Professor at Purdue University in the Department of Food and Nutrition from 1978 to 1982. In 1982, he joined the faculty at the University of Arizona in the Department of Family and Community Medicine, Nutrition section, and is a Research Professor.

Dr. Watson's major research interest involves effects of dietary materials on disease resistance, biochemical changes, and immune functions. This includes understanding the effects of alcohol and drug abuse on immunosuppression associated with AIDS.

He has published 250 research papers and review chapters, and edited 25 books. He is a member of several national and international nutrition, immunology, and cancer societies.

CONTRIBUTORS

Martin J. Berg, M.S., Research Scientist, Center for Neurochemistry, The Nathan Kline Institute for Psychiatric Research, Orangeburg, New York

Cheryl F. Dreyfus, Ph.D., Associate Professor, Department of Neuroscience and Cell Biology, University of Medicine and Dentistry of New Jersey, Robert Wood Johnson Medical School, Piscataway, New Jersey

Meredith M. Garcia, Ph.D., Postdoctoral Fellow, Department of Anatomy, Tulane Medical School, New Orleans, Louisiana

David J. Greenblatt, M.D., Chief, Division of Clinical Pharmacology, New England Medical Center Hospital; Professor, Departments of Psychiatry, Pharmacology, and Medicine, Tufts University School of Medicine, Boston, Massachusetts

Richard E. Harlan, Ph.D., Associate Professor, Department of Anatomy, Tulane Medical School, New Orleans, Louisiana

John A. Harvey, Ph.D., Chief, Division of Behavioral Neurobiology; Professor, Department of Pharmacology, The Medical College of Pennsylvania, Philadelphia, Pennsylvania

Cecilia J. Hillard, Ph.D., Assistant Professor, Department of Pharmacology and Toxicology, Medical College of Wisconsin, Milwaukee, Wisconsin

Sari Izenwasser, Ph.D., Research Associate, Department of Pharmacology, Uniformed Services University, Bethesda, Maryland

Conan Kornetsky, Ph.D., Laboratory of Behavioral Pharmacology, Boston University School of Medicine, Boston, Massachusetts

Charles Ksir, Ph.D., Professor, Department of Psychology, University of Wyoming, Larmie, Wyoming

P. Y. Law, Ph.D., Department of Pharmacology, School of Medicine, University of Minnesota, Minneapolis, Minnesota

H. H. Loh, Department of Pharmacology, School of Medicine, University of Minnesota, Minneapolis, Minnesota

Neville Marks, Ph.D., Associate Professor, Department of Psychiatry, New York University; Assistant Director, Center for Neurochemistry, The Nathan Kline Institute for Psychiatric Research, Orangeburg, New York

Lawrence G. Miller, M.D., Associate Director, Division of Clinical Pharmacology; Assistant Professor, Departments of Pharmacology and Psychiatry, Tufts University School of Medicine, and New England Medical Center Hospital, Boston, Massachusetts

Henry C. Nipper, Ph.D., Associate Professor, Department of Pathology, Creighton University; Scientific Director, Clinical Chemistry and Toxicology, Department of Pathology, Saint Joseph Hospital, Omaha, Nebraska

S. Michael Owens, Ph.D., Associate Professor, Department of Pharmacology and Toxicology, University of Arkansas for Medical Sciences, Little Rock, Arkansas

Graham A. Patrick, Ph.D., Associate Professor, Department of Pharmacology and Toxicology, Virginia Commonwealth University, Richmond, Virginia

Paul V. Polishuk, M.S., Scientific Associate, Department of Pathology, Creighton University School of Medicine, Omaha, Nebraska

Hervé C. Porchet, M.D., Associate Director, Corporate Clinical Pharmacology, ARES SERONO, Geneva, Switzerland

George V. Rebec, Ph.D., Professor and Director, Program in Neural Science, Department of Psychology, Indiana University, Bloomington, Indiana

Anthony G. Romano, Ph.D., Research Assistant Professor, Department of Pharmacology, The Medical College of Pennsylvania, Philadelphia, Pennsylvania

Marlene A. Wilson, Ph.D., Assistant Professor, Department of Pharmacology, University of South Carolina School of Medicine, Columbia, South Carolina

Gerald J. Yutrzenka, Ph.D., Associate Professor, Department of Physiology and Pharmacology, University of South Dakota School of Medicine, Vermillion, South Dakota

Mark A. Zorbas, Ph.D., Postdoctoral Fellow, Department of Pharmacology, Baylor College of Medicine, Houston, Texas

TABLE OF CONTENTS

Chapter 1

BRAIN-STIMULATION REWARD: A METHOD FOR ASSESSING THE NEUROCHEMICAL BASES OF DRUG-INDUCED EUPHORIA

Sari Izenwasser and Conan Kornetsky

TABLE OF CONTENTS

I. BACKGROUND

Since 1954, when Olds and Milner[1] first discovered that animals would work to receive electrical stimulation to discrete brain regions, much interest has been paid to this finding. The technique whereby animals self-administer electrical stimulation to a particular brain area, which has been called by various names including intracranial self-stimulation (ICSS) and brain-stimulation reward (BSR), has been used to assess the abuse potential of drugs. It is now widely believed that drugs are abused because of the reinforcement or ''euphoria'' they produce. Since the electrical stimulation itself appears to be reinforcing in certain brain regions but not in others, it is likely that those regions to which animals will deliver electrical stimulation are involved in the production of euphoria and thus in the effects produced by abused drugs. In fact, it has been reported that electrical stimulation of the ventral tegmental area increases responding on the d-amphetamine-appropriate lever in animals trained to discriminate 1.0 mg/kg amphetamine from saline.[2] This finding shows that the stimulation itself has amphetamine-like stimulus properties. Additionally, ICSS seems to activate the same neuronal substrates that mediate the reinforcing effects of other reinforcers, such as food and water.[3] Thus, although the brain-stimulation model does not mimic the behavior associated with human drug abuse so closely as do other models, such as self-administration, it does allow for direct access to the brain regions and neurochemical bases underlying the reinforcing effects of abused drugs.

There have been many comprehensive reviews of self-stimulation behavior. Some have focused on the technique itself and the difficulties associated with interpretation of the results;[4,5] others have evaluated the technique as a model for drug-induced euphoria; and some have focused on a particular class of drugs or neuronal pathway in the brain.[3,6-11] This chapter will focus on the use of brain stimulation to determine the underlying neurochemical bases of drug-induced reinforcement.

There are a number of different models that use electrical brain stimulation to study reinforcement. The earliest method used the rate of lever pressing as the dependent variable.[1] In this paradigm, a simple press of a lever resulted in the delivery of an intracranial electrical stimulation. In general with this paradigm, as the stimulation intensity increases, so does the response rate (number of lever presses) up to an asymptotic level. At high intensities, response rates tend to decrease, thus giving an inverted-U-shaped curve for response rate as a function of intensity. A simple rate-of-response method is generally used: animals are trained to self-stimulate at a less than maximal intensity, and a drug is considered to be reinforcing if response rate increases and dysphoric if response rate decreases. This method, although one of the easiest to use, does not differentiate between the effects of a drug on the reinforcing value of the stimulation and the direct effect of the drug on response rate itself. Since many drugs have either stimulant or sedative effects, it is difficult to interpret what a change in response rate for self-stimulation means. Another problem with a simple rate-of-response method is that a change in response rate measured at a single intensity is difficult to interpret unless one knows which side of the inverted-U-shaped intensity-rate function one is on. In response to these problems, a number of more complex behavioral paradigms have been developed.

Some methods have used shuttle boxes[12,13] in which the animal must move from one side of the box to the other in order to receive stimulation. Others include the curve-shift rate-frequency model:[14] the stimulation intensity is kept constant while the frequency changes, and the response rate is determined as a function of frequency. Following drug administration, a shift to the right of this rate-frequency curve is interpreted as a decrease, and a shift to the left as an increase in the reinforcement value of the stimulation. Other methods include post-reinforcement pause[15] and On-Off, in which the duration of self-stimulation is measured.[16] For a more complete review of these and other methods used to measure ICSS see References 4, 5, and 17.

Rather than rate of response, threshold-of-stimulation intensity is used in a number of methods as a dependent variable (see, for example, References 15, 18-20). Some of these employ two levers, one of which the rat presses to receive stimulation that decreases in intensity over time. Pressing the second lever resets the intensity level to maximum, and the threshold is determined to be the intensity at which the animal presses the reset lever.[18] In one rate-independent threshold paradigm, animals make a discrete response to receive stimulation at varying intensity levels, and the intensity below which an animal will not respond is determined (threshold of reinforcement).[5,19] A drug is considered to be reinforcing if it causes the threshold to decrease, that is, if the animal will respond to receive stimulation that it previously did not find reinforcing. Abused substances from many different drug categories lower the reinforcement threshold. Figure 1 shows the threshold-lowering effects of cocaine (Figure 1A) and morphine (Figure 1B). Conversely, an increase in threshold following drug administration suggests that the drug produces dysphoria: because of the limited physical requirements of the task, it is easier for the animal to respond, and it is thus possible to distinguish changes in reinforcement value from performance deficits produced by the drug (for a review of these methods see Reference 5).

ICSS can also be used to measure the depression and anhedonia that are characteristic symptoms of cocaine withdrawal. When ICSS thresholds are measured during a withdrawal period following prolonged access to cocaine, they are elevated compared to pre-drug baseline levels, and the magnitude and duration of the threshold elevation is proportional to the amount of cocaine consumed (Figure 2). Since the effect is the opposite of the threshold lowering observed following acute cocaine administration, it is thought to reflect the anhedonia or depression associated with cocaine withdrawal.[21] Similar results are seen following termination of chronic administration of amphetamine[22,23] or morphine.[24]

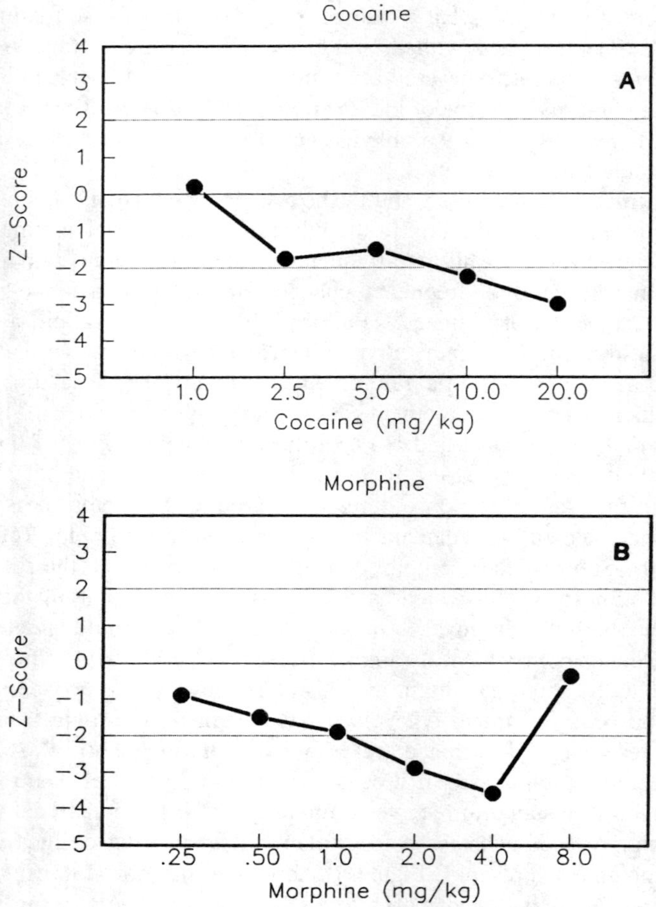

FIGURE 1. Cocaine (A) and morphine (B) each dose-dependently lower the threshold for self-stimulation to the medial forebrain bundle. Data are represented as mean standard scores (z-scores) of changes in the reward threshold from pre- to post-cocaine or morphine. A z-score of ± 2 indicates the 95% confidence limits. (A. From Kornetsky, C. and Espositio, R. U., *Brain Res.*, 209, 496, 1981. With permission. B. Courtesy of Dr. Conan Kornetsky, Laboratory of Behavioral Pharmacology, Boston University School of Medicine, Boston, 1991.)

In addition to predicting the abuse potential of drugs, the self-stimulation model of drug-induced euphoria can also be used to determine the underlying neurochemical bases of reinforcement and to identify the drugs that produce it. In 1956, Olds et al.[25] reported that chlorpromazine and reserpine each decreased lever pressing for electrical brain stimulation, theirs being the first report of the effect of a drug on brain-stimulation reward. This study was

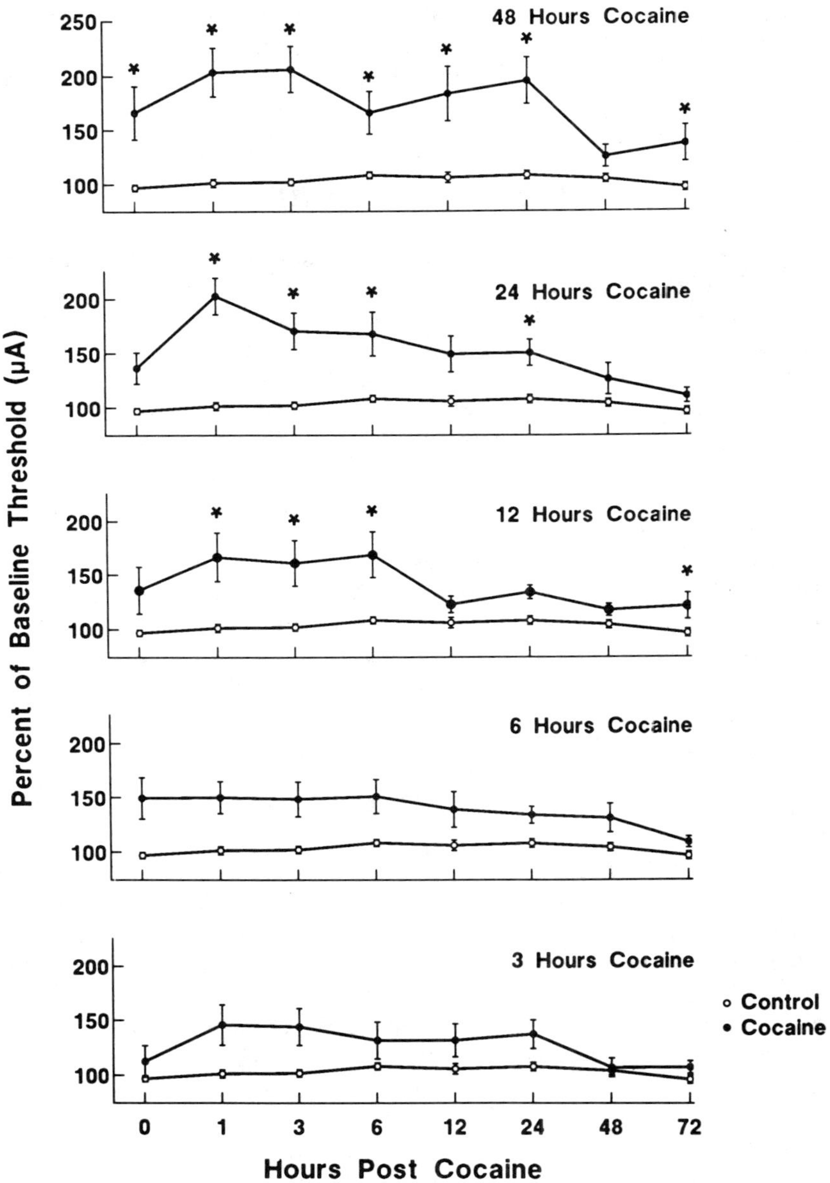

FIGURE 2. Intracranial self-stimulation thresholds were increased for up to 72 h after cessation of cocaine self-administration. The magnitude and duration of the increase appears to be related to the number of hours of cocaine self-administration. (Reprinted by permission of Elsevier Science Publishing Co. Inc., from Postcocaine anhedonia: An animal model of cocaine withdrawal, by Athina Markou and George F. Koob, *Neuropsychopharmacology*, 4, 17–16, 1991. Copyright 1991 by the American College of Neuropsychopharmacology.)

important in that it showed that self-stimulation behavior could be modified by drug administration and so suggested that the technique might be a useful tool in studying drug effects. Since then much work has been done in attempting to determine which neurotransmitter or neuropeptide system is critical for reinforcement.

II. NOREPINEPHRINE

Early theories suggested that norepinephrine was the primary neurotransmitter involved in mediating reinforcement processes.[26] This belief stemmed from the observation that there is a close correlation between ascending noradrenergic projections and brain areas that support ICSS. However, as pointed out by German and Bowden[27] in 1974, norepinephrine is widely distributed throughout the brain, and it is not surprising to find that noradrenergic tracts pass through these regions. Furthermore, much of the evidence supporting a major role for norepinephrine came from early findings that drugs such as cocaine,[28] d-amphetamine,[26] and MAO inhibitors[29] all increase ICSS behavior and that, conversely, drugs that deplete catecholamine stores, block receptors, or inhibit synthesis all decrease responding for ICSS.[30-34] It was later realized, however, that the compounds used in these early studies influence dopaminergic as well as noradrenergic activity. Furthermore, since lesions of the major noradrenergic tracts have no effect on ICSS,[35-37] it is clear that norepinephrine is not a necessary component for the production of euphoria by drugs. Based on these findings, Crow in 1972, modified the catecholamine hypothesis of reward to include dopamine.[38]

III. DOPAMINE

Most of the evidence about the neurochemical bases underlying drug reinforcement suggests that dopamine plays a major role. There is a large overlap between the brain regions that contain dopamine cell bodies and those that support self-stimulation.[39] The pathways that appear to be most highly involved in reinforcement are the mesolimbic/mesocortical dopamine pathways. High rates of self-stimulation are supported in the medial forebrain bundle (MFB), through which these dopaminergic projections pass; the ventral tegmental area (VTA), where the mesolimbic dopamine pathway originates;[38,40] and the nucleus accumbens (NAcc), which is a major site of innervation by mesolimbic dopamine. Because of the high rate of response for stimulation in the MFB, this site has been used for placement of electrodes in the majority of studies reported. The nigrostriatal dopamine pathway, however, which has cell bodies in the substantia nigra and terminals in the caudate nucleus (striatum), does not appear to be highly involved in the production of drug-induced euphoria.[3] Lesions of this pathway do not appear to have an effect on self-stimulation behavior. Rather, a loss of dopaminergic innervation in this pathway results in Parkinson's symptoms such as akinesia and catalepsy.

In addition to the differences in response rate for stimulation in one brain region vs. another, there is much neurochemical and pharmacological evidence to support a finding that mesolimbic dopamine plays a role in mediating the reinforcing effects of abused drugs. The effect of ICSS on dopaminergic activity and the interaction of compounds that act either directly or indirectly as selective dopamine agonists or antagonists with ICSS have been extensively studied.

Using 2-[^{14}C]-deoxyglucose autoradiography, Porrino et al. (1990) showed that the brain regions that exhibit the most significant changes in local cerebral metabolic rate of glucose (LCMRglu) following MFB stimulation are the nucleus accumbens, prefrontal cortex, olfactory tubercle, and lateral septum.[41] Stimulation of the VTA also increases LCMRglu in the prefrontal cortex, nucleus accumbens, and lateral septum.[42,43] A similar pattern of glucose utilization is seen following cocaine administration, significant increases being seen in the olfactory tubercle, medial prefrontal cortex, and substantia nigra. When a reinforcing dose of either cocaine or morphine is combined with MFB stimulation, further increases in activity are seen in the olfactory tubercle (Plate 1)* suggesting that this structure may be important for the reinforcing effects of morphine, cocaine, and brain-stimulation reward.[44]

The development of *in vivo* procedures that can monitor dopamine overflow has allowed a more direct measure of dopamine release following MFB stimulation.[45,46] Using *in vivo* voltammetry, Kuhr et al. (1987) have shown that dopaminergic neurons are activated by MFB stimulation.[47] *In vivo* microdialysis studies have also shown that there is enhanced dopamine release during ICSS to the MFB.[48] With *in vivo* microdialysis, Di Chiara and Imperato (1988) showed that many drugs abused by humans (opiates, ethanol, nicotine, amphetamine, and cocaine) increase extracellular dopamine in the NAcc, whereas drugs that are dysphoric (e.g., the κ agonists U50, 488, tifluadom, and bremazocine) reduce dopamine release.[49] Drugs such as diphenhydramine (an antihistamine), imipramine (an antidepressant), and atropine (a muscarinic cholinergic agonist), which are not abused by humans, have no effect on the concentration of synaptic dopamine in the NAcc. Blaha and Phillips (1990) used *in vivo* electrochemistry to show that both experimenter-administered stimulation and self-stimulation of the VTA produce increases in dopamine oxidation in the NAcc and that the magnitude of these increases is a function of current intensity.[50] Furthermore, increases in response rates for ICSS following systemic administration of nomifensine or GBR 12909 (dopamine uptake inhibitors) are accompanied by significant increases in the amplitude of the chronoamperometric signal. Conversely, desipramine, a noradrenergic uptake inhibitor, has no effect either on response rate for stimulation or on dopamine oxidation.

* Plate 1 follows page 22.

A. DOPAMINE AGONISTS

The effects of cocaine and amphetamine on ICSS have been extensively studied. Both are indirect dopamine agonists that increase synaptic dopamine although via different mechanisms. Cocaine inhibits the reuptake of dopamine into presynaptic terminals.[51,52] The major action of amphetamine is to stimulate dopamine release;[53,54] it enhances the response rate[26,55-58] and lowers the stimulation threshold[57,59-63] for brain stimulation. Studies of ICSS in a shuttle box have also shown enhanced stimulation time following injection of methamphetamine[33] or d-amphetamine.[64]

Similarly, cocaine lowers the threshold for stimulation to the MFB.[65-67] Cocaine also increases the rate and duration of prefrontal cortex self-stimulation.[68] The same doses of cocaine that lower the reinforcement threshold lead to increases in the threshold for detection of the stimulation, which is measured using a similar paradigm.[69] Thus, the threshold lowerings are clearly not attributable to an increase in the animal's ability to perceive the stimulation, and the increased sensitivity for brain-stimulation reward is not merely a function of general stimulation of the animal by the drug. In an attempt to measure the neurochemical correlates of ICSS directly, Phillips et al. (1989) recently reported that dopamine levels in the MFB are increased following stimulation of the VTA, but only when the current intensity exceeds the threshold value for self-stimulation behavior.[71] Additionally, cocaine and GBR 12909, a highly selective dopamine uptake inhibitor, facilitate self-stimulation and concomitantly increase extracellular dopamine. The pattern of the increase in extracellular dopamine matches the response pattern for self-stimulation. Similar increases in extracellular dopamine have been measured in microdialysis samples from the NAcc following stimulation of the MFB.[72]

ICSS threshold lowerings have been seen not only in conjunction with cocaine and amphetamine, but also in conjunction with other indirect dopamine agonists such as amfonelic acid[73,74] and methylphenidate.[75]

Apomorphine, a direct agonist that acts at dopamine receptors, has yielded conflicting results on ICSS: it has been reported to increase,[64] or to decrease,[6] or to have no effect on brain-stimulation behavior.[5] One factor that has to be considered is the dose of apomorphine tested. An increase in ICSS is seen with low doses,[64] whereas decreases are seen when high doses are used.[6] However, Carey (1982) reported that low response rates are suppressed while high rates are unaffected by apomorphine.[76] What is clear is that the effects of apomorphine on self-stimulation are rate dependent and that it is difficult to separate out the competing motoric effects of this drug. Apomorphine is reported to be a partial agonist at D_1 dopamine receptor sites as well as a full agonist at D_2 sites;[77] thus it is likely that at low doses it acts predominantly via D_2 receptors, whereas at higher doses it also acts via D_1 receptors. The interactions between D_1 and D_2 dopamine receptors are complex, and the roles that each plays in mediating drug-induced reinforcement are themselves unclear; it is difficult to interpret the effects of a drug which acts nonselectively. Yet another factor should be considered in interpreting the effects of apomorphine and other dopaminergic agents that act similarly: at high doses they have emetic properties that are likely to contribute to their effects on responding for rewarding brain stimulation.

B. DOPAMINE ANTAGONISTS

There have been many reports showing that dopamine antagonists attenuate or abolish the reinforcing effects of brain stimulation. Separating the drug effects on the reinforcement properties of the stimulation from performance deficits has not, however, been easy. Neuroleptic drugs such as pimozide, haloperidol, and *cis*-flupenthixol attenuate ICSS by lowering response rate or increasing threshold (see, for example, References 57 and 78 to 82). Because normal dopamine function is required for normal motor function, neuroleptics, which interfere with dopamine transmission, produce deficits in operant performance that interfere with the ability of the animal to respond in many behavioral paradigms. However, a number of well-designed behavioral studies attempting to separate the motor debilitating effects from the anhedonic effects produced by neuroleptics have suggested that these drugs do reduce the reinforcement value of reinforcers such as food, water, and brain stimulation.[83] Furthermore, although pimozide increases the reinforcement threshold, it has no effect on the stimulation-detection threshold. Thus, it does not change the animal's ability to perceive and respond to the stimulation.[84]

The efficacy of neuroleptic drugs in suppressing brain stimulation appears to be correlated with their affinity for D_2, but not D_1, dopamine receptors.[85] There are, however, reports suggesting that D_1 receptors are also involved in reinforcement. Systemic administration of SCH 23390, a selective D_1 receptor antagonist, reduces response rate for self-stimulation.[86] Furthermore, administration of SCH 23390 into the NAcc produces a decrease in responding for stimulation to the VTA,[87] whereas sulpiride, a D_2 antagonist, has no effect.

IV. ENDOGENOUS OPIOID PEPTIDES

The effects of opiate drugs, which have been widely abused for centuries, have been extensively studied in relation to ICSS. Systemic administration of morphine increases response rate[88-90] and lowers threshold[19,91,92] for ICSS to the mesolimbic dopamine system. Acute systemic morphine injection does not, however, affect threshold for stimulation to the substantia nigra.[92] Other opioid drugs such as heroin[93] and the mixed opioid agonist/antagonists nalbuphine,[94] pentazocine,[95] and buprenorphine[96] also lower the self-stimulation threshold.

Whether the reinforcing effects of opioid drugs are mediated through opiate receptors and endogenous opioids or through changes in dopaminergic activity is not completely understood. The dopaminergic brain regions that support ICSS also have significant levels of enkephalins[97-100] and opioid binding sites.[101,102] Thus, there has been much interest in the role that endogenous opioids may play in mediating ICSS and hence drug reinforcement. The compounds most often used in these studies have been the opiate antagonists naloxone and naltrexone (for a more complete review see Reference 11). If endogenous opioid peptides mediate drug reinforcement, it is likely that blockade of opiate receptors should alter ICSS behavior. As expected, naloxone blocks the increases in self-stimulation behavior that are produced by morphine.[11] The effects of systemic naloxone alone, however, have been variable. Naloxone has no effect on the reinforcement threshold when rate-independent procedures are used.[103-105] When response rate is used as the dependent variable, there are reports of either no change[62,106,107] or a slight decrease[108-110] in response rate for ICSS to the MFB. Schaeffer (1988) and colleagues have shown, however, that naloxone has a greater effect on a fixed-ratio than on a continuous-reinforcement schedule for ICSS.[11] They suggest that naloxone may have a greater effect when the dopamine system already has been highly activated. Diprenorphine, an opioid receptor antagonist that is less selective for μ-opioid receptors than naloxone and binds equally well to μ-opioid, κ-opioid, and δ-opioid receptor subtypes, increases the self-stimulation response rate.[111,112] These findings suggest that perhaps κ or δ receptors are more involved in opioid reinforcement than are μ receptors. However, neither of the κ receptor agonists cyclazocine and ethylketocyclazocine nor the κ receptor antagonist U50-488H has any effect on ICSS thresholds.[113] Injection of DTLET, a δ-opioid receptor agonist, into either the lateral ventricle or the nucleus accumbens increases response rate for self-stimulation to the MFB.[114] An increase is also seen after intraventricular injection of amphetamine or apomorphine, though intraventricular administration of morphine or DAGO (a μ-opioid receptor agonist) decreases response rate. Thus, it appears that the opioid receptor subtypes may have different and possibly opposite effects on reinforcement. Since diprenorphine binds equally well to μ-opioid, κ-opioid, and δ-opioid receptor subtypes, and naloxone binds preferentially to μ-opioid receptors, this may explain the differences in the effects of these two drugs when they are administered systemically. Furthermore, since diprenorphine is known to have agonist as well as antagonist actions at opioid receptors, it is also possible that the increase in ICSS behavior is due to its opioid agonist actions.

The combination of an opiate analgesic with a stimulant is known on the street as a ''speedball'' and is more reinforcing than either drug separately. When morphine and d-amphetamine are administered concomitantly, they produce an even greater change in threshold for ICSS than either drug alone.[63] Since amphetamine stimulates noradrenergic as well as dopaminergic activity, this potentiation of morphine's effect could be modulated by either system. When a low dose of amfonelic acid (a dopamine releaser and uptake inhibitor) is co-administered with morphine, the effect is potentiated to an even greater magnitude than when amphetamine is used. Nisoxetine (a selective norepinephrine uptake inhibitor), however, when administered concomitantly with morphine, does not potentiate the threshold lowerings seen with morphine alone.[73] Thus, it is most likely that the increase in dopaminergic activity modulates the potentiation of morphine's reinforcing effect by amphetamine.

Naloxone blocks the rate-increasing effects of amphetamine,[107] ethanol,[115] and chlordiazepoxide.[115] It also blocks the threshold lowering produced by cocaine,[66] amphetamine,[62] pentobarbital,[105] or phencyclidine.[116]

Stimulation of the MFB increases preproenkephalin mRNA in the rat striatum.[117] This effect is also seen after administration of either methamphetamine or quinpirole (a D_2-dopamine receptor agonist) but not after SKF 38393 (a D_1-dopamine receptor agonist). The interactions between dopamine and endogenous opioids are complex, and as yet it is still unclear which, if either, of these systems is more important in the production of reinforcement by drugs. It is likely, though, that both systems play important roles.

V. NICOTINE

Assessing the reinforcing effects of nicotine by using brain stimulation has been somewhat more difficult than by using more direct-acting dopaminergic agents. Nicotine has been reported to increase, decrease, or have no effect on ICSS. As with many other classes of compounds, it has been difficult to separate the locomotor-stimulating effects of nicotine from its more direct effect on the reward value of the stimulation. Pradhan and Bowling (1971) showed that nicotine facilitates low rates of responding but depresses high rates of responding for self stimulation.[118] Nicotine lowers the threshold for rewarding stimulation, and this effect is blocked by the nicotinic receptor antagonist mecamylamine but not by naloxone.[119] Other studies have shown that in addition to depending on the pre-drug response rate, the effect of nicotine on ICSS depends on the dose of nicotine[118,120] and on the time after injection that the animals are tested.[121,122] Corrigall and Coen (1989) have also reported difficulties in assessing the reinforcing effects of nicotine using self-administration as a model of reinforcement.[123] They show that the unit dose of drug administered is critical and that it is difficult to establish convincing evidence for nicotine reinforcement unless a limited-access schedule of self-administration is used.

Whether the reinforcing effects of nicotine, as assessed with ICSS, are modulated by cholinergic activity is not known. In addition to stimulating the release of acetylcholine, nicotine stimulates dopamine release both *in vitro*[124,125] and *in vivo*[126] and inhibits dopamine reuptake *in vitro*.[127] Interestingly, the effect of nicotine on dopamine release is greater in the nucleus accumbens than in the caudate-putamen.[126,128] Thus it is possible that nicotine's reinforcing effects are attributable to its indirect dopamine agonist actions rather than to a direct mediation of reward by acetylcholine.

There have been other studies aimed at studying the role of acetylcholine in mediating ICSS. Domino and Olds (1968) showed that physostigmine leads to an increase in brain acetylcholine levels and a decrease in self-stimulation response rate.[129] Conversely, atropine and scopolamine, both selective muscarinic cholinergic antagonists, raise the threshold for ICSS.[5,130] Like nicotine, however, acetylcholine is known to interact directly with dopaminergic neurons to stimulate dopamine release.[131] Thus it is possible that acetylcholine may modulate reinforcement indirectly via modulation of dopaminergic activity.

VI. GABA AND BENZODIAZEPINES

The effects of anxiolytic drugs on ICSS have been extensively studied. Meprobamate,[132] chlordiazepoxide,[115,133-135] diazepam,[136,137] pentobarbital,[105,132,135,138] and clonazepam[135] all increase ICSS behavior when the appropriate conditions and doses are used.

The GABA antagonists bicuculline and picrotoxin were found by Kent and Fedinets (1976) each to decrease response rate for self-stimulation.[139] When a rate-independent threshold procedure was used, picrotoxin lowered the threshold for VTA stimulation, an effect that was not blocked by naloxone.[140] Porrino and Coons, however, reported the opposite effect, that picrotoxin increases the self-stimulation threshold.[141] The animals in this study apparently exhibited marked stereotypical behaviors that may have interfered with their ability to respond to the stimulation. Using a two-lever reset method to determine ICSS threshold, Zarevics and Setler found an increase in threshold with picrotoxin and a decrease with the GABA-mimetic muscimol.[142] Since there is a tonic inhibition by GABA of dopamine neurons, blockade of this inhibition by a GABA antagonist results in increased dopaminergic activity.

VII. Δ^9-TETRAHYDROCANNABINOL

Both the main ingredient in marijuana, Δ^9-tetrahydrocannabinol (Δ^9-THC), and the cannabinoid levonantradol have been reported to have no effect on ICSS thresholds for stimulation to the MFB.[143] Recently, it has been shown that Δ^9-THC does facilitate self-stimulation to the MFB, but only in Lewis rats.[144] The mechanism by which Δ^9-THC enhances self-stimulation is not known, nor is it known why this enhancement occurs only in Lewis rats. Like many of the other drugs that augment ICSS, however, Δ^9-THC increases extracellular dopamine by inhibiting dopamine uptake[145] and increasing dopamine release[146] *in vitro*. *In vivo* microdialysis studies have also shown that Δ^9-THC increases extracellular dopamine in both the caudate putamen[147] and nucleus accumbens.[148] Thus, as with other classes of drugs, it is likely that Δ^9-THC is producing its reinforcing effect by activating dopamine pathways.

VIII. ETHANOL

The effects of ethanol on ICSS have been mixed. Low doses of ethanol enhance response rates for self-stimulation, and this effect is blocked by naloxone.[115] There is no change either in response rate or in threshold for ICSS following intragastric injection[149] or following intraperitoneal injection of ethanol except at debilitating doses.[150] When animals are trained to drink ethanol immediately before the testing session, however, the threshold for ICSS is lowered and the response rate increases.[151] Conversely, a chronic alcohol diet leads to a decrease in the rate of lever pressing for ICSS.[152]

IX. OTHER BRAIN SITES

An area of research not discussed in detail in this chapter is the comparison of the effects of drugs on self-stimulation of different brain regions. Most studies of self-stimulation behavior have used mesolimbic dopaminergic regions such as the ventral tegmental area, medial forebrain bundle, and nucleus accumbens, because these are areas that employ a high rate of responding for self-stimulation and that have been reported to be important in mediating reinforcement. However, there is evidence that there are differences in drug effects depending on the brain region used. Nazzaro et al. (1981) have shown, for example, that acute morphine administration lowers the self-stimulation threshold in rats with electrodes in the ventral tegmental area, but not in the substantia nigra.[92] Kornetsky and colleagues have shown that stimulation to some brain regions is aversive and that animals will work to turn the stimulation off. When electrodes are implanted into brain regions that are along pain pathways, brain stimulation can be used to measure the analgesic effects of drugs. Analgesic drugs, such as morphine, will raise the threshold of stimulation intensity that animals will tolerate.[19,153] To have behavioral models of analgesia and reinforcement that employ similar paradigms allows a direct comparison of the supraspinal analgesic effects and the reinforcing effects of compounds. Since a major class of abused drugs is made up of opioid analgesics, it becomes possible to separate the underlying neurochemical bases of the analgesia and reinforcement produced by these compounds.[73,113,154] Using electrodes in different brain regions can contribute to an understanding of the roles that different pathways may play in modulating drug effects.

X. SUMMARY

Brain-stimulation reward paradigms can be used to assess the underlying neurochemical bases of reinforcement. Much of the evidence, from many different classes of drugs, suggests that dopamine plays a major role in the mediation of drug reinforcement. As noted above, however, it is likely that one or more neuropeptides also may be involved in producing drug-induced euphoria.

Self-stimulation paradigms, especially when used in conjunction with other pharmacological and behavioral tests, can clearly contribute to a further understanding of how drugs are working and which brain pathways are involved in their effects. For example, as cited above, the combination of brain stimulation with *in vivo* microdialysis, 2-[^{14}C]deoxyglucose autoradiography, or drug-discrimination techniques has begun to give a clear picture of the neurochemical changes produced by abused drugs and has shown that the stimulation itself has actions very similar to those of abused drugs such as opiates and stimulants.

REFERENCES

1. **Olds, J. and Milner, P.,** Positive reinforcement produced by electrical stimulation of septal area and of other regions of rat brain, *J. Comp. Physiol. Psychol.,* 47, 419, 1954.
2. **Druhan, J. P., Fibiger, H. C., and Phillips, A. G.,** Amphetamine-like stimulus properties produced by electrical stimulation of reward sites in the ventral tegmental area, *Behav. Brain Res.,* 38, 175, 1990.
3. **Wise, R. A.,** Catecholamine theories of reward: a critical review, *Brain Res.,* 152, 215, 1978.
4. **Liebman, J. M.,** Discriminating between reward and performance: a critical review of intracranial self-stimulation methodology, *Neurosci. Biobehav. Rev.,* 7, 45, 1983.
5. **Kornetsky, C. and Bain, G.,** Brain-stimulation reward: a model for drug-induced euphoria, in *Modern Methods in Pharmacology,* Vol. 6, 1990, 211.
6. **Wauquier, A.,** The influence of psychoactive drugs on brain self-stimulation in rats: a review, in *Brain-Stimulation Reward,* Wauquier, A. and Rolls, E. T., Eds., Elsevier, New York, 1976, 123.
7. **Fibiger, H. C.,** Drugs and reinforcement mechanisms: a critical review of the catecholamine theory, *Annu. Rev. Pharmacol. Toxicol.,* 18, 37, 1978.
8. **Wise, R. A. and Bozarth, M.A.,** Brain substrates for reinforcement and drug self-administration, *Prog. Neuropsychopharmacol.,* 5, 467, 1981.
9. **Liebman, J. M.,** Anxiety, anxiolytics and brain stimulation reinforcement, *Neurosci. Biobehav. Rev.,* 9, 75, 1985.
10. **Bozarth, M. A. and Wise, R. A.,** Neural substrates of opiate reinforcement, *Prog. Neuropharmacol. Biol. Psychiatr.,* 7, 569, 1983.
11. **Schaefer, G. J.,** Opiate antagonists and rewarding brain stimulation, *Neurosci. Biobehav. Rev.,* 12, 1, 1988.
12. **Atrens, D. M., Ljunberg, T., and Ungerstedt, U.,** Modulation of reward and aversion processes in the rat diencephalon by neuroleptics: differential effects of clozapine and haloperidol, *Psychopharmacologia,* 49, 97, 1976.
13. **Levitt, R. A., Baltzer, J. H., Evers, T. M., Stilwell, D. J., and Furby, J. E.,** Morphine and shuttle-box self-stimulation in the rat: a model for euphoria, *Psychopharmacologia,* 54, 307, 1977.
14. **Edmonds, D. E. and Gallistel, C. R.,** Parametric analysis of brain stimulation reward in the rat. III. Effect of performance variables on the reward summation function, *J. Comp. Physiol. Psychol.,* 87, 876, 1974.
15. **Huston, J. P. and Mills, A. W.,** Threshold of reinforcing brain stimulation, *Commun. Behav. Biol.,* 5, 331, 1971.

16. **Liebman, J. M., Hall, N., and Prowse, J.,** Effects of various catecholamine receptor antagonists, muscle relaxation and physical hindrance on shuttlebox self-stimulation, *Pharmacol. Biochem. Behav.,* 16, 785, 1982.
17. **Stellar, J. R. and Rice, M. B.,** Pharmacological basis of intracranial self-stimulation reward, in *The Neuropharmacological Basis of Reward,* Liebman, J. M. and Cooper, S. J., Eds., Clarendon Press, Oxford, 1989, 14.
18. **Stein, L. and Ray, O. S.,** Brain stimulation reward "thresholds" self determined in the rat, *Psychopharmacologia,* 1, 251, 1960.
19. **Marcus, R. and Kornetsky, C.,** Negative and positive intracranial reinforcement thresholds: effects of morphine, *Psychopharmacologia,* 38, 1, 1974.
20. **Zarevics, P. and Setler, P. E.,** Effects of GABAergic drugs on brain stimulation reward as assessed by a 'threshold' method, *Brain Res.,* 215, 201, 1982.
21. **Markou, A. and Koob, G. F.,** Postcocaine anhedonia: an animal model of cocaine withdrawal, *Neuropsychopharmacology,* 4, 17, 1991.
22. **Leith, N. J. and Barrett, R. J.,** Amphetamine and the reward system: evidence for tolerance and post-drug depression, *Psychopharmacologia,* 46, 19, 1976.
23. **Cassens, G., Actor, C., Kling, M., and Schildkraut, J. J.,** Amphetamine withdrawal: effects on threshold of intracranial reinforcement, *Psychopharmacologia,* 73, 318, 1981.
24. **Schaefer, G. J. and Michael, R. P.,** Changes in response rates and reinforcement thresholds for intracranial self-stimulation during morphine withdrawal, *Pharmacol. Biochem. Behav.,* 25, 1263, 1986.
25. **Olds, J., Killam, K. F., and Bach-y-Rita, P.,** Self-stimulation of the brain used as a screening method for tranquilizing drugs, *Science,* 124, 265, 1956.
26. **Stein, L.,** Self-stimulation of the brain and the central stimulant action of amphetamine, *Fed. Proc.,* 23, 836, 1964.
27. **German, D. C. and Bowden, D. M.,** Catecholamine systems as the neural substrate for intracranial self-stimulation: a hypothesis, *Brain Res.,* 73, 381, 1974.
28. **Crow, T. J.,** Enhancement by cocaine of intracranial self-stimulation in the rat, *Life Sci.,* 9, 375, 1970.
29. **Poschel, B. P. H.,** Mapping of rat brain for self-stimulation under monoamine oxidase blockade, *Physiol. Behav.,* 4, 325, 1969.
30. **Stein, L.,** Effects and interactions of imipramine, chlorpromazine, reserpine, and amphetamine on self-stimulation: possible neurophysiological basis of depression, in *Recent Advances in Biological Psychiatry,* Vol. 4, Wortis, J., Ed., Plenum Press, New York, 1962, 288.
31. **Esposito, R. U., Faulkner, W., and Kornetsky, C.,** Specific modulation of brain stimulation reward by haloperidol, *Pharmacol. Biochem. Behav.,* 10, 937, 1979.
32. **Esposito, R. U., Perry, W., and Kornetsky, C.,** Chlorpromazine and brain-stimulation reward: potentiation of effects by naloxone, *Pharmacol. Biochem. Behav.,* 15, 903, 1981.
33. **Poschel, B. P. H. and Ninteman, F. W.,** Psychotropic drug effects on self-stimulation of the brain: a control for motor output, *Psychol. Rep.,* 19, 79, 1966.
34. **Black, W. C. and Cooper, B. R.,** Reduction of electrically rewarded behavior by interference with monoamine synthesis, *Physiol. Behav.,* 5, 1405, 1970.
35. **Clavier, R. M. and Routtenberg, A.,** Brainstem self-stimulation attenuated by lesions of medial forebrain bundle but not by lesions of locus coeruleus or the caudate ventral norepinephrine bundle, *Brain Res.,* 101, 251, 1976.
36. **Van der Kooy, D., Fibiger, H. C., and Phillips, A. G.,** Monoamine involvement in hippocampal self-stimulation, *Brain Res.,* 136, 119, 1977.
37. **Phillips, A. G., Van der Kooy, D., and Fibiger, H. C.,** Maintenance of intracranial self-stimulation in hippocampus and olfactory bulb following regional depletion of noradrenaline, *Neurosci. Lett.,* 4, 77, 1977.
38. **Crow, T. J.,** Catecholamine-containing neurons and electrical self-stimulation. I. A review of some data, *Psychol. Med.,* 2, 414, 1972.

39. **Corbett, D. and Wise, R.,** Intracranial self-stimulation in relation to the ascending dopamine systems of the midbrain: a movable electrode mapping study, *Brain Res.,* 185, 1, 1980.
40. **Routtenberg, A. and Malsbury, C.,** Brainstem pathways of reward, *J. Comp. Physiol. Psychol.,* 68, 22, 1969.
41. **Porrino, L. J., Huston-Lyons, D., Bain, G., Sokoloff, L., and Kornetsky, C.,** The distribution of changes in local cerebral energy metabolism associated with brain-stimulation reward to the medial forebrain bundle of the rat, *Brain Res.,* 511, 1, 1990.
42. **Esposito, R. U., Porrino, L. J., Seeger, T. F., Crane, A. M., Everist, H. D., and Pert, A.,** Changes in local cerebral glucose utilization during rewarding brain stimulation, *Proc. Natl. Acad. Sci.,* 81, 635, 1984.
43. **Porrino, L. J., Esposito, R. U., Seeger, T. F., Crane, A. M., Pert, A., and Sokoloff, L.,** Metabolic mapping of the brain during rewarding self-stimulation, *Science,* 224, 306, 1984.
44. **Kornetsky, C., Huston-Lyons, D., and Porrino, L. J.,** The role of the olfactory tubercle in the effects of cocaine, morphine and brain-stimulation reward, *Brain Res.,* 541, 75, 1991.
45. **Ungerstedt, U.,** Measurement of neurotransmitter release by intracranial dialysis, in *Measurement of Neurotransmitter Release in Vivo,* Marsden, C. A., Ed., John Wiley & Sons, New York, 1984, 81.
46. **Wightman, R. M., Kuhr, W. G., and Ewing, A. G.,** Voltametric detection of dopamine release in the rat corpus striatum, *Ann. NY Acad. Sci.,* 473, 92, 1986.
47. **Kuhr, W. G., Wightman, R. M., and Rebec, G. V.,** Dopaminergic neurons: simultaneous measurement of dopamine release and single-unit activity during stimulation of the medial forebrain bundle, *Brain Res.,* 418, 122, 1987.
48. **Nakahara, D., Ozaki, N., Miura, Y., Miura, H., and Nagatsu, T.,** Increased dopamine and serotonin metabolism in rat nucleus accumbens produced by intracranial self-stimulation of medial forebrain bundle as measured by *in vivo* microdialysis, *Brain Res.,* 495, 178, 1989.
49. **Di Chiara, G. and Imperato, A.,** Drugs abused by humans preferentially increase synaptic dopamine concentrations in the mesolimbic system of freely moving rats, *Proc. Natl. Acad. Sci.,* 85, 5274, 1988.
50. **Blaha, C. D. and Phillips, A. G.,** Application of *in vivo* electrochemistry to the measurement of changes in dopamine release during intracranial self-stimulation, *J. Neurosci. Methods,* 34, 125, 1990.
51. **Heikkila, R. E., Orlansky, H., and Cohen, G.,** Studies on the distinction between uptake inhibition and release of ^{3}H-dopamine in rat brain tissue slices, *Biochem. Pharmacol.,* 24, 847, 1975.
52. **Izenwasser, S., Werling, L. L., and Cox, B. M.,** Comparison of the effects of cocaine and other inhibitors of dopamine uptake in rat striatum, nucleus accumbens, olfactory tubercle, and medial prefrontal cortex, *Brain Res.,* 520, 303, 1990.
53. **Uretsky, N. J. and Snodgrass, S. R.,** Studies on the mechanism of stimulation of dopamine synthesis by amphetamine in striatal slices, *J. Pharmacol. Exp. Ther.,* 202, 565, 1977.
54. **Schwarz, R. D., Uretsky, N. J., and Bianchine, J. R.,** The relationship between the stimulation of dopamine synthesis and release produced by amphetamine and high potassium in striatal slices, *J. Neurochem.,* 35, 1120, 1980.
55. **Crow, T. J.,** Mode of enhancement of self stimulation in rats by methamphetamine, *Nature,* 224, 709, 1969.
56. **Wise, C. D. and Stein, L.,** Amphetamine: facilitation of behavior by augmented release of norepinephrine from the medial forebrain bundle, in *International Symposium on Amphetamines and Related Compounds,* Costa, E. and Garattini, S., Eds., Raven Press, New York, 1970, 463.

57. **Zarevics, P. and Setler, P. E.,** Simultaneous rate-independent and rate-dependent assessment of intracranial self-stimulation: evidence for the direct involvement of dopamine in brain reinforcement mechanisms, *Brain Res.,* 169, 499, 1979.
58. **Greenshaw, A. J., Sanger, D. J., and Blackman, D. E.,** Effects of d-amphetamine and of β-phenylethylamine on fixed interval responding maintained by self-regulated lateral hypothalamic stimulation in rats, *Pharmacol. Biochem. Behav.,* 23, 519, 1983.
59. **Cassens, G. P. and Mills, A. W.,** Lithium and amphetamine: opposite effects on threshold of intracranial reinforcement, *Psychopharmacologia,* 30, 283, 1973.
60. **Wauquier, A. and Niemegeers, C. J. E.,** Intracranial self-stimulation in rats as a function of various stimulus parameters. IV. The influence of amphetamine on medial forebrain bundle stimulation with monopolar electrodes, *Psychopharmacologia,* 34, 265, 1974.
61. **Schaefer, G. J. and Holtzman, S. G.,** Free-operant and autotitration brain self-stimulation procedures in the rat: a comparison of drug effects, *Pharmacol. Biochem. Behav.,* 10, 127, 1979.
62. **Esposito, R. U., Perry, W., and Kornetsky, C.,** Effects of d-amphetamine and naloxone on brain stimulation reward, *Psychopharmacologia,* 69, 187, 1980.
63. **Hubner, C. B., Bain, G. T., and Kornetsky, C.,** The combined effect of morphine and d-amphetamine on the threshold for brain-stimulation reward, *Pharmacol. Biochem. Behav.,* 28, 311, 1987.
64. **Liebman, J. M. and Butcher, L. L.,** Comparative involvement of dopamine and noradrenaline in rate-free self-stimulation in substantia nigra, lateral hypothalamus, and mesencephalic gray, *Naunyn-Schmiedeberg's Arch. Pharmacol.,* 284, 167, 1974.
65. **Wauquier, A. and Niemegeers, C. J. E.,** Intracranial self-stimulation in rats as a function of various stimulus parameters. V. The influence of cocaine on medial forebrain bundle stimulation with monopolar electrodes, *Psychopharmacologia,* 38, 201, 1974.
66. **Esposito, R. U., Motola, A. H., and Kornetsky, C.,** Cocaine: acute effects on reinforcement thresholds for self-stimulation behavior to the medial forebrain bundle, *Pharmacol. Biochem. Behav.,* 8, 437, 1978.
67. **Frank, R. A., Martz, S., and Pommering, T.,** The effect of chronic cocaine on self-stimulation train-duration thresholds, *Pharmacol. Biochem. Behav.,* 29, 755, 1988.
68. **Moody, C. A. and Frank, R. A.,** Cocaine facilitates prefrontal cortex self-stimulation, *Pharmacol. Biochem. Behav.,* 35, 743, 1990.
69. **Kornetsky, C. and Esposito, R. U.,** Reward and detection thresholds for brain stimulation: dissociative effects of cocaine, *Brain Res.,* 209, 496, 1981.
70. **Schaefer, G. J. and Michael, R. P.,** An analysis of the effects of amphetamine on brain self-stimulation behavior, *Behav. Brain Res.,* 29, 93, 1988.
71. **Phillips, A. G., Blaha, C. D., and Fibiger, H. C.,** Neurochemical correlates of brain-stimulation reward measured by *ex vivo* and *in vivo* analyses, *Neurosci. Biobehav. Rev.,* 13, 99, 1989.
72. **Nakahara, D., Ozaki, N., Kapoor, V., and Nagatsu, T.,** The effect of uptake inhibition on dopamine release from the nucleus accumbens of rats during self- or forced stimulation of the medial forebrain bundle: a microdialysis study, *Neurosci. Lett.,* 104, 136, 1989.
73. **Izenwasser, S. and Kornetsky, C.,** The effect of amfonelic acid or nisoxetine in combination with morphine on brain-stimulation reward. *Pharmacol. Biochem. Behav.,* 32, 983, 1989.
74. **Knapp, C. M. and Kornetsky, C.,** The effects of amfonelic acid alone and in combination with naloxone on brain-stimulation reward. *Pharmacol. Biochem. Behav.,* 32, 977, 1989.
75. **Leith, N. J. and Barrett, R. J.,** Self-stimulation and amphetamine: Tolerance to *d* and *l* isomers and cross tolerance to cocaine and methylphenidate, *Psychopharmacologia,* 74, 23, 1981.
76. **Carey, R. J.,** Rate dependent inhibition of self-stimulation by apomorphine, *Pharmacol. Biochem. Behav.,* 16, 859, 1982.

77. **Creese, I., Hamblin, M. W., Leff, S. E., and Sibley, D. R.,** CNS dopamine receptors, in *Handbook of Psychopharmacology: Biochemical Studies of CNS Receptors,* Iversen, L. L., Iversen, S. D., and Snyder, S. H., Eds., Plenum Press, New York, 1983, 81.
78. **Fouriezos, G., Hansson, P., and Wise, R. A.,** Neuroleptic-induced attenuation of brain-stimulation reward in rats, *J. Comp. Physiol. Psychol.,* 92, 661, 1978.
79. **Fouriezos, G. and Wise, R. A.,** Pimozide-induced extinction of intracranial self-stimulation: response patterns rule out motor or performance deficits, *Brain Res.,* 103, 377, 1976.
80. **Ettenberg, A. and White, N.,** Pimozide attenuates conditioned taste preferences induced by self-stimulation in rats, *Pharmacol. Biochem. Behav.,* 15, 915, 1981.
81. **Gallistel, C. R., Boytim, M., Gomita, Y., and Klebanoff, L.,** Does pimozide block the reinforcing effect of brain stimulation, *Pharmacol. Biochem. Behav.,* 17, 769, 1982.
82. **Stellar, J. R., Kelly, A., and Corbett, D.,** Effects of peripheral and central dopamine blockade on lateral hypothalamic stimulation reward: evidence for both reward and motor deficits, *Pharmacol. Biochem. Behav.,* 18, 433, 1983.
83. **Ettenberg, A.,** Dopamine, neuroleptics and reinforced behavior, *Neurosci. Biobehav. Rev.,* 13, 105, 1989.
84. **Bird, M. and Kornetsky, C.,** Dissociation of the attentional and motivational effects of pimozide on the threshold for rewarding brain stimulation, *Neuropsychopharmacology,* 3, 33, 1990.
85. **Gallistel, C. R. and Davis, A. J.,** Affinity for the dopamine D_2 receptor predicts neuroleptic potency in blocking the reinforcing effect of MFB stimulation, *Pharmacol. Biochem. Behav.,* 19, 867, 1983.
86. **Nakajima, S. and McKenzie, G. M.,** Reduction of the rewarding effect of brain stimulation by a blockade of dopamine D_1 receptor with SCH 23390, *Pharmacol. Biochem. Behav.,* 24, 919, 1986.
87. **Kurumiya, S. and Nakajima, S.,** Dopamine D_1 receptors in the nucleus accumbens: involvement in the reinforcing effect of tegmental stimulation, *Brain Res.,* 448, 1, 1988.
88. **Adams, W. J., Lorens, S. A., and Mitchell, C. L.,** Morphine enhances lateral hypothalamic self-stimulation in the rat, *Proc. Soc. Exp. Biol.,* 140, 770, 1972.
89. **Lorens, S. A. and Mitchell, C. L.,** Influence of morphine on lateral hypothalamic self-stimulation in the rat, *Psychopharmacologia,* 32, 271, 1973.
90. **Lorens, S. A.,** Comparison of the effects of morphine on hypothalamic and medial frontal cortex self-stimulation, *Psychopharmacologia,* 48, 217, 1976.
91. **Esposito, R. U. and Kornetsky, C.,** Morphine lowering of self-stimulation thresholds: lack of tolerance with long-term administration, *Science,* 195, 189, 1977.
92. **Nazzaro, J. M., Seeger, T. F., and Gardner, E. L.,** Morphine differentially affects ventral tegmental and substantia nigra brain reward thresholds, *Pharmacol. Biochem. Behav.,* 14, 325, 1981.
93. **Hubner, C. B. and Kornetsky, C.,** The effects of heroin on brain-stimulation reward, *Soc. Neurosci. Abstr.,* 10, 1106, 1984.
94. **Unterwald, E. M. and Kornetsky, C.,** Effects of nalbuphine alone and in combination with tripelennamine on rewarding brain stimulation thresholds in the rat, *Pharmacol. Biochem. Behav.,* 25, 629, 1986.
95. **Unterwald, E. M. and Kornetsky, C.,** Effects of concomitant pentazocine and tripelennamine on brain-stimulation reward, *Pharmacol. Biochem. Behav.,* 21, 961, 1984.
96. **Hubner, C. B. and Kornetsky, C.,** The reinforcing properties of the mixed agonist-antagonist buprenorphine as assessed by brain-stimulation reward, *Pharmacol. Biochem. Behav.,* 30, 195, 1988.
97. **Hong, J. S., Yang, H. Y. T., Tratta, W., and Costa, E.,** Determination of metenkephalin in discrete regions of rat brain, *Brain Res.,* 134, 383, 1977.
98. **Sar, M., Stumpf, W. E., Miller, R. J., Chang, K.-J., and Cuatrecasas, P.,** Immunohistochemical localization of enkephalin in rat brain and spinal cord, *J. Comp. Neurol.,* 182, 17, 1978.

99. **Johnson, R. P., Sar, M., and Stumpf, W. E.,** A topographic localization of enkephalin on the dopamine neurons of the rat substantia nigra and ventral tegmental area demonstrated by combined histofluorescence-immunocytochemistry, *Brain Res.*, 194, 566, 1980.
100. **Wamsley, J. K., Young, W. S., and Kuhar, M. J.,** Immunohistochemical localization of enkephalin in rat forebrain, *Brain Res.*, 190, 153, 1980.
101. **Pert, C. B., Kuhar, M. J., and Snyder, S. H.,** Opiate receptor: autoradiographic localization in rat brain, *Proc. Natl. Acad. Sci.*, 19, 3729, 1976.
102. **Pollard, H., Llorens-Cortes, C., and Schwartz, J. C.,** Enkephalin receptors on dopaminergic neurons in rat striatum, *Nature*, 268, 745, 1977.
103. **Stilwell, D. J., Levitt, R. A., Horn, C. A., Irvin, M. D., Gross, K., Parsons, D. S., Scott, R. H., and Bradley, E. L.,** Naloxone and shuttlebox self-stimulation in the rat, *Pharmacol. Biochem. Behav.*, 13, 739, 1980.
104. **Perry, W., Esposito, R. U., and Kornetsky, C.,** Effects of chronic naloxone treatment on brain stimulation reward, *Pharmacol. Biochem. Behav.*, 14, 247, 1981.
105. **Seeger, T. F., Carlson, K. R., and Nazzaro, J. M.,** Pentobarbital induces a naloxone-reversible decrease in mesolimbic self-stimulation threshold, *Pharmacol. Biochem. Behav.*, 15, 583, 1981.
106. **Wauquier, A., Niemegeers, C. J. E., and Lal, H.,** Differential antagonism by naloxone of inhibitory effects of haloperidol and morphine on brain self-stimulation, *Psychopharmacologia*, 37, 303, 1974.
107. **Holtzman, S. G.,** Comparison of the effects of morphine, pentazocine, cyclazocine and amphetamine on intracranial self-stimulation in the rat. *Psychopharmacologia*, 46, 223, 1976.
108. **Stapleton, J. M., Merriman, V. J., Coogle, C. L., Gelbard, S. D., and Reid, L. D.,** Naloxone reduces pressing for intracranial stimulation of sites in the periaqueductal gray area, accumbens nucleus, substantia nigra, and lateral hypothalamus, *Physiol. Psychol.*, 7, 427, 1979.
109. **Schaefer, G. J. and Michael, R. P.,** Threshold differences for naloxone and naltrexone in the hypothalamus and midbrain using fixed ratio brain self-stimulation in rats, *Psychopharmacologia*, 74, 17, 1981.
110. **Franklin, K. B. J. and Robertson, A.,** Effects and interactions of naloxone and amphetamine on self-stimulation of the prefrontal cortex and dorsal tegmentum, *Pharmacol. Biochem. Behav.*, 16, 433, 1982.
111. **Pollerberg, G. E., Costa, T., Shearman, G. T., Herz, A., and Reid, L. D.,** Opioid antinociception and positive reinforcement are mediated by different types of opioid receptors, *Life Sci.*, 33, 1549, 1983.
112. **La Gasse, J. M., Wetstein, L. M., Czirr, S. A., and Reid, L. D.,** Morphine and diprenorphine together potentiate responding for positively reinforcing brain stimulation, *Psychobiology*, 15, 261, 1987.
113. **Unterwald, E. M., Sasson, S., and Kornetsky, C.,** Evaluation of supraspinal analgesic activity and abuse liability of EKC, *Eur. J. Pharmacol.*, 133, 275, 1987.
114. **De Witte, P., Heidbreder, C., and Roques, B. P.,** Kelatorphan, a potent enkephalinase inhibitor, and opioid receptor agonists DAGO and DTLET, differentially modulate self-stimulation behaviour depending on the site of administration, *Neuropharmacology*, 28, 667, 1989.
115. **Lorens, S. A. and Sainati, S. M.,** Naloxone blocks the excitatory effect of ethanol and chlordiazepoxide on lateral hypothalamic self-stimulation behavior, *Life Sci.*, 23, 1359, 1978.
116. **Kornetsky, C. and Markowitz, R. A.,** Phencyclidine and naloxone: effects on sensitivity to aversive and rewarding stimulation in the rat, in *PCP-Historical and Current Perspectives (Phencyclidine-Opiate Interactions)*, Domino, E. F., Ed., NPP, Ann Arbor, MI, 1981, 321.

117. **Bannon, M. J., Kelland, M., and Chiodo, L. A.,** Medial forebrain bundle stimulation or D-2 dopamine receptor activation increases preproenkephalin mRNA in rat striatum, *J. Neurochem.*, 52, 859, 1989.
118. **Pradhan, S. N. and Bowling, C.,** Effects of nicotine on self-stimulation in rats, *J. Pharmacol. Exp. Ther.*, 176, 229, 1971.
119. **Lyons, D. J., Bain, G. T., and Kornetsky, C.,** The effects of nicotine on brain-stimulation reward, *Soc. Neurosci. Abstr.*, 14, 765, 1988.
120. **Newman, L. M.,** Effects of cholinergic agonists and antagonists on intracranial self-stimulation behavior in the rat, *J. Comp. Physiol. Pharmacol.*, 79, 394, 1972.
121. **Olds, J. and Domino, E. F.,** Comparison of muscarinic and nicotinic agonists on self-stimulation behavior, *J. Pharmacol. Exp. Ther.*, 166, 189, 1969.
122. **Clarke, P. B. S. and Kumar, R.,** Effects of nicotine and d-amphetamine on intracranial self-stimulation in a shuttle box test in rats, *Psychopharmacologia,* 84, 109, 1984.
123. **Corrigall, W. A. and Coen, K. M.,** Nicotine maintains robust self-administration in rats on a limited access schedule, *Psychopharmacologia,* 99, 473, 1989.
124. **Westfall, T. C.,** Effect of nicotine and other drugs on the release of [^{3}H]-norepinephrine and [^{14}C]-dopamine in rat brain striatum and hypothalamus slices, *Neuropharmacology,* 13, 693, 1974.
125. **Giorguieff, M. F., Le Floc'h, M. L., Glowinski, J., and Besson, M. J.,** Involvement of cholinergic presynaptic receptors of nicotinic and muscarinic types in the control of the spontaneous release of dopamine from striatal dopaminergic terminals in the rat, *J. Pharmacol. Exp. Ther.*, 200, 535, 1977.
126. **Imperato, A., Mulas, A., and di Chiara, G.,** Nicotine preferentially stimulates dopamine release in the limbic system of freely moving rats, *Eur. J. Pharmacol.*, 132, 337, 1986.
127. **Izenwasser, S., Jacocks, H. M., Rosenberger, J. G., and Cox, B. M.,** Nicotine indirectly inhibits [^{3}H]-dopamine uptake at concentrations that do not directly promote [^{3}H]-dopamine release in rat striatum, *J. Neurochem.*, 56, 603, 1991.
128. **Grenhoff, J. and Svensson, T. H.,** Selective stimulation of limbic dopamine activity by nicotine, *Acta. Physiol. Scand.*, 133, 595, 1988.
129. **Domino, E. F. and Olds, M. E.,** Cholinergic inhibition of self-stimulation behavior, *J. Pharmacol. Exp. Ther.*, 164, 202, 1968.
130. **Kofman, O., McGlynn, S. M., Olmstead, M. C., and Yeomans, J. S.,** Differential effects of atropine, procaine and dopamine in the rat ventral tegmentum on lateral hypothalamic rewarding brain stimulation, *Behav. Brain Res.*, 38, 55, 1990.
131. **Giorguieff, M. F., Le Floc'h, M. L., Westfall, T. C., Glowinski, J., and Besson, M. J.,** Nicotinic effect of acetylcholine on the release of newly synthesized [^{3}H]-dopamine in rat striatal slices and cat caudate nucleus, *Brain Res.*, 106, 117, 1976.
132. **Olds, J. and Travis, M. P.,** Effects of chlorpromazine, meprobamate, pentobarbital and morphine on self-stimulation, *J. Pharmacol. Exp. Ther.*, 128, 397, 1960.
133. **Domino, E. F. and Olds, M. E.,** Effects of d-amphetamine, scopolamine, chlordiazepoxide and diphenylhydantoin on self-stimulation behavior and brain acetylcholine, *Psychopharmacologia,* 23, 1, 1972.
134. **Panskepp, J., Gandelman, R., and Trowill, J.,** Modulation of hypothalamic self-stimulation and escape behavior by chlordiazepoxide, *Physiol. Behav.*, 5, 965, 1970.
135. **Herberg, L. J. and Williams, S. F.,** Anti-conflict and depressant effects by GABA agonists and antagonists, benzodiazepines and non-gabaergic anticonvulsants on self-stimulation and locomotor activity, *Pharmacol. Biochem. Behav.*, 19, 625, 1983.
136. **Caudarella, M., Campbell, K. A., and Milgram, N. W.,** Differential effects of diazepam (valium) on brain stimulation reward sites, *Pharmacol. Biochem. Behav.*, 16, 17, 1982.
137. **Ichimaru, Y., Moriyama, M., and Gomita, Y.,** Effects of antianxiety and antipsychotic drugs on DRL responding for brain stimulation, *Life Sci.*, 32, 437, 1983.

138. **Mogenson, G. J.,** Effects of sodium pentobarbital on brain self-stimulation, *J. Comp. Physiol. Psychol.*, 58, 461, 1964.
139. **Kent, E. W. and Fedinets, P.,** Effects of GABA blockade on lateral hypothalamic self-stimulation, *Brain Res.*, 107, 628, 1976.
140. **Nazzaro, J. and Gardner, E. L.,** GABA antagonism lowers self-stimulation thresholds in the ventral tegmental area, *Brain Res.*, 189, 279, 1980.
141. **Porrino, L. J. and Coons, E.,** Effects of GABA receptor blockade on stimulation-induced feeding and self-stimulation, *Pharmacol. Biochem. Behav.*, 12, 125, 1980.
142. **Zarevics, P. and Setler, P. E.,** Effects of GABAergic drugs on brain stimulation reward as assessed by a "threshold" method, *Brain Res.*, 215, 201, 1981.
143. **Kucharski, L. T., Williams, J., and Kornetsky, C.,** The effects of levonantradol on rewarding brain stimulation thresholds in the rat, *Pharmacol. Biochem. Behav.*, 19, 149, 1983.
144. **Gardner, E. L., Paredes, W., Smith, D., Donner, A., Milling, C., Cohen, D., and Morrison, D.,** Facilitation of brain stimulation reward by Δ^9-tetrahydrocannabinol, *Psychopharmacologia*, 96, 142, 1988.
145. **Banerjee, S. P., Snyder, S. H., and Mechoulam, R.,** Cannabinoids: influences on neurotransmitter uptake in rat brain synaptosomes, *J. Pharmacol. Exp. Ther.*, 194, 74, 1975.
146. **Poddar, M. K. and Dewey, W. L.,** Effects of cannabinoids on catecholamine uptake and release in hypothalamic and striatal synaptosomes, *J. Pharmacol. Exp. Ther.*, 214, 63, 1980.
147. **Ng Cheong Ton, J. M., Gerhardt, G. A., Friedemann, M., Etgen, A. M., Rose, G. M., Sharpless, N. S., and Gardner, E. L.,** The effects of Δ^9-tetrahydrocannabinol on potassium-evoked release of dopamine in the rat caudate nucleus: an *in vivo* electrochemical and *in vivo* microdialysis study, *Brain Res.*, 451, 59, 1988.
148. **Chen, J., Paredes, W., and Gardner, E. L.,** *In vivo* brain microdialysis studies of Δ^9-tetrahydrocannabinol on presynaptic dopamine efflux in nucleus accumbens of the Lewis rat, *Soc. Neurosci. Abstr.*, 15, 1096, 1989.
149. **Schaefer, G. J. and Michael, R. P.,** Ethanol and current thresholds for brain self-stimulation in the lateral hypothalamus of the rat, *Alcohol*, 4, 209, 1987.
150. **Kornetsky, C., Bain, G. T., Unterwald, E. M., and Lewis, M. J.,** Brain stimulation reward: effects of ethanol, *Alcoholism: Clin. Exp. Res.*, 12, 609, 1988.
151. **Moolten, M. and Kornetsky, C.,** Oral self-administration of ethanol and not experimenter-administered ethanol facilitates rewarding electrical brain stimulation, *Alcohol*, 7, 221, 1990.
152. **Schaefer, G. J. and Michael, R. P.,** Effects of an alcohol diet on locomotor activity and the acquisition of brain self-stimulation in rats, *Alcohol*, 8, 71, 1991.
153. **Wheeling, H. S., Sasson, S., and Kornetsky, C.,** Tolerance to the effects of morphine on escape from reticular formation stimulation, *Subst. Alcohol Actions (Misuse)*, 2, 107, 1981.
154. **Izenwasser, S. and Kornetsky, C.,** Potentiation of morphine analgesia by d-amphetamine is mediated by norepinephrine and not dopamine, *Pain*, 33, 363, 1988.

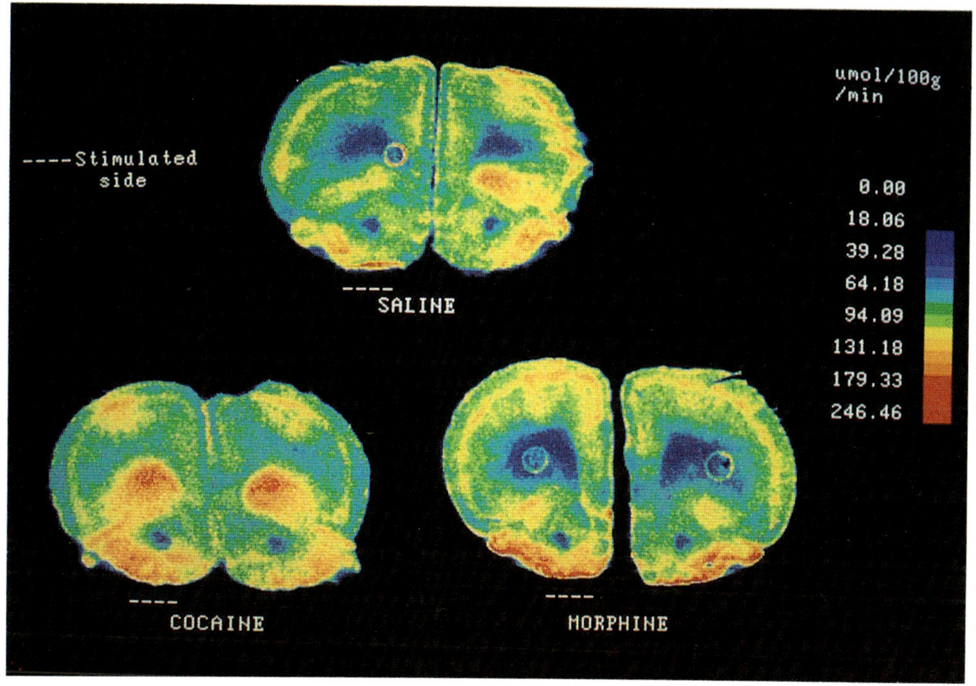

PLATE 1. 2-[^{14}C]Deoxyglucose autoradiograms of brain sections at the level of the olfactory tubercle of rats receiving electrical stimulation to the medial forebrain bundle plus saline, 10 mg/kg cocaine, or 4 mg/kg morphine. Site of stimulation was on the left. Note that the optical density, which is a reflection of the relative rate of glucose utilization, is increased in the olfactory tubercle by stimulation alone and that concomitant administration of either morphine or cocaine leads to further increases in this region. (Modified from Kornetsky, C., Huston-Lyons, D., and Porrino, L. J., *Brain Res.*, 541, 75, 1991.)

Chapter 2

DRUG EFFECTS ON SENSORIMOTOR INTEGRATION IN A MODEL SYSTEM

Anthony G. Romano and John A. Harvey

TABLE OF CONTENTS

I. INTRODUCTION

The nictitating membrane (NM) reflex of the rabbit has proven to be an excellent model system for assessing drug effects on sensory processing, sensorimotor gating, and associative learning.[1,2] Historically, the rabbit NM preparation was first used in behavioral investigations of classical conditioning.[3] Because of the wealth of behavioral data generated and the relative simplicity of the preparation, subsequent studies have used the NM preparation to examine the neurophysiological bases of associative learning (see assorted chapters in Reference 4). The data generated by these studies have provided information concerning the anatomy and physiology both of the NM reflex and of the learned NM response. In the pages to follow, some of this background information will be reviewed, and it will be shown how it has contributed to understanding of drug effects on sensorimotor gating and associative learning.

II. THE NM REFLEX

The anatomy and physiology of the rabbit's NM reflex is relatively simple.[5] Tactile stimulation of the cornea or electrical stimulation of the skin just lateral to the eye, for example, activate inputs to the spinal trigeminal nucleus, pars oralis, from which second-order trigeminal neurons project to retractor bulbi motoneurons in the accessory abducens nucleus either directly or indirectly via the reticular formation.[6-8] Activation of the accessory abducens nucleus leads to contraction of the retractor bulbi muscle via the sixth nerve. Contraction of retractor bulbi pulls the eye back into the orbit, and the force of this mechanical action squeezes the NM out over the cornea. Extension of the NM has been shown to be caused solely by retraction of the eyeball when the reflex is elicited by a puff of air directed at the cornea; a shock-elicited reflex engages retractor bulbi as well as other extraocular muscles and muscles innervated by the facial nerve.[9-11] Retraction of the NM involves both passive and active components.[12,13] The passive component consists of relaxation of the retractor bulbi muscle, thus allowing the eyeball to return to its resting position in the orbit and forcing a return of the NM into the nasal canthus. At the same time, motoneurons in the vicinity of the oculomotor nucleus are activated, causing contraction of the levator palpebrae muscle. A portion of this muscle inserts directly on the NM. Thus, contraction of the levator palpebrae muscle facilitates the return of the NM into the nasal canthus. Autonomic inputs from the superior cervical ganglion may also contribute to NM retraction,[14] but the autonomic influence in the rabbit is minor in comparison with the autonomic influence in the cat.[15]

A. CLASSICAL CONDITIONING OF THE NM RESPONSE

Two typical classical conditioning procedures are diagrammed in Figure 1.

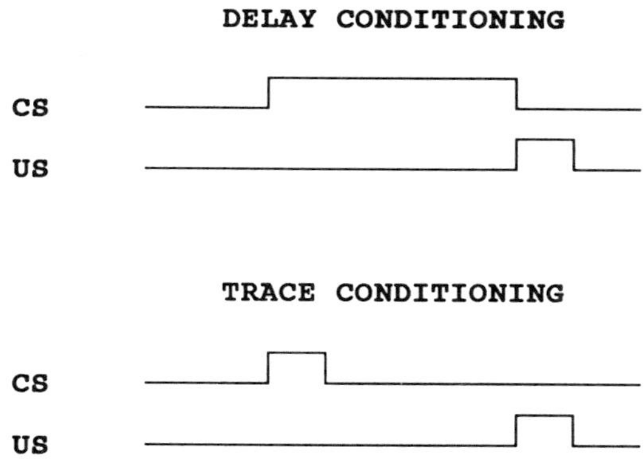

FIGURE 1. Schematic representation of the temporal relationship between the CS and US for delay vs. trace conditioning procedures.

One stimulus, the conditioned stimulus (CS), is paired with another stimulus, the unconditioned stimulus (US), in a strict temporal order, and changes in the NM response system are monitored. The US is a motivationally significant event and reliably elicits an unconditioned response (UR), i.e., the NM reflex. By contrast, the CS is motivationally neutral and thus should elicit only random and infrequent responses. As a consequence of repeated CS-US pairings, the CS comes to elicit an NM response that anticipates the US. This response is known as a conditioned response (CR), and, because it occurs in the same response system activated by the US, it often resembles the UR.

The two conditioning procedures diagrammed in Figure 1 differ both in their implementation and in their effects on the rate of learning. In delay conditioning, the CS is continuously present until the occurrence of the US and often overlaps the US. The delay conditioning procedure tends to be used far more often than the trace conditioning procedure. In trace conditioning, a temporal gap separates the offset of the CS from the onset of the US. The delay procedure produces a faster rate of learning and higher asymptotic level of conditioned responding (CR) than does the trace procedure.[16]

The rate and degree of learning may be indexed in several ways. For the NM response, most investigators choose to measure the frequency of CRs and express that frequency as a percentage. However, both the latency and amplitude of the CR have also been used as measures of learning. All three measures are highly correlated with one another and show reliable changes as a function of CS-US pairings. Thus, CR frequency increases and CR latency decreases over the course of training. If CS-alone test trials are interspersed among CS-US trials, the complete topography of the CR, including CR amplitude, can be measured in the absence of contamination by the UR. The change in CR amplitude mirrors the change in CR frequency; both increase over the course of training.

The classical conditioning procedure introduced by Pavlov is often confused with other associative learning paradigms. Such confusion stems from a focus on only one aspect of Pavlov's procedure, pairing the CS with the US and recording changes in the response to the CS. There are two aspects of Pavlov's procedure that are often overlooked. First, the US reliably elicited a quantifiable reflex response in one or more motor systems. Second, after repeated CS-US pairings, the change in response to the CS also occurred in one or more of the motor systems activated by the US. Thus, the CR and the UR shared a common response system.

There are clear advantages in requiring that the CR and UR be measured in a common response system, especially in the context of drug effects. The most important advantage arises out of the distinction between learning effects and performance effects. Because drugs and other manipulations can have global effects on behavior, an alteration in the frequency or topography of the CR can be attributed either to an effect on the inferred learning process or to an effect on the performance of the NM response. If the performance of the NM response, i.e., the UR, is affected, then obviously an uncontaminated drug effect on learning as reflected by alterations in the CR cannot be ascertained.

B. NONASSOCIATIVE DETERMINANTS OF RESPONDING

Even under the best circumstances, not all NM responses that anticipate the US can be considered to be CRs. Most response systems, including the NM system, show some baseline rate of responding. Some of these baseline responses will happen to fall within the CS-US interval and will be falsely identified as CRs. Other sources of nonassociative responding may also contribute to an overestimate of the frequency of CRs. These nonassociative responses include alpha responses and pseudo-CRs.

Alpha responses are reflexively elicited by the CS in the target response system under investigation. Although the alpha response can be completely habituated by repeated CS-alone presentations, subsequent US presentations may reinstate and augment the alpha response, a process called *sensitization*. It should be noted that there is very little evidence that alpha responding occurs in the NM response system.

A pseudo-CR is a response to an originally neutral CS that occurs as the result of prior presentations of the US. The mechanism underlying pseudoconditioning is unclear, but it appears to involve generalization from the US to the CS. Although nonassociative responses cannot be completely eliminated, an estimate of their frequency, relative to the frequency of CRs, may be obtained by employing a control group that receives only unpaired presentations of the CS and US. A summative measure of the frequency of all possible nonassociative responses may be obtained by examining responding during the unpaired CS presentations.

The use of such a control group can aid in determining the behavioral mechanism(s) whereby a drug apparently affects associative learning. For example, the hallucinogens d-lysergic acid diethylamide (LSD), d,1-2,5-dimethoxy-4-methylamphetamine (DOM), and methylenedioxyamphetamine (MDA) all produce an enhancement in the rate of CR acquisition.[17,18] However, in addition to enhancing CR acquisition, DOM produces an increase in nonassociative responding to the CS during unpaired presentations of the CS and US, as LSD and MDA do not. Thus, DOM's enhancement of NM responding during paired CS-US presentations cannot be attributed to an effect solely on associative learning.

C. REFLEX MODIFICATION

Before the detailed discussion of drug effects on classical conditioning of the NM response that will follow, a consideration of unlearned phenomena using the NM preparation is in order. These phenomena are best classified as instances of sensorimotor gating. If a reflex-eliciting stimulus such as a corneal air puff or a periorbital shock is preceded by an auditory stimulus, the amplitude and latency of the NM reflex may be altered.[19-21] This alteration in reflex topography by a prior stimulus event, termed reflex modification, has been studied in a number of reflex systems, including the acoustic startle reflex of the rat, the human knee jerk and eyeblink reflexes, and the light-elicited startle reflex of the pigeon (see Reference 22 for a review).

Because the reflex modification paradigm used with the rabbit NM preparation is formally similar to the classical conditioning paradigm, the auditory stimulus will be referred to as a conditioned stimulus (CS) and the reflex-eliciting stimulus as an unconditioned stimulus (US). It should be noted, moreover, that, despite the formal similarity between the two paradigms, a key feature of reflex modification is that it occurs prior to any evidence of learning and is present during the first pairing of the CS and US.[19]

The direction and magnitude of NM reflex modification varies as a function of CS-US interval, CS intensity, and US intensity.[19] The CS-US interval has been the variable most commonly manipulated. In general, at CS-US intervals less than about 80 msec, the amplitude of the NM reflex is attenuated. This reflex inhibition is replaced by reflex facilitation at longer CS-US intervals during which the amplitude of the NM reflex is greatly augmented. Reflex facilitation has a fairly long time-course. Other things being equal, reflex facilitation can be obtained at CS-US intervals ranging from 100 to 8000 msec.[19,20] The magnitude of reflex facilitation at a given CS-US interval can be altered by manipulating either CS intensity or US intensity. Weaker CSs produce less facilitation than stronger CSs at all CS-US intervals (see Figure 1 in Reference 19). By contrast, weaker USs tend to produce more facilitation than stronger USs (see Figure 4 in Reference 19). The latter phenomenon is probably due to a ceiling effect, for a strong US by itself will elicit a near maximal reflex that leaves little room for further increases in amplitude when that US is preceded by a reflex facilitating CS.

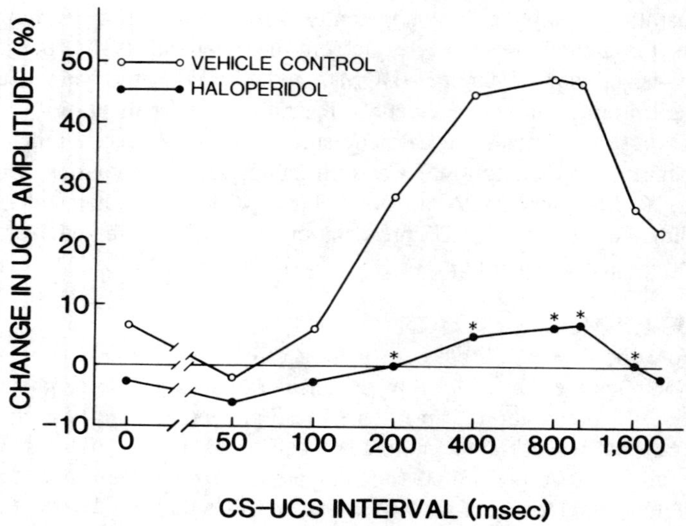

FIGURE 2. Effect of haloperidol (250 μg/kg iv) on modification of the NM reflex by an auditory stimulus at various CS-US (tone-shock) intervals. For intervals greater than 800 msec, the trace conditioning procedure was used. The percentage change in reflex amplitude is relative to reflex amplitude during shock-alone presentations. (From Harvey, J. A. and Gormezano, I., *J. Pharmacol. Exp. Ther.*, 218, 712, 1981. With permission.)

III. DRUG EFFECTS ON NM REFLEX FACILITATION

Several drugs have been shown to affect both the magnitude and the time-course of NM reflex facilitation. Parenthetically, it should be noted that drug effects on reflex modification of the acoustic startle reflex of the rat have been the subject of several recent investigations by Davis, Mansbach, and Swerdlow and their colleagues (see, for example, References 23-25). Their studies are beyond the scope of the present chapter but should be consulted by those interested in using modification of the acoustic startle response of the rat as a model for assessing drug effects on sensorimotor gating.

Harvey and Gormezano[26] examined the effects of haloperidol (250 μg/kg i.v.) on reflex facilitation at nine CS-US intervals ranging from 0 to 2000 ms, measured from CS onset to US onset. The reflex-eliciting stimulus was a 100-ms, 1-mA periorbital shock. The CS was a 90-dB 800-ms tone. Each block of ten trials consisted of one US alone and the nine different CS-US intervals presented in random order. This admixture of CS-US intervals within subjects precluded the development of any overt signs of learning even though more than 1700 CS-US pairings had been given. The reflex modification function is shown in Figure 2. At the shortest CS-US intervals (0, 50, and 100 ms), control animals and haloperidol-injected animals showed essentially identical reflex amplitudes and some evidence of reflex inhibition at the 50-ms interval. Reflex facilitation was evident in control animals beginning at 100 ms and increased to asymptotic levels over the next 300 ms. Facilitation began to decline in control animals at 1000 msec although significant facilitation was still evident at the 2000-msec CS-US interval. Haloperidol-injected animals showed a significant impairment in reflex facilitation at all CS-US intervals. The decline in the magnitude of reflex facilitation in haloperidol-injected animals was accompanied by a shortening of the time-course; facilitation was evident only between 400 and 1000 msec following CS onset. Although reflex facilitation was depressed following haloperidol treatment, the amplitude of the reflex in the absence of the CS was unaffected. As mentioned above, an increase in US intensity will also dampen the reflex facilitation function.[19] However, that effect is always accompanied by an increase in the baseline amplitude of the reflex during presentations of US alone. Thus, haloperidol's effects on reflex facilitation are not equivalent to an increase in US intensity. Haloperidol's effects more closely resemble the effect of decreasing CS intensity; less facilitation was obtained at all CS-US intervals, but baseline amplitude was unchanged.

Further evidence that haloperidol affected reflex facilitation by altering sensory processing of the CS comes from classical conditioning studies. Harvey and Gormezano[26] also reported that haloperidol retarded CR acquisition when either a tone or a flashing light was used as the CS. In a follow-up experiment, drug-naive animals were first given 10 d of acquisition training to a 90-dB tone CS. The CS-intensity threshold for eliciting CRs was determined in subsequent test sessions when animals were injected with either vehicle or haloperidol. Significantly fewer CRs were observed under haloperidol at CS intensities between 45 and 65 dB. Haloperidol produced an increase in the CS-intensity threshold of about 12 dB. A similar CS-intensity effect produced by haloperidol has recently been reported by Sears and Steinmetz.[27]

IV. DRUG EFFECTS ON NM REFLEX FACILITATION AND CR ACQUISITION

As mentioned above, the CS-US interval is a potent determinant of the magnitude of reflex facilitation. In the rabbit NM preparation, the CS-US interval is also a potent determinant of the rate of CR acquisition.[16,28,29] Several investigators have noted that there is a close relationship between the magnitude of reflex facilitation and the rate of CR acquisition as a function of CS-US interval.[19,21,30] However, Weisz and Walts[20] recently argued that the close relationship between the two phenomena exists only when a relatively brief (100 to 125 ms) auditory stimulus is used. In other words, the two phenomena are closely related only when trace conditioning procedures are used.

Harvey et al.[30] directly examined the relationship between reflex facili-

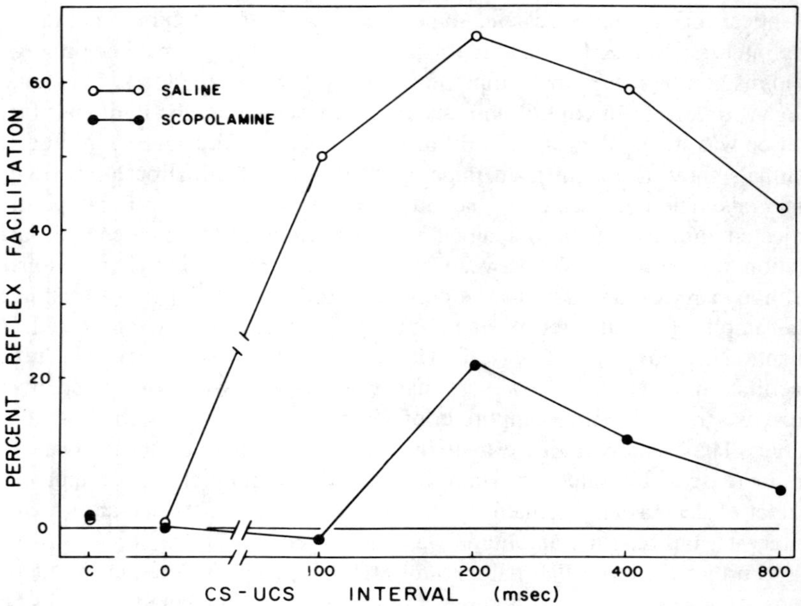

FIGURE 3. Effect of scopolamine (0.4 mg/kg iv) on tone-induced NM reflex facilitation. Separate groups of animals were trained at each of the CS-US intervals. Trace conditioning procedures were used beginning with the 200-ms CS-US interval. (From Harvey, J. A. et al., *J. Neurosci.*, 5, 596, 1985. With permission.)

tation and CR acquisition using trace conditioning procedures. Various CS-US intervals were examined in scopolamine- vs. vehicle-injected animals. Separate groups of animals were given pairings of a 100-ms tone CS with a periorbital shock US at CS-US intervals of 0, 100, 200, 400, or 800 ms. Shock-alone trials were interspersed among the CS-US trials in order to assess the baseline amplitude of the reflex. Half the animals in each group were given either scopolamine (0.4 mg/kg) or saline injections prior to each of ten daily sessions. Reflex facilitation was measured on the first day of training when there was no evidence of conditioned responding. The reflex facilitation data are shown in Figure 3. For vehicle-injected animals, the relationship between magnitude of reflex facilitation and CS-US interval took the form of an inverted U-shaped function. Reflex facilitation was absent at the 0-ms interval, but it increased over time, reached a peak at the 200-ms CS-US interval, and then declined over the longer intervals. A similar shape for the reflex facilitation function was seen in scopolamine-injected animals. However, scopolamine significantly reduced the magnitude of reflex facilitation and shortened its time-course. Facilitation was maximal at the 200-ms CS-US interval, but the reflex was significantly less than in the saline controls. Furthermore, little or no facilitation was observed at the 100- or 800-ms

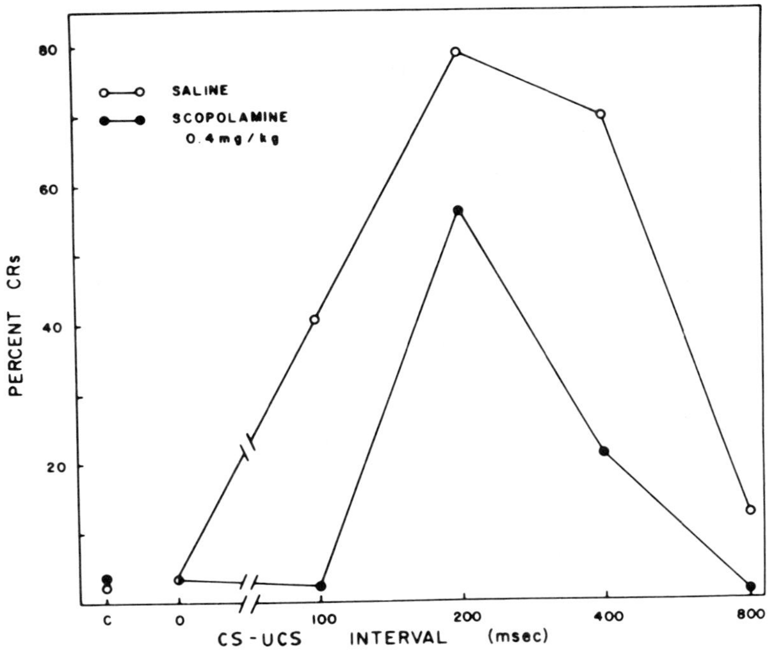

FIGURE 4. Mean percentages of CRs as a function of CS-US interval for the saline and scopolamine treated groups depicted in Figure 3. (From Harvey, J. A. et al., *J. Neurosci.*, 5, 596, 1985. With permission.)

intervals. Finally, scopolamine had no effect on the amplitude of the reflex on shock-alone trials, indicating that scopolamine's effect on reflex facilitation was probably owing to an alteration in CS processing.

CR began to emerge during the remaining training sessions. Regardless of drug condition, the function relating the percentage of CRs to the CS-US interval was strikingly similar to the reflex facilitation function. As shown in Figure 4, both scopolamine- and saline-injected animals showed an inverted U-shaped function with the percentage of CRs reaching a maximum at the 200-ms CS-US interval. Furthermore, scopolamine significantly impaired the acquisition of conditioned responding at all CS-US intervals. Harvey et al.[30] determined the correlation between magnitude of reflex facilitation and the subsequent percentage of CRs within each drug treatment. For saline-treated animals, the correlation was +0.86; for scopolamine-treated animals, the correlation was +0.95.

A number of other studies have also shown that scopolamine retards the acquisition of rabbit NM conditioning.[31-34] In two of those studies, the CS intensity threshold for eliciting CRs was evaluated in scopolamine- vs. vehicle-treated animals. Both Harvey et al.[31] and Moore et al.[32] found a significant

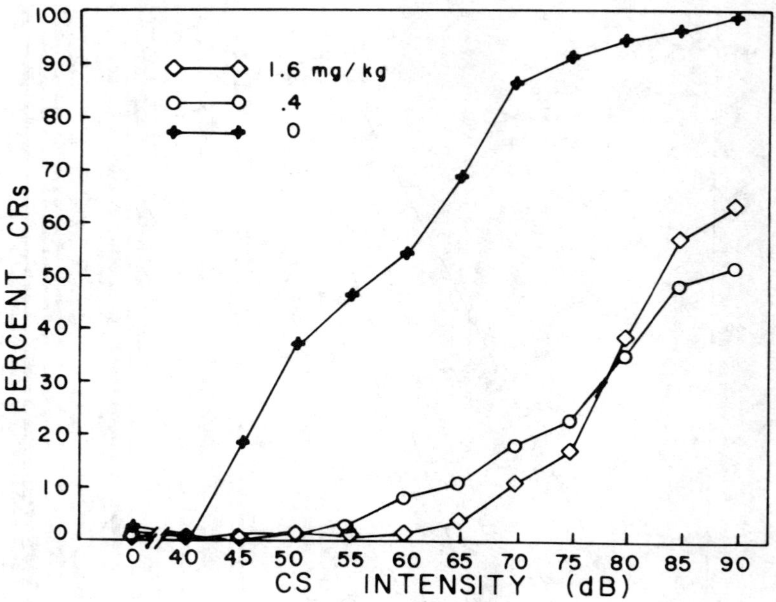

FIGURE 5. Mean percentages of CRs for saline- and scopolamine-treated animals as a function of CS intensity. All animals had received 10 d of acquisition training to the 90-dB tone CS at a time during which no drug or vehicle was injected. (From Harvey, J. A. et al., *J. Pharmacol. Exp. Ther.*, 225, 42, 1983. With permission.)

elevation in the CS intensity threshold following scopolamine treatment. The data shown in Figure 5 are reproduced from Harvey et al.[31] Note that the 0.4 mg/kg dose of scopolamine is the same as that used in the reflex facilitation study. Note also that scopolamine depressed the frequency of CRs at virtually every CS intensity. Thus, scopolamine is similar to haloperidol in that both drugs depress reflex facilitation and retard CR acquisition by decreasing sensory processing of the CS.

Although the magnitude of reflex facilitation across CS-US intervals is a good predictor of the rate of CR acquisition within a drug treatment, it is not always a good predictor of CR acquisition between drug treatments. LSD has been shown to enhance the rate of CR acquisition and decrease the CS intensity threshold for eliciting CRs.[35,36] One would expect, therefore, that reflex facilitation would also be enhanced under LSD. However, Harvey et al.[37] found that this was not the case. The experimental design was similar to the one used to examine the reflex-facilitation — CR-acquisition function under scopolamine. Separate groups of animals were trained at CS-US intervals of 0, 100, 200, 400, and 800 ms. Each group was subdivided into LSD-treated (12.9 μg/kg) animals and vehicle-treated animals. Reflex facilitation was determined during the first training session, prior to the occurrence

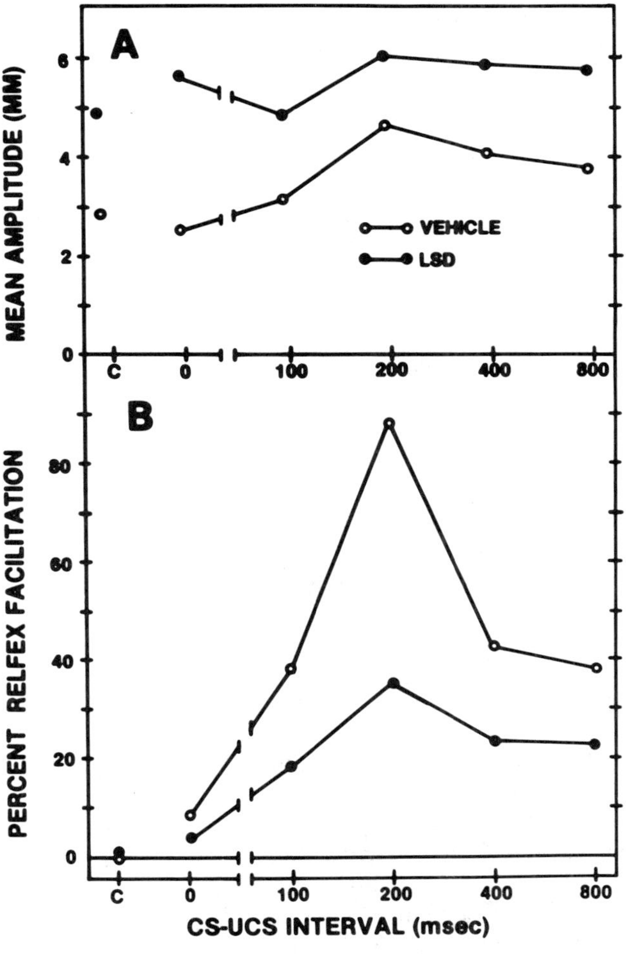

FIGURE 6. Effect of LSD (12.9 μg/kg iv) on tone-induced NM reflex facilitation. Separate groups of animals were trained at each of the CS-US intervals. Trace conditioning procedures were used beginning with the 200-ms CS-US interval. Panel A shows the mean amplitude of the reflex on shock-alone trials (points above C) and on CS-US trials; panel B depicts NM reflex amplitude on CS-US trials as a percentage change from reflex amplitude on shock-alone trials. (From Harvey, J. A. et al., *Pharmacol. Biochem. Behav.*, 30, 427, 1988. With permission.)

of CR acquisition. The results are shown in Figure 6. Both treatments yielded an inverted U-shaped function relating reflex facilitation to CS-US interval with a peak at 200 ms. However, as Panel B indicates, the percentage of reflex facilitation was significantly depressed in LSD-treated animals. This depression was attributed to a decrease in the intensity threshold for eliciting

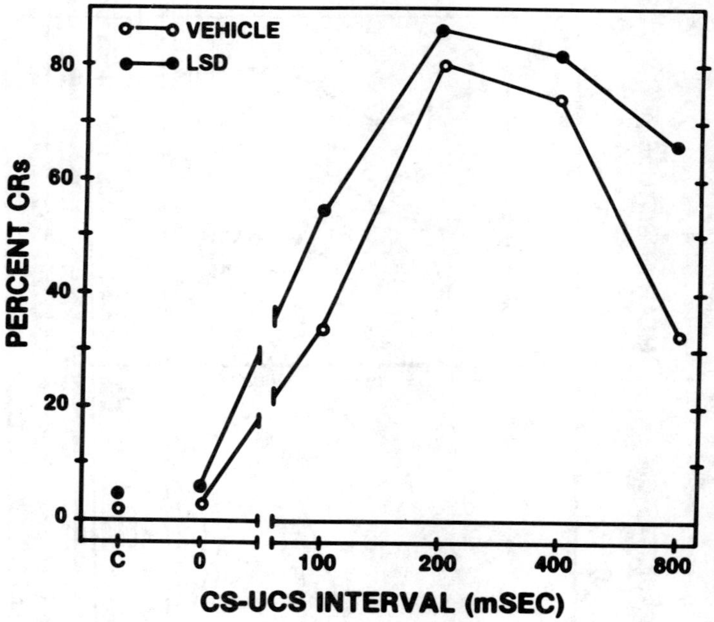

FIGURE 7. Mean percentages of CRs as a function of CS-US interval for the vehicle- and LSD-treated groups depicted in Figure 6. (From Harvey, J. A. et al., *Pharmacol. Biochem. Behav.*, 30, 427, 1988. With permission.)

the reflex following LSD treatment. As mentioned above, strong USs produce less facilitation than weak USs.[19] Thus, LSD's effects on reflex facilitation were equivalent to an increase in US intensity. This is most clearly illustrated in Panel A of Figure 6. The baseline amplitude of the reflex on shock-alone trials was approximately 2 mm higher for LSD-treated animals.

Although reflex facilitation was depressed by LSD, Figure 7 shows that CR acquisition was enhanced. This last result can be attributed to an effective increase in US intensity coupled with an effective increase in CS intensity. In untreated animals, the rate of CR acquisition is positively correlated with increases both in US intensity and in CS intensity.[28,38] Finally, within treatments, the magnitude of reflex facilitation was highly correlated with the subsequent percentage of CRs; LSD-treated animals showed a correlation of +0.94, and vehicle-treated animals showed a correlation of +0.85. Thus, when trace conditioning procedures are employed, the magnitude of reflex facilitation over a range of CS-US intervals is highly predictive of the subsequent rate of CR acquisition. Moreover, this relationship is little affected by drug treatments even though such treatments may have apparently opposing effects on reflex facilitation and rate of learning.

In summary, both haloperidol and scopolamine have been shown to retard

CR acquisition, decrease tone-induced facilitation of the NM reflex, and increase the CS-intensity threshold for eliciting CRs. By contrast, LSD has been shown to enhance CR acquisition, decrease tone-induced facilitation of the NM reflex, and decrease the CS-intensity and US-intensity thresholds for eliciting CRs and URs, respectively. These findings have led to the view that many drugs alter the ability of stimuli to activate premotor blink areas of the reticular formation and thus block or enhance the ability of such stimuli to enter into associative learning and, once learning has occurred, to block the ability of the CS to elicit CRs.[1]

V. SUMMARY

As the studies cited indicate, the rabbit NM preparation has proven to be an extremely sensitive model for assessing drug effects on sensorimotor integration as reflected by reflex facilitation and CR acquisition. Although this discussion has been restricted to only two phenomena arising from the use of the NM preparation in the rabbit, the preparation has also proven to be useful in investigations of drug effects on a variety of other "higher-order" learning phenomena. An extensive list of these phenomena and associated drug effects, including the effects of drugs of abuse, may be found in Schindler and Harvey.[2]

ACKNOWLEDGMENT

Preparation of this manuscript was supported by grant DA04944 from the National Institute on Drug Abuse.

REFERENCES

1. **Harvey, J. A.,** Effects of drugs on associative learning, in *Psychopharmacology: The Third Generation of Progress,* Meltzer, H. Y., Ed., Raven Press, New York, 1987, 1485.
2. **Schindler, C. W. and Harvey, J. A.,** Use of classical conditioning procedures in behavioral pharmacology, *Drug Dev. Res.,* 20, 169, 1990.
3. **Gormezano, I., Kehoe, E. J., and Marshall, B. S.,** Twenty years of classical conditioning research with the rabbit, in *Progress in Psychobiology and Physiological Psychology,* Sprague, J. M. and Epstein, A. N., Eds., Academic Press, New York, 1983, 197.
4. **Gormezano, I., Prokasy, W. F., and Thompson, R. F.,** *Classical Conditioning,* 3rd ed., Lawrence Erlbaum Associates, Hillsdale, NJ, 1987.
5. **Prince, J. H.,** *The Rabbit in Eye Research,* Charles C Thomas, Springfield, IL, 1964.
6. **Harvey, J. A., Land, T., and McMaster, S. E.,** Anatomical study of the rabbit's corneal-VIth nerve reflex: connections between cornea, trigeminal sensory complex, and the abducens and accessory abducens nuclei, *Brain Res.,* 301, 307, 1984.

7. **Holstege, G., Tan, J., van Hamm, J. J., and Graveland, G. A.,** Anatomical observations on the afferent projections to the retractor bulbi motoneuronal cell group and other pathways possibly related to the blink reflex in the cat, *Brain Res.*, 374, 321, 1986.
8. **Guégan, M. and Horcholle-Bossavit, G.,** Reflex control of the retractor bulbi muscle in the cat, *Pflügers Archiv.*, 389, 143, 1981.
9. **Berthier, N. E.,** The role of the extraocular muscles in the rabbit nictitating membrane response: a reexamination, *Behav. Brain Res.*, 14, 81, 1984.
10. **Berthier, N. E. and Moore, J. W.,** Role of extraocular muscles in the rabbit *(Oryctolagus cuniculus)* nictitating membrane response, *Physiol. Behav.*, 24, 931, 1980.
11. **Marek, G. J., McMaster, S. E., Gormezano, I., and Harvey, J. A.,** The role of the accessory abducens nucleus in the rabbit nictitating membrane response, *Brain Res.*, 299, 215, 1984.
12. **Coburn, K. L.,** Separate channels mediate prestimulus facilitation and suppression of rabbit nictitating membrane reflex, *Behav. Neurosci.*, 100, 576, 1986.
13. **Harrison, T. A. and Cegavske, C. F.,** Role of the levator palpebrae superioris (LPS) muscle in effecting nictitating membrane movement in the rabbit, *Physiol. Behav.*, 26, 159, 1981.
14. **Jacobowitz, D., Johnson, P., Kitchner, I., and Koelle, G. B.,** The effect of hemicholinium (HC-3) on sympathetic transmission at the nictitating membrane of the rabbit, *Br. J. Pharmacol.*, 25, 527, 1965.
15. **Cegavske, C. F., Thompson, R. F., Patterson, M. M., and Gormezano, I.,** Mechanisms of efferent neuronal control of the reflex nictitating membrane response in rabbit *(Oryctolagus cuniculus), J. Comp. Physiol. Psychol.*, 90, 411, 1976.
16. **Schneiderman, N.,** Interstimulus interval function of the nictitating membrane response of the rabbit under delay versus trace conditioning, *J. Comp. Physiol. Psychol.*, 62, 397, 1966.
17. **Harvey, J. A., Gormezano, I., and Cool, V.,** Effects of d-lysergic acid diethylamide, d-2-bromolysergic acid diethylamide, dl-2,5-dimethoxy-4-methylamphetamine and d-amphetamine on classical conditioning of the rabbit nictitating membrane response, *J. Pharmacol. Exp. Ther.*, 221, 289, 1982.
18. **Romano, A. G., Bormann, N. M., and Harvey, J. A.,** A unique enhancement of associative learning produced by methylenedioxyamphetamine, *Behav. Pharmacol.*, 2, 225, 1991.
19. **Ison, J. R. and Leonard, D. W.,** Effects of auditory stimuli on the amplitude of the nictitating membrane reflex of the rabbit *(Oryctolagus cuniculus), J. Comp. Physiol. Psychol.*, 75, 157, 1971.
20. **Weisz, D. J. and Walts, C.,** Reflex facilitation of the rabbit nictitating membrane response by an auditory stimulus as a function of interstimulus interval, *Behav. Neurosci.*, 104, 11, 1990.
21. **Young, R. A., Cegavske, C. F., and Thompson, R. F.,** Tone-induced changes in excitability of abducens motoneurons and of the reflex path of nictitating membrane response in rabbit *(Oryctolagus cuniculus), J. Comp. Physiol. Psychol.*, 90, 424, 1976.
22. **Hoffman, H. S. and Ison, J. R.,** Reflex modification in the domain of startle. I. Some empirical findings and their implications for how the nervous system processes sensory input, *Psychol. Rev.*, 87, 175, 1980.
23. **Davis, M., Mansbach, R. S., Swerdlow, N. R., Campeau, S., Braff, D. L., and Geyer, M. A.,** Apomorphine disrupts the inhibition of acoustic startle induced by weak prepulses in rats, *Psychopharmacology,* 102, 1, 1990.
24. **Mansbach, R. S. and Geyer, M. A.,** Effects of phencyclidine and phencyclidine biologs on sensorimotor gating in the rat, *Neuropsychopharmacology,* 2, 299, 1989.
25. **Swerdlow, N. R., Mansbach, R. S., Geyer, M. A., Pulvirenti, L., Koob, G. F., and Braff, D. L.,** Amphetamine disruption of prepulse inhibition of acoustic startle is reversed by depletion of mesolimbic dopamine, *Psychopharmacology,* 100, 413, 1990.

26. **Harvey, J. A. and Gormezano, I.,** Effects of haloperidol and pimozide on classical conditioning of the rabbit nictitating membrane response, *J. Pharmacol. Exp. Ther.*, 218, 712, 1981.
27. **Sears, L. L. and Steinmetz, J. E.,** Haloperidol impairs classically conditioned nictitating membrane responses and conditioning-related cerebellar interpositus nucleus activity in rabbits, *Pharmacol. Biochem. Behav.*, 36, 821, 1990.
28. **Smith, M. C.,** CS-US interval and US intensity in classical conditioning of the rabbit's nictitating membrane response, *J. Comp. Physiol. Psychol.*, 66, 679, 1968.
29. **Smith, M. C., Coleman, S. R., and Gormezano, I.,** Classical conditioning of the rabbit's nictitating membrane response at backward, simultaneous, and forward CS-US intervals, *J. Comp. Physiol. Psychol.*, 69, 226, 1969.
30. **Harvey, J. A., Gormezano, I., and Cool-Hauser, V. A.,** Relationship between heterosynaptic reflex facilitation and acquisition of the nictitating membrane response in control and scopolamine-injected rabbits, *J. Neurosci.*, 5, 596, 1985.
31. **Harvey, J. A., Gormezano, I., and Cool-Hauser, V. A.,** Effects of scopolamine and methylscopolamine on classical conditioning of the rabbit nictitating membrane response, *J. Pharmacol. Exp. Ther.*, 225, 42, 1983.
32. **Moore, J. W., Goodell, N. A., and Solomon, P. R.,** Central cholinergic blockade by scopolamine and habituation, classical conditioning, and latent inhibition of the rabbit's nictitating membrane response, *Physiol. Psychol.*, 4, 395, 1976.
33. **Salvatierra, A. T. and Berry, S. D.,** Scopolamine disruption of septo-hippocampal activity and classical conditioning, *Behav. Neurosci.*, 103, 715, 1989.
34. **Solomon, P. R., Solomon, S. E., Vander Schaaf, E., and Perry, H. E.,** Altered activity in the hippocampus is more detrimental to classical conditioning than removing the structure, *Science*, 220, 329, 1983.
35. **Gormezano, I. and Harvey, J. A.,** Sensory and associative effects of LSD in classical conditioning of rabbit *(Oryctolagus cuniculus)* nictitating membrane response, *J. Comp. Physiol. Psychol.*, 94, 641, 1980.
36. **Siegel, S. and Freedman, D. X.,** Effects of LSD-25 on classical trace conditioning, *Pharmacol. Biochem. Behav.*, 30, 427, 1988.
37. **Harvey, J. A., Gormezano, I., Cool-Hauser, V. A., and Schindler, C. W.,** Effects of LSD on classical conditioning as a function of CS-UCS interval: relationship to reflex facilitation, *Pharmacol. Biochem. Behav.*, 30, 433, 1988.
38. **Scavio, M. J. and Gormezano, I.,** CS intensity effects on rabbit nictitating membrane conditioning, extinction and generalization, *Pavlovian J. Biol. Sci.*, 9, 25, 1974.

Chapter 3

SECOND MESSENGERS IN THE TRANSDUCTION OF MULTIPLE OPIOID RECEPTOR SIGNALS

P. Y. Law and H. H. Loh

TABLE OF CONTENTS

I. INTRODUCTION

Since the initial identification of stereoselective opioid receptors in neural tissues[1-3] and the subsequent demonstration of the existence of multiple opioid receptor types (see reviews in References 4 to 6), the cellular events associated with activation of the receptors have been under intensive investigation. Although the commonly observed primary effects of opioid agonist on sensitive neurons are inhibition of electrical excitability and reduction in transmitter release, elucidation of cellular machineries involved in the expression of these opioid effects has been difficult. The presence of multiple opioid peptide gene products, even in model systems such as clonal cell lines, in addition to multiple opioid receptors, has made studies of the molecular mechanisms of opioid actions a continuing challenge.

The task could be simplified if the cellular effectors regulated by various opioid receptor types were identified. Given the advent of molecular cloning techniques, cloning of the opioid receptor and subsequent protein sequence deduction from the nucleotide sequence might illuminate the second messengers that could be regulated by opioid receptors. In this connection it should be noted that, with the exception of the IGF-II receptor,[7,8] all known membrane receptors whose functions are transduced by the family of G-proteins have seven α-helical structures that putatively span the membrane and form the ligand-binding domain (see reviews in References 9 to 12).

Attempts to clone the various opioid receptor types have yet to yield a clear picture of the functions of the receptors, however. An opioid ligand-binding protein that has been cloned by Loh and colleagues[13] has no obvious hydrophobic domain for membrane anchoring, and the deduced protein sequence suggests homology with the neural cell adhesion molecule (NCAM) and belongs to the immunoglobulin superfamily of proteins. Which second messenger system is transducing the signal of this protein remains to be investigated. At present, therefore, which effector is involved in a specific opioid receptor function can be deduced only from biochemical and electrophysiological studies. Therefore, the question of which effector system is participating in the cellular expression of a specific opioid receptor type will be addressed in this chapter. Whether or not the interaction between the receptor and effector is direct will be discussed.

II. THE ROLE OF G-PROTEINS IN OPIOID RECEPTOR FUNCTIONS

Findings of many biochemical studies support the view that multiple opioid receptors mediate their functions via a family of GTP-binding proteins, G-proteins. This conclusion is based on the observation that opioid receptor-agonist interactions follow the diagnostic tests for the involvement of G-proteins in the transduction of receptor signals, as outlined by Birnbaumer et al.[14] The main features of the tests are (1) receptor stimulation of effector functions is dependent on GTP or GTP analogues; (2) agonists stimulate a low K_m GTPase; (3) agonists stimulate the release of GDP which originates from the GTP molecule prebound to the membranes; (4) agonists stimulate the binding of GTP analogues and activate the G-proteins; and (5) agonist affinity for the receptor is decreased in the presence of GTP, GDP or their metabolically stable analogues, and the coupling of the receptor with the G-protein would generate a high affinity state for the agonist. For various multiple opioid receptor types, μ-, δ-, and κ-opioid receptors, agonist-receptor interactions fulfill all or partially the criteria of the diagnostic tests. A detailed examination will be made of evidence in support of the view that G-proteins are the transducer of opioid receptors, but first an examination will be made of the basic mechanism of G-protein action which is the foundation of the diagnostic tests.

A. G-PROTEIN MECHANISMS

It is not necessary here to discuss the function of G-proteins in detail because there are numerous excellent review articles that summarize findings in the field of G-proteins and discuss the roles of various G-proteins in the signal transduction process.[14-18] However, the properties of G-proteins that give rise to the criteria of the diagnostic tests will be summarized.

From molecular cloning and protein purification studies, the structures of G-proteins involved in signal transduction have been elucidated. Basically, these signal-transduction G-proteins belong to the family of proteins that propagates signals by GDP/GTP exchange and that terminates signals by the activation of endogenous GTPases.[18] Typically, the signal-transduction G-protein consists of an α-subunit with molecular mass of 39 to 50 kDa and possessing the GTPase, a β-subunit of 35 to 36 kDa and a γ-subunit of 6 to 10 kDa. There are three different forms of β-subunits[19-24] and two or three different forms of γ-subunit.[25-27] The complexity and the diversity of the G-proteins lies in the α-subunits.

Thus far, 14 different α-subunits of G-proteins have been identified either by moleculer cloning or by purification experiments. These 14 α-subunits are (1) four $G_{s\alpha}$s, the α-subunits of G-proteins that have been demonstrated to stimulate adenylate cyclase and Ca^{2+} channels[28-32] and are substrate for ADP-ribosylation by the toxin of Vibrio cholera (cholera toxin, CTX)[32] (these four different $G_{s\alpha}$s have been shown to be different splicing products of the same gene[33]); (2) three $G_{i\alpha}$s, the α-subunits of G-proteins that are involved in the inhibition of adenylate cyclase, regulation of K^+-channels, and phospholipases and are substrates for ADP-ribosylation by the bacterial toxin from Bordetella pertussis (PTX)[34-38] (the homology in protein sequences within these $G_{i\alpha}$s is ≥90%[35]); (3) two $G_{o\alpha}$s, the α-subunits of G-proteins that regulate Ca^{2+} channels and phospholipase C and are PTX substrates[39-43] (again these two $G_{o\alpha}$s are splicing products of the same gene[44]); (4) one G_{olf}, the α-subunit of the G-protein that is involved in smell transduction and is substrate for PTX[45]); (5) one Tr, the α-subunit of transduction that is found in rods and is substrate for both CTX and PTX[46]); (6) one Tc, the α-subunit of transduction that is found in cones and is substrate for both CTX and PTX[47]); (7) one G_q, a PTX-insensitive α-subunit that has been demonstrated recently to regulate the polyphosphoinositide-specific phospholipase C;[48] and (8) one $G_{z/x}$, the α-subunit of a G-protein cloned from a brain library and expressed only in CNS and spleen (it is not a PTX substrate,[49,50] and its functions are unknown).

The sequences of these G_α-subunits are highly homologous, especially within a given class. For example, protein sequences deduced from the cDNA sequences of the three $G_{i\alpha}$s reveal that $G_{i\alpha 2}$ is 88% homologous to $G_{i\alpha 1}$ and 85% identical to $G_{i\alpha 3}$. $G_{i\alpha 3}$ is 94% identical to $G_{i\alpha 1}$.[35] The sequence of 18 amino acids proposed to be the GTP binding site is identical in these three proteins. There are additional homologous regions among these Gα-subunits that also share with the bacterial elongation factor EFTu and the *ras* oncogene product p21.[51]

Despite the similarities between various G_αs, it is possible to have significant different multiple forms of G-proteins that contain the same G_α but have different β- and γ-subunits. Because there is evidence to suggest that β- and γ-subunits of G-proteins can modulate receptor-mediated activation of G-proteins[52,53] and that β- and γ-subunits can directly regulate the adenylate cyclase of K^+-channel activity,[54-56] multiple forms of the G-proteins could result in multiple cellular responses to the activation of the receptor.

During agonist activation of receptor, the G-protein is believed to undergo several transformations in order to transduce and amplify the receptor signal. Prior to receptor activation, the G-protein exists in the heterotrimer associated form with GDP occupying the nucleotide binding site at the α-subunit. The α-subunits associate with the membrane via the association of βγ-subunits, which may be isoprenylated.[57,58] Upon interaction with receptor-agonist complexes, there is a dissociation of bound GDP and an association of GTP. The association of GTP causes the dissociation of the heterotrimers of the G-proteins into α-subunits and βγ-subunits. Signals could thus be propagated by the interaction of either α-subunits or βγ-subunits with the effector molecules. Termination of the signal is accomplished by the activation of the endogenous GTPase in the α-subunits and reformation of the heterotrimers.

Under this scheme, it is the role of receptor-ligand complexes to promote the dissociation of the bound GDP and the association of the GTP. PTX catalyses the ADP-ribosylation of the carboxyl-terminus cysteine moiety of the α-subunits of G_is and G_os, thus preventing promotion of GDP dissociation by the receptor-agonist complexes. CTX catalyzes the ADP-ribosylation of the arginine moiety at residue 170—202 of the various G_αs, thus inactivating the endogenous GTPase and preventing the termination of agonist signals.

Based on this mechanism of G-proteins, one can formulate several criteria for diagnosing whether a membrane receptor mediates its functions via G-proteins. Because activation of the G-proteins requires the association of GTP, the agonist cannot have any effect in the absence of GTP. Agonist should also promote the dissociation of bound GDP prior to the binding of GTP. The binding of agonist to the receptor should result in the increase of high affinity GTPase activity because this enzymatic activity signifies the termination of signal transduction. From the thermodynamic basis of equilibrium constants and the assumption that the agonist promotes the interaction between receptor and effector, it can be deduced that when the receptor is interacting with the G-protein α-subunits, it should have a higher affinity for the agonist than when the receptor is dissociated from the G_{α}s. Hence, if opioid receptor-agonist interactions exhibit such characteristics, probably the opioid receptor signals are being transduced by this family of G-proteins.

B. TRANSDUCTION OF THE SIGNALS OF MULTIPLE OPIOID RECEPTORS VIA G-PROTEINS

As discussed in the previous section, if G-proteins are the signal transducers of opioid receptors, then opioid agonist-receptor interaction should exhibit the characteristics summarized in the diagnostic tests. For δ-opioid receptor, the agonist-receptor interaction and the subsequent effector activation fulfill all the criteria mentioned above. Both in brain striatal membranes[59,60] and in the clonal cell line model neuroblastoma × glioma NG108-15 hybrid cells,[61] δ-opioid agonist inhibition of adenylate cyclase activity requires the presence of GTP. The δ-opioid agonist stimulates a low K_m GTPase in NG108-15 hybrid cells,[62,63] and the agonist stimulation is eliminated upon prior chronic treatment of the hybrid cells with agonist.[63] Agonist promotes the binding of GTP_{γ}s and decreases the concentration of GTP required to inhibit adenylate cyclase activity.[64] GTP and its analogue, GppNHp, decrease agonist affinity for δ-opioid receptor both in brain and in NG108-15 cells.[65,66] Under conditions in which δ-opioid agonists inhibit adenylate cyclase activities, δ-opioid receptors exist in multiple affinity states.[67,68] Finally, δ-opioid receptor regulation of adenylate cyclase activities,[69] GTPase activities,[70] and agonist high affinity for receptor[64,71] were all attenuated by pretreating the hybrid cells with PTX. As discussed above, PTX is a toxin having ADP-ribose transferase activities and has been reported to catalyse the ADP-ribosylation by NAD^+ of the cysteine moiety near the carboxyl terminal of a group of G-protein α-subunits, which led to subsequent loss of agonist activities. Therefore, δ-opioid receptor probably transduces its signal via the G-proteins family.

There is ample evidence that μ- and κ-opioid receptor types also transduce their signals via the G-protein family. Both μ- and κ-opioid agonist binding can be attenuated by guanine nucleotides.[72,73] The size of detergent-solubilized receptors can be altered by the addition of GTP,[74] suggesting that the receptors are interacting with other membrane proteins. Both μ- and κ-agonists can stimulate a low K_m GTPase,[75,76] while a putative homogeneous μ-opioid receptor potentiates the binding of GTPγs to two different G-proteins in a reconstituted system.[77] A κ-opioid receptor preparation in guinea pig cerebellum reconstituted with various purified G-proteins stimulates the GDP/GTP exchange of two selective G-proteins.[78] In clonal cell line models human neuroblastoma SK-N-SH[79] and human neuroblastoma SH-SY-5Y[80] and in tumor cells μ-opioid agonists[81] inhibit adenylate cyclase activities which are GTP-dependent. κ-Agonist has been observed to inhibit brain adenylate cyclase in a GTP-dependent manner also.[76,82] Furthermore, these μ- and κ-opioid agonist activities can be blocked by PTX pretreatment.[80,81] κ-Agonist modulation of Ca^{2+} channels *in locus ceruleus*[83] and in spinal cord dorsal root ganglion cocultures[84] could be attenuated by PTX treatment. Thus, although all the criteria for testing the involvement of G-protein in transducing the μ- and κ-opioid receptor signals were not observed in a single model system, from all the available data, both μ- and κ-opioid receptors appear to transduce their signals via the G-protein family.

C. POSSIBLE MULTIPLE CELLULAR RESPONSES BY THE ACTIVATION OF OPIOID RECEPTORS

If indeed the various opioid receptors do transduce their signals via various G-proteins, then the probable cellular effectors activated by the opioid receptors might be deduced from the identities of the G-proteins with which the receptors interact. However, the functional identities of various G-proteins have not been fully established. For G_s and T_r or T_c, their functions have been established clearly to be stimulating adenylate cyclase and phosphodiesterase activities, respectively.[32,46,47] As for various $G_{i\alpha}$s and G_os, their exact functions remain the subject of investigations. Birnbaumer and co-workers have suggested, based on work employing both purified G-proteins and recombinant G_α-subunits, that $G_{i\alpha 3}$ may be the α-subunit of the G-protein involved in the regulation of K^+-channels and that $G_{i\alpha 1}$ may mediate stimulation of phospholipase C hydrolysis of polyphosphoinositides.[37,85] On the basis of work using antibodies and reconstitution experiments,[39,86] G_o appears to be involved in receptor regulation of Ca^{2+} channels. On the basis of work with peptide antibodies developed against $G_{i\alpha 2}$, Milligan and co-workers have suggested that this G-protein is responsible for receptor mediated inhibition of adenylate cyclase activities.[87]

Thus, it appears that the functions of various $G_{i\alpha}$s and G_os are fairly well defined. However, these G-proteins have overlapping functions. The G_αs may regulate more than one effector's functions. In addition to stimulating adenylate cyclase activity, $G_{s\alpha}$ has been demonstrated to be involved in the direct regulation of voltage-gated dihydropyridine-sensitive Ca^{2+} channels.[31] The regulation of the same Ca^{2+} channels by the same $G_{s\alpha}$ might be accomplished by the phosphorylation of the channels by cAMP-dependent protein kinase which is activated by elevated levels of cAMP. The G_o could mediate receptor inhibition of voltage-dependent Ca^{2+} channels[39,86] and could stimulate phosphoinositide-specific phospholipase C; these functions of G_o were demonstrated in Xenopus oocytes and in reconstitution experiments with membranes pretreated with PTX.[43] Furthermore, the same effector molecule could be regulated by multiple G-proteins; bacterially expressed $G_{i\alpha3}$ stimulates the atrial "muscarinic" K^+-channel, and the bacterially expressed $G_{i\alpha1}$ and $G_{i\alpha2}$ have the same potencies and efficacies as those of $G_{i\alpha3}$ in the same assay.[85] Thus, if all the components involved are expressed in the same cells, activation of G-protein coupled receptor, such as opioid receptor types, would elicit multiple cellular responses to the receptor agonist. It would not be surprising to observe that one opioid receptor type could elicit multiple cellular responses in the same cell. Because of the "sharing" of the same pool of G-proteins among various receptor types,[88] it would not be surprising to observe also that the same effector could be activated by different opioid receptor types expressed in the same neuronal cell. Therefore, the reported data on effectors activated by multiple opioid receptor types are reviewed below, the three major types reported on being examined individually.

III. THE EFFECTORS INVOLVED IN MULTIPLE OPIOID RECEPTOR ACTIONS

A. δ-OPIOID RECEPTOR

The original prototypic agonist for δ-opioid receptor has been defined to be enkephalin and its metabolic stable analog D-Ala[2], D-Leu[5]-enkephalin (DADLE). However, DADLE has since been demonstrated to have similar potency at both μ- and δ-opioid receptors.[89] Therefore, it is imperative to keep in mind when examining the literature on the possible regulation of a specific second messenger by the δ-opioid receptor that, unless a homogenous preparation of receptor is present in the system, DADLE may not be sufficiently selective for δ-opioid receptors. In heterogeneous systems, a more δ-opioid selective agonist, such as 2-D-penicillamine,5-D-penicillamine-enkephalin (DPDPE) must be used to define the effects of δ-opioid agonist.

As discussed in Section II.B. above, there is ample evidence to support the notion that δ-opioid receptor transduces its signals via the G-protein family, in particular the G-proteins that are substrates for PTX-catalyzed ADP-ribosylation. Hence, treatment of neural tissues, model cell lines, or *in vitro* cultures of neural tissues with PTX normally eliminate the ability of the δ-opioid agonist to regulate second-messenger systems such as adenylate cyclase, Ca^{2+}-channels, and K^+-channels. The α-opioid agonist receptor interaction can be modulated by guanine nucleotides.[65,66] Target inactivation analysis reveals that the α-opioid receptor exists in membrane as complexes with molecular weight 100 to 200 kDa.[90-92] The size of the complexes can be reduced by the inclusion of Na^+ and GTP analogs.[92] The interaction between δ-opioid receptor and G-proteins appears to be strong,[93] as illustrated by Costa et al., who found negative intrinsic activities of antagonist in the model cell line neuroblastoma × glioma NG108-15 hybrid cells.[94] This model cell line, among others (murine neuroblastoma N18TG2, N4TG2, N1E115 and human neuroblastoma SH-SY-5Y), has been the key in studies of opioid-receptor interaction with effectors.

In the early 1970s, Collier and co-workers reported that morphine-inhibited prostaglandin E_1 (PGI_1) stimulated production of cAMP by brain striatal adenylate cyclase and proposed it as a basis for the opiate drug actions.[95] However, this report of opiate inhibition of adenylate cyclase was controversial. Reports of studies in which inhibition and no inhibition of striatal adenylate cyclase activities have been observed[96-98] are equally numerous. The controversy partially was resolved by the observation that δ-opioid agonists, namely, the enkephalin pentapeptide and its analogues, attenuate PGE_1-stimulated adenylate cyclase activities[99] and production of intracellular cAMP level induced by PGE_1[100] in the clonal cell line, NG108-15 hybrid cell. This opioid inhibition was subsequently demonstrated by Blume et al. to be dependent on both GTP and Na^+.[61]

The absolute requirement of GTP of the δ-opioid receptor activities in the hybrid cells was then found to extend to the striatal adenylate cyclase activity. Law et al.[59] and subsequently Copper et al.[60] demonstrated that δ-opioid agonists inhibit basal striatal adenylate cyclase activity in a dose-dependent manner and that this inhibition is dependent on both GTP and Na^+. Na^+-dependence could be observed only after the complete removal of all Na^+ from reagents[61] and the concentration of Na^+ required correlated with the intracellular Na^+-concentration (10 to 20 m*M*)[59,61] and not the extracellular concentration. This GTP- and Na^+-dependence of agonist activities could be observed also in agonist receptor binding assays. In both brain and cell membrane preparations, δ-agonist binding to receptor is attenuated by the presence of GTP or its analogues and Na^+.[65,66] The reduction is owing to a decrease in agonist affinities resulting from the formation of multiple affinity states of δ-opioid receptor.[67,68] The multiple affinity states of the δ-opioid receptors reflect the interaction of the binding sites with the G-proteins, for the percentage of receptor in high affinity states can be attenuated (but not eliminated as reported by Hsiu et al.[71]) by prior treatment of the membranes with PTX.[64]

Since the initial observations, δ-opioid agonist inhibition of adenylate cyclase activities has been reported for other brain regions (locus coereleus[101] and cortex[102]) and other neuroblastoma cell lines (N18TG2[103] and N1E115[104]). Inhibition of adenylate cyclase activity was observed also in *in vitro* cultures of spinal-cord ganglion explants.[105] In all cases in which δ-opioid receptor-mediated inhibition of adenylate cyclase was observed, the inhibition was GTP-dependent, and in the cases in which PTX-treatment was carried out, the inhibition could be attenuated by this toxin. The involvement of a transducer/amplifier (G-protein) in the δ-opioid receptor inhibition of adenylate cyclase was also implicated by the observation that inactivation of >90% of δ-opioid binding sites in NG108-15 cells with chlornaltraxamine (an opioid receptor covalent label that is an antagonist) did not alter the maximal inhibition level of δ-opioid agonists.[106] This observation, together with the observations of PTX pretreatment and the finding that opioid agonist inhibition of adenylate cyclase activities in NG108-15 hybrid cells correlated with the high affinity state of the agonists,[67,68] clearly suggests involvement of G-proteins in δ-opioid receptor inhibition of adenylate cyclase activity.

It is of interest to note that the magnitude of δ-opioid receptor inhibition of brain adenylate cyclase can be modified by low pH treatment.[107-111] Low pH treatment (pH 4.5) increases opioid inhibition of adenylate cyclase by blocking G_s-stimulated adenylate cyclase,[109,110] increasing guanine nucleotide regulation of agonist binding,[108] and modifying the guanine nucleotide exchange between G_i and G_s.[111] These data further support the notion that δ-opioid receptor inhibition of adenylate cyclase is mediated via G_i. Low pH treatment attenuated the G_s activity, and thus magnified the G_i activity, which directly mediated opioid receptor function in this system. The interplay between G_s and G_i on membrane adenylate cyclase could be observed also by pretreatment of membrane with PTX. This toxin, by eliminating the receptor/G_i input into the control of adenylate cyclase, allowed the overexpression of the G_s-stimulated activities above the control level.[112] This G_s stimulation is presumably a tonic input that is overriden by the G_i activity.

In addition to the direct G_i-mediated inhibition of adenylate cyclase activity, δ-opioid receptors may also regulate neuronal adenylate cyclase by controlling the plasma membrane concentrations of calmodulin. It has been established that in neural tissues, there are two pools of adenylate cyclase activity: one is sensitive to calmodulin, and one is not. In membranes prepared from brain hippocampus, δ-opioid agonists have been reported to inhibit calmodulin-sensitive adenylate cyclase activity.[113] δ-Opioid agonists in NG108-15 cells[114] and morphine in brain tissues[115] alter the distribution of calmodulin bound to the plasma membrane. Since Ca^{2+}/calmodulin activates adenylate cyclase, such action of δ-opioid agonist might increase the intracellular cAMP level. Such cellular redistribution of calmodulin could be the result of opioid-receptor-induced dissociation of the G-protein heterotrimers, for it has been reported recently that dissociated βγ-subunits of G-proteins can interact with calmodulin molecules.[116] In any case, what functions these βγ-subunit-calmodulin complexes have in the expression of opioid reception activation remains to be elucidated.

As discussed in Section II. above, one of the criteria for receptor/G-protein interaction is the ability of agonist to stimulate the high-affinity GTPase activities. The ability of δ-opioid receptor to stimulate this GTPase has been documented with brain membrane[117] and membranes isolated from NG108-15 hybrid cells.[118] Since this GTPase activity turns off receptor activation, treatments such as PTX[70] or reagents such as those reactive to sulfhydryl groups, which also attenuate opioid agonist inhibition of adenylate cyclase activities, might also modify opioid agonist stimulation of GTPase activity.[119]

However, caution must be taken because measured GTPase activity reflects the hydrolysis of GTP bound to multiple G_α-subunits. There are reports to suggest that a single receptor type can couple to multiple G_α-subunits simultaneously. For example, m_2-muscarinic receptors expressed in CHO cells could inhibit adenylate cyclase activity (which is PTX sensitive) and also stimulate phospholipase C (which is PTX insensitive).[120-122] Therefore, one might not be able to correlate agonist-stimulated GTPase activity with the observed second-messenger response. Such is the case of δ-opioid agonist stimulation of low K_m GTPase and the inhibition of adenylate cyclase. Costa and co-workers have reported a difference in the desensitization of receptor-stimulated GTPase and that of inhibition of adenylate cyclase during chronic agonist treatment.[123] This observation suggests that δ-opioid receptors may interact with multiple G-proteins and induce multiple cellular responses.

The presence of multiple PTX substrates in neural membranes has been well documented. In neuroblastoma × glioma NG108-15 cells, there are at least four different PTX substrates: $G_{i\alpha 2}$ $G_{i\alpha 3}$, and two $G_{o\alpha}$s.[124] These two $G_{o\alpha}$s, separable only by two-dimensional electrophoresis, are two different splicing products of same gene.[44] Which one of these PTX substrates mediates δ-opioid receptor inhibition of adenylate cyclase could not be answered by PTX catalyzed ADP-ribosylation of the α-subunit. Although PTX catalyzes the ADP-ribosylation of only the heterotrimer and not the dissociated α-subunits of G-proteins,[14-18] because of the relatively high concentrations of PTX substrates in neural tissues (≥1%[14-18]) as compared to opioid receptor (<0.005% density in NG108-15 cells), it is extremely difficult to identify the G-protein involved in δ-opioid receptor function by PTX labeling. However, it is possible to identify the G-proteins involved in opioid action by cholera toxin (CTX) catalyzed ADP-ribosylation of the α-subunit. Normally, CTX catalyzes the ADP-ribosylation of only Gsα, at the Arg(186/187/201/202) of the various Gsα-subunits.[32] The same Arg is also present in all the PTX substrates ($G_{i1/2/3}$: Arg 180/170/179; G_o, Arg 179) and is not ADP-ribosylated by CTX. But when the G-protein interacts with the receptor, CTX can then catalyze the ADP-ribosylation of the $G_{i\alpha}$- and $G_{o\alpha}$-subunits, in a process that is agonist concentration-dependent. Milligan and his co-workers[125] and Klinz and Costa[126] have demonstrated that δ-opioid agonists promote the CTX-catalyzed ADP-ribosylation of PTX substrates in NG108-15 cells, which is naloxone reversible. Prior treatment with PTX eliminated this effect.[125] Thus, CTX-catalyzed ADP-ribosylation of the PTX substrates probably reflects the interaction of the receptor with the specific G-protein(s) that are transducing the signals of the δ-opioid receptor. However, because the SDS-PAGE systems used by the two groups could not resolve all the PTX substrates, it was not possible in the studies cited to distinguish which PTX substrates is involved in the opioid action.

In an attempt to settle this issue, we have recently utilized the urea gradient/SDS-PAGE system reported by Birnbaumer and co-workers[14] and

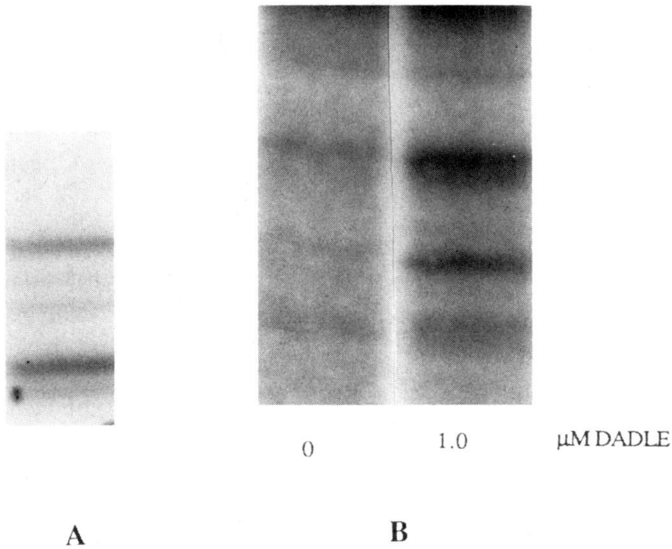

FIGURE 1. Effect of DADLE on the CTX labeling of various PTX substrates expressed in NG108-15 cells. The multiple PTX substrates in NG108-15 cells, were ADP-ribosylated with [^{32}P]-NAD$^+$ and separated by an urea gradient/SDS PAGE system as described[127] (Panel A). The ability of CTX to ADP-ribosylate these PTX substrates in the presence or absence of 1 μM of DADLE was carried out, and the products were separated on a similar urea gradient/SDS PAGE system (Panel B).

have resolved the PTX substrates in NG108-15 cells in a single electrophoretic dimension.[127] Instead of the expected four different PTX substrates, five different PTX substrates in NG108-15 cells were observed. (Figure 1). Through Western analysis using peptide antisera developed against specific G_{α}-subunits, it was concluded that NG108-15 hybrid cells express one $G_{i\alpha 2}$-, two $G_{i\alpha 3}$-, and two $G_{o\alpha}$-subunits. Furthermore, by carrying out the CTX-catalyzed ADP-ribosylation experiments in the presence of δ-opioid agonists, it was demonstrated that δ-opioid receptors could couple to $G_{i\alpha 2}$, one of the $G_{i\alpha 3}$s, and one of the $G_{o\alpha}$s simultaneously, with similar concentrations of receptor occupancy (Figure 1). Similar observations were reported by Offermann et al.[128] by using a photoreactive GTP analog (α-[^{32}P]-GTP azidoanilide). The opioid agonist D-Ala^2,D-Leu^5-enkephalin (DADLE) promoted the labeling of a 39 K protein ($G_{o\alpha}$), 40 K protein ($G_{i\alpha 2}$), and a 41 K protein (unknown); the process was DADLE concentration-dependent and antagonized by naloxone. Thus, these independent studies indicate that δ-opioid receptors in NG108-15 cells can interact with multiple G-proteins, and perhaps have multiple effect.

Although in many other studies, $G_{i\alpha 2}$ has been implicated in the regulation of adenylate cyclase and $G_{i\alpha3}$ in the regulation of K^+-channels,[37] the exact function of each $G_{i\alpha}$ is not well understood. Furthermore, recombinant $G_{i\alpha1,2}$, $G_{i\alpha1,3}$ are equally potent in regulating the atrial K^+-channels.[85] Thus, which of these PTX substrates is involved in δ-opioid agonist regulation of adenylate cyclase remains to be determined.

One possible approach to resolving this is reported by Milligan and coworkers. By developing specific peptide antibodies against the various $G_{i\alpha}$-subunits, this group demonstrated attenuation of opioid-agonist-stimulated low K_m GTPase activity,[87] and subsequently agonist inhibition of adenylate cyclase activity with $G_{i\alpha2}$ antibodies. Other peptide antibodies that recognize other Gα-subunits did not modify opioid-agonist inhibition of adenylate cyclase activities. In view of the authors' recent observations[127] and those reported by Offermann et al.,[128] in which δ-opioid receptors interacted with multiple G-proteins simultaneously, it is surprising to observe that antibodies specific to a single Gα-subunit ($G_{i\alpha2}$) could completely abolish opioid-receptor-stimulated GTPase. However, because of the inherent cross-reactivity of the antibodies against $G_{i\alpha2}$ to other $G_{i\alpha}$-subunits (notably with $G_{i\alpha3}$), the final identification of $G_{i\alpha2}$ as the G-protein for the inhibition of adenylate cyclase activity remains to be confirmed.

Although not so well characterized as δ-opioid agonist inhibition of adenylate cyclase activity, δ-opioid receptor regulation of other cellular effectors has been observed. A notable example is the ability of δ-opioid agonist to modulate voltage-dependent calcium channels in NG108-15 hybrid cells.[39,129] By differentiating the NG108-15 hybrid cells by prolonged elevation of intracellular cAMP levels, several groups have observed the induction of more than one type of voltage-sensitive calcium channels.[130,131] Moreover, after differentiation, DADLE can inhibit the voltage-sensitive Ca^{2+} channels.[39,129] This δ-opioid receptor regulation of Ca^{2+} channels appears to be direct, for there is no reported evidence to suggest opioid receptor regulation of phospholipase C and the generation of the second messengers inositol triphosphates and diacylglycerol, as in the case of bradykinin action in these hybrid cells.[132] These two second messengers can activate the Ca^{2+}-dependent protein kinase, which in turn can modulate Ca^{2+} channel activity.

Although the specific Ca^{2+} channels regulated by the δ-opioid receptor were not investigated, this opioid inhibition could be blocked by pretreatment with PTX, suggesting the involvement of G_i/G_o. Addition of either the $G_{o\alpha}$-subunits or G_o heterotrimer restored the DADLE inhibition of the Ca^{2+} channels.[39] This observation together with the failure of muscarinic agonist (which inhibits adenylate cyclase activity in this cell line) to inhibit the same Ca^{2+} channels led the authors to conclude that this specific δ-opioid receptor effect in NG108-15 cells was mediated by G_o. However, the G_o preparations used were not subjected to high-resolution chromatography, such as the MonoQ column, normally used by others to resolve the various G_α-subunits. Similar preparations of G_o from the brain have been demonstrated to contain heterogeneous PTX substrates.[133] In addition, the report that $G_{o\alpha}$ has a higher potency than the heterotrimers for restoring δ-opioid receptor activities contradicts the generally accepted theory of G-protein actions. Experimental evidence suggested that the dissociated G_α-subunits would not associate with the receptors.[14-18] Therefore, the designation of G_o as the G-protein that mediates the δ-opioid receptor inhibition of Ca^{2+} channels remains tentative. Nevertheless, recent data reported by Milligan and co-workers[86] using G_o-specific antibodies to attenuate the α_2-adrenergic receptor-mediated inhibition of Ca^{2+} channels in the differentiated NG108-15 cells lend credence to the reconstitution studies discussed above.

The possible interaction of δ-opioid receptor with Ca^{2+} channels has been reported in dorsal-root ganglion (DRG) neurons.[84,134-136] Acute exposure of the DRG neurons to DADLE markedly shortened the duration of the calcium component of the action potentials of these cells. This opioid effect could be blocked by naloxone or pretreatment with PTX,[136] suggesting the direct involvement of the opioid receptor and the transduction of its signal via G-proteins. However, since changes in levels of other second messengers, such as a decrease in intracellular cAMP also occurred in response to opioid receptor activation in these neurons,[105] it could not be determined whether the electrophysiological effect represent a direct interaction of the δ-opioid receptor with Ca^{2+} channels. Because μ- or κ-opioid receptor types are also present in these DRG neurons, and their agonists elicit similar effects, the participation of the δ-opioid receptor as inferred by the use of DADLE as prototypic ligand is speculative.

Another mechanism by which opioid ligands may reduce transmitter release is by hyperpolarization of the sensitive neurons, which would involve the opioid receptor controlling K^+ channels. δ-Opioid receptor-mediated hyperpolarization of neurons has been reported peripherally in the submucosal neurons of the guinea pig ileum,[137,138] centrally with the locus coeruleus,[138-142] and the pyramidal cells of hippocampus.[143-147] In the pyramidal cells of the hippocampus, opioid agonists excite the cells by inhibiting the interneuronal inhibitory inputs.[148,149] This inhibition is associated with an increase in the membrane conductance, and the extrapolated reversal potential was more negative than the resting potential.[149] A direct involvement of K^+ channels was not demonstrated; however, in the cases of guinea pig ileum and locus coeruleus, hyperpolarization is associated with an increase in potassium conductance as verified by intracellular recordings.[137,138] This increase in potassium conductance results in the reduction of the excitability of these neurons and thus can account for the inhibitory effect of the opioid ligands. This inward rectifying potassium conductance was activated by both μ- and δ-opioid receptor in locus coeruleus[138] and by δ-opioid receptor in the guinea pig submucous neurons.[150] Even though in both tissues, δ-opioid receptors can regulate the intracellular cAMP level, application of forskolin or dibutyrl cAMP did not alter the opioid responses, which suggests a direct regulation of the K^+ channels by receptor.[138,151] The participation of G-proteins in this δ-opioid receptor action was demonstrated by intracellular injection of $GTP_{\gamma}s$, which prolonged the hyperpolarization after removal of the agonist.[138] Injection of PTX or pretreatment of the ileum with PTX attenuated δ-opioid receptor regulation of the K^+-channels. The identity of the G-protein involved in this opioid effect has not been examined. On the basis of other studies, such as those with atrial cells, it is speculated that a G_k may exist. But recombinant $G_{i\alpha3}$ and also $G_{i\alpha2}$ or $G_{i\alpha1}$ in equal potency could regulate atrial K^+ channels.[85] One could conclude that G-protein such as $G_{i\alpha2}$ might not be involved in this δ-opioid receptor action because that activation of protein kinase C with phorbol esters did not mimic or attenuate the opioid activation of K^+ conductance.[138] Treatment of NG108-15 cells with phorbol ester attenuated δ-opioid receptor inhibition of adenylate cyclase activity,[152] an effect probably owing to the phosphorylation of the $G_{i\alpha2}$-subunit.[153] However, the final identification of the G-protein involved in this δ-opioid receptor function remains to be demonstrated.

Two other cellular effectors that have been reported to be controlled by the δ-opioid receptor are guanylate cyclase[154,155] and the Na^+/H^+ antiporter.[156] Evidence for the control of these two effectors by opioid ligands has been controversial. Earlier reports, which have been refuted by the same authors, suggested that opioid could increase the production of cGMP and accelerate the Na^+/H^+ exchange in the hybrid cells, thus increasing the intracellular pH. Using a pH-sensitive fluorescent probe, 2,7-biscarboxyethyl-5(6)-carboxyfluorescein, Isom et al.[156] reported that receptors that inhibit adenylate cyclasein NG108-15 cells accelerated the Na^+/H^+ exchange and that this receptor effect was not PTX sensitive, unlike inhibition of adenylate cyclase. These observations are very attractive, because the control of intracellular cGMP levels is related to the intracellular pH. For example, in pinealocytes, it has been reported that the blockade of Na^+/H^+ antiporter with amiloride inhibitors decreases the intracellular cGMP level in a dose-dependent manner.[157] α_1-Adrenergic agonists in this preparation accelerated the Na^+/H^+ exchange and increased intracellular pH. The same agonists stimulated the accumulation of intracellular cGMP. This same mechanism could underlie the controversial δ-opioid agonist increases in cGMP levels and intracellular pH. Since the initial report, the ability of opioid agonist to increase intracellular cGMP levels has been reported in the hybrid cells in the presence of prostaglandin $E_{2\alpha}$.[154] Opioid stimulation of cGMP level also has been reported in neostriatal slices.[158] The inconsistent observations of the Na^+/H^+ exchanger could be owing to the heterogeneity of NG108-15 cell preparations used by the authors, for by using single-cell fluorescence, it is possible to observe a naloxone-reversible δ-opioid agonist-induced increase in the Na^+/H^+.[245] Whether these δ-opioid receptor effectors are mediated by G-proteins that are PTX substrates, or with G-proteins that are not PTX substrates, as reported previously, remains to be determined.

B. μ-OPIOID RECEPTOR

The classical prototypic ligand for the μ-opioid receptor is morphine, as defined by Martin et al.[159] However, since the initial classification, the relative nonselectivity of morphine toward the various multiple opioid receptors has been documented (e.g., see Reference 89). Therefore, numerous reports in the literature using morphine as the ligand to define the involvement of the μ-opioid receptor as the regulator of a specific second messenger might not be valid. Nevertheless, it has been accepted that the involvement of G-proteins in the transduction of μ-opioid receptor signals is implicated by the ability of various guanine nucleotide analogs to decrease the agonist affinities for the receptor.[70,160-162] A direct demonstration of the interaction between G-proteins and μ-opioid receptors has been reported by Ueda and co-workers.[77,163] In their reports, a putatively homogenous preparation of μ-opioid binding sites could stimulate the low K_m GTPase activities of G_o and G_i preparations in a reconstituted system. The specificity of the μ-opioid receptor interaction with signal transduction G-proteins was demonstrated by the inability of the purified μ-opioid receptor to increase the GTPase activity of *ras* in the same reconstituted system.[77] This finding is not surprising, since the GTPase activity of *ras* is activated only in the presence of GAP.[18] Nevertheless, the ability of the purified μ-opioid receptor to interact with and stimulate the low K_m GTPase of the purified G-protein preparations suggests that the transduction of μ-opioid receptor signals, similar to that of δ-opioid receptor signals, is mediated via the same G-protein family.

Unlike the δ-opioid receptor, which is present in multiple murine neuroblastoma cell lines, the μ-opioid receptor is absent in all but one known clonal cell line, which has made investigation of second messengers involved in its action more difficult. However, in many instances, the μ-opioid receptor is regulating the same sets of second messengers as those regulated by the δ-opioid receptor. Although there may be some brain regional distribution differences in these two opioid receptor types, the final effects of the μ- and δ-opioid agonists are similar. For example, supraspinal analgesia has been reported to be mediated via both μ- and δ-opioid receptors.[164] As a matter of fact, it has been proposed that μ- and δ-opioid receptors exist in a complex.[165] Thus, it is reasonable to believe that the μ-opioid receptor is interacting with the same pools of G-proteins, and hence the same second messengers are being regulated.

One second messenger regulated by the μ-opioid receptor is the production of intracellular cAMP. The best system for illustrating μ-opioid receptor regulation of adenylate cyclase activity is in the clonal cell lines of human neuroblastoma SH-SY-5Y cells. This cell line, similar to the human neuroblastoma cell line SK-N-SH,[166,167] contains both the μ- and δ-opioid receptor types.[80,168] With the use of the δ-opioid receptor selective ligand, [^{3}H]-2-D-penicillamine,5-D-penicillamine-enkephalin (^{3}H-DPDPE) or the μ-opioid receptor selective ligands, [^{3}H]-D-Ala^2-*N*-methyl-Phe4-Gly(ol)^5-enkephalin ([^{3}H]-

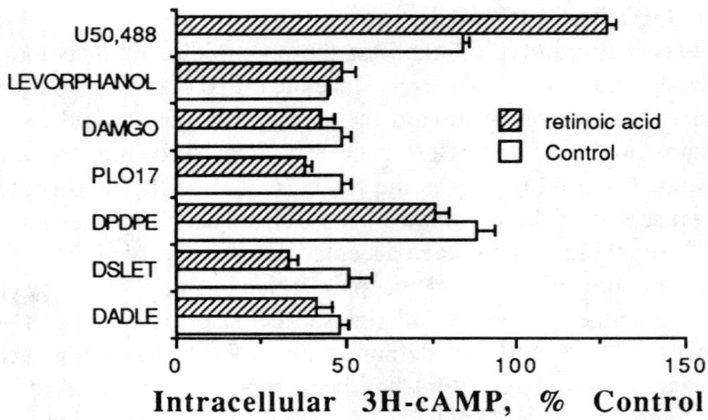

FIGURE 2. Ability of various opioid agonists to inhibit forskolin-stimulated production intracellular [^{3}H]-cAMP level in human neuroblastoma SH-SY-5Y cells treated with or without 10 μ*M* retinoic acid for 6 d.

DAGO) or ^{3}H-D-Phe-Cys-Tyr-D-Trp-Orn-Thr-Pen-Thr-NH2 ([^{3}H]-CTOP),[80,169] a specific μ-opioid binding site, which has twice the density as δ-opioid binding sites, was demonstrated in the SH-SY-5Y cells.[80] Unlike the case reported with the SK-N-SH cells,[166,167] the presence of the μ-opioid receptor is observed in SH-SY-5Y cells without culturing of the cells with differentiating agents such as retinoic acid, NGF, or dibutyrl cAMP. Furthermore, these μ-opioid binding sites are coupled to adenylate cyclase activity. μ-Opioid agonists can inhibit basal and PGE_1-stimulated adenylate cyclase in SH-SY-5Y cells. However, the inhibition of adenylate cyclase is much less than that observed in other murine neuroblastoma (25% vs. 50% inhibition observed in the NG108-15 cells). The relatively low level of inhibition might be due to partial agonist properties in some of the μ-opioid receptor ligands used. However this is unlikely because DPDPE possesses only fractional activities in the SH-SY-5Y cells and because the putative δ-opioid selective antagonist ICI-174864 did not antagonize the actions of several μ-opioid agonists.[80]

The activity of this μ-opioid receptor could be better observed by measurement of intracellular cAMP level. As shown in Figure 2, when the intracellular cAMP level was determined by the method of prelabeling the intracellular ATP pools with [^{3}H]-adenine, the authors observed a similar μ-opioid agonist inhibition of the forskolin-stimulated increase in the intracellular ^{3}H-cAMP. With or without culturing the SH-SY-5Y cells in the presence of 10 μ*M* retinoic acid for 4 days, only μ-opioid receptor selective ligands inhibited the production of cAMP. The δ-opioid agonist DPDPE and the κ-opioid receptor selective agonist U50,488, did not attenuate the intracellular cAMP production. Moreoever, the level of cAMP production was further decreased to 50% by ligands (Figure 2). Thus, most likely, the μ-opioid receptor in the SH-SY-5Y cells regulates the adenylate cyclase activity. Because the presence of GTP is required for the observed opioid inhibition in the adenylate cyclase assays[80] and because opioid-agonist binding in both the SH-SY-5Y cells and the SK-N-SH cells can be modulated by guanine nucleotides,[80,167] the transduction of this μ-opioid receptor signal via the G-protein family is inferred. However, which G-protein is involved has not been determiend yet.

μ-Opioid receptor-mediated inhibition of adenylate cyclase activity has also been reported in the primary mouse tumor 7315c. Frey and Kebabian reported that this tumor contains a homogenous population of μ-opioid receptors, activation of which inhibited prolactin release and adenylate cyclase.[170] A μ-opioid selective ligand, such as FK33824, was seen to be more potent than DADLE in the adenylate cyclase inhibition. The opioid inhibition required GTP and metabolically stable analogues of GTP-inhibited adenylate cyclase activity in these tumor cells in the absence of an agonist. As with other opioid receptors that couple to adenylate cyclase, chronic exposure of 7315c cells to morphine resulted in the uncoupling of the μ-opioid receptor from the G-protein.[171] Hence, it is likely that the observed μ-opioid receptor-mediated inhibition of adenylate cyclase activity in these tumor cells is mediated via the G-protein family.

Inhibition of adenylate cyclase activity by μ-opioid agonist has been reported with neural tissues also. In three different brain regions — locus coeruleus,[101] striatum,[102,173-175] and periaqueductal gray area[172] — μ-opioid agonist inhibition of adenylate cyclase has been reported. In the striatum, the μ-opioid agonist DAMGO inhibited dopamine-stimulated adenylate cyclase activity.[173-175] In the locus coeruleus, findings showed morphine inhibition of adenylate cyclase *in vitro,* which was GTP-dependent and reversible by naloxone.[101] Interestingly, in this locus coeruleus membrane preparation, findings also showed a persistent decrease in adenylate cyclase after *in vivo* morphine treatment[101] and an increase both in the cAMP-dependent protein kinase[176] and in ADP-ribosylation of $G_{i\alpha}$- and $G_{o\alpha}$-subunits[177] after chronic treatment. These observations suggest the interaction of the μ-opioid receptor with the α-subunits of these two G-proteins, which extend a tonic control of the adenylate cyclase. Such a notion has precedent in NG108-15 hybrid cells, where removal of tonic control by $G_{i\alpha}$-subunits via PTX treatment resulted in an increase in overall adenylate cyclase activity.[112] However, in the hybrid cells, which have a much higher opioid receptor density than any brain region, measurement of adenylate cyclase activity in the presence of naloxone did not result in any observable increase in enzymatic activity. Thus, the reported results with locus coeruleus membranes might suggest other cellular responses to μ-opioid receptor activation in addition to the alteration of intracellular cAMP.

In rat neostriatal membrane or slices, μ-opioid agonists have been reported to inhibit adenylate cyclase activity[102] and to stimulate low K_m GTPase activity.[178,179] As discussed in an earlier section, striatal adenylate cyclase and GTPase activities are regulated by the δ-opioid receptor type. However, Schoffelmeer et al.[173-175] and Kelly and Nahorski[180] reported that μ-opioid receptor also participated in the regulation of the intracellular cAMP level, particularly if the intracellular cAMP level was elevated by the stimulation of D_1-dopamine receptors. In another study the efflux of cAMP and the release of the neurotransmitter (norepinephrine) were blocked by the activation of μ-opioid receptors.[181] In all these assays, activation of δ-opioid receptors produced similar responses. Interestingly, when the δ-opioid receptors were irreversibly blocked by fentanyl isothiocyanate, naloxone did not reverse the remaining μ-opioid responses, leading to the speculation that the μ- and δ-opioid receptors existed in a complex.[174] However, one must be careful in the interpretation of such data, for fentanyl isothiocyanate is not specific to δ-opioid receptor and is not an antagonist. With a reactive group such as thioisothiocyanate, alkylation of sites other than the δ-opioid receptor binding sites would occur. Furthermore, since naloxone is a competitive antagonist for μ-opioid receptor, it is difficult to visualize why naloxone could not antagonize the DAMGO inhibition of adenylate cyclase in membranes or slices pretreated with fentenyl isothiocyanate.

On the other hand, μ-opioid receptor mediated inhibition of adenylate cyclase activity in periaqueductal grey (PAG) of rat brain is well characterized. That region of the midbrain has been implicated in both endogenous pain control and opioid analgesia. The receptor in question has been implicated to have μ-opioid characteristics. Autoradiographic studies suggest the opioid binding sites in periaqueductal grey are of the μ-opioid type.[182] Radioligand binding assays have also demonstrated that the opioid binding site is this type.[183] Furthermore, this μ-opioid receptor type can stimulate low K_m GTPase activity, an effect blocked by naloxone or PTX pretreatment of the membrane.[184] Subsequently, G-protein-mediated inhibition of adenylate cyclase activity was observed in the presence of μ-opioid agonist.[172] Thus, in the periaqueductal gray region, the μ-opioid receptor regulates adenylate cyclase. Again, the specific G-proteins involved in this μ-opioid receptor action have not been determined.

Again, as with the δ-opioid receptor, the mechanism for μ-opioid receptor-mediated inhibition of transmitter release could be demonstrated to be activation of K^+ channels, thus hyperpolarizing the membrane and reducing the excitability of the neurons. The μ-opioid receptor activation of the K^+ channels was demonstrated by voltage clamp studies with neurons of the rat locus coeruleus and guinea pig submucous plexus.[141,185-187] Initial findings were reported for studies in which extracellular recording electrodes were used to measure the depression of action-potential discharges in the myenteric neurons after administration of morphine.[188-191] Because the inhibition persisted after all neural transmission was blocked, Dingledine and Goldstein proposed that opioid agonist induce hyperpolization.[192] This hypothesis was verified by a later study using intracellular recordings, carried out by North and co-workers in which hyperpolarization of myenteric neurons by normorphine,[193] a μ-opioid receptor agonist, and an increase in potassium conductance[185,186] were found. These findings were extended later to the locus coeruleus with more opioid-receptor selective ligands, such as DAMGO for μ-opioid receptor and DPDPE for δ-opioid receptor.[138] As in experiments with myenteric plexus, the activation of μ-opioid receptor or δ-opioid receptor increased the K^+ conductance. Patch clamp studies with the cell-attached configuration suggested direct interaction of the receptor with the K^+ channels.[194] The opening of the K^+ channels by opioid agonist could be potentiated by $GTP_\gamma s$, suggesting the involvement of G-proteins.

A close association of Ca^{2+} and opioid receptor activity, especially that of μ-opioid receptor, has been proposed and documented.[195] Attenuation of normorphine or morphine action in guinea pig ileum preparations by increasing extracellular Ca^{2+} levels was reported by several groups.[196-198] Conversely, potentiation of DAMGO effects in the ileum were observed when extracellular Ca^{2+} was lowered.[199] Furthermore, the hyperpolarization of myenteric S neurons by μ-agonists was regulated similarly by the extracellular Ca^{2+} level.[185] Although these results and others indicate a close relationship between Ca^{2+} and μ-opioid receptor activities, there is no evidence to suggest μ-opioid receptor regulation of the Ca^{2+}-channels. The exact mechanism of this Ca^{2+} effect on μ-opioid receptor action requires further elucidation.

Finally, as discussed in Section II above, association of μ-opioid receptor with multiple G-proteins has been implicated by reconstitution studies. Through the use of a putatively purified μ-opioid receptor preparation, reconstitution of the opioid receptor with the $G_{i\alpha 2}$ and $G_{o\alpha}$s was observed.[77,163] Successful reconstitution was monitored by the ability of μ-opioid agonist to stimulate the low K_m GTPase activity. Reconstitution with *ras* protein did not result in the ability of opioid agonist to stimulate the GTPase activity. Similar reconstitution experiments have been reported with the μ-opioid receptor from cortex.[200] In their report, Remmers and Medzihradsky used high pH to remove G-protein activity in the donor membrane (cortex); the membranes were fused to the acceptor membranes (C6 glioma cells); and μ-opioid receptor activities were restored. These studies, together with others, clearly suggest a direct interaction between μ-opioid receptors and G-proteins. Since $G_{i\alpha 2}$ and $G_{o\alpha}$s have been implicated in the regulation of effectors other than adenylate cyclase and K^+-channels,[14-18] it is probable that μ-opioid receptors can also regulate other cellular effectors.

C. κ-OPIOID RECEPTOR

When the κ-opioid receptor was initially defined by Martin and co-workers, the ligand used was ketocyclazocine.[159] Since then, it has been demonstrated that ketocyclazocine or its ethyl derivative does not selectively distinguish the three major opioid receptor types, μ- δ-, and κ-opioid.[89] The ligands of choice today for κ-opioid receptor are U50,488 and U69,593, which have high selectivity toward this type of receptor.[201] The κ-opioid receptor selectivity of these drugs has been demonstrated by their diuretic properties in contrast to the antidiuretic properties of μ-opioid agonist.[202] The putative endogenous ligand for the κ-opioid receptor has been proposed to be dynorphin.[203-205] However, because the various dynorphins have affinities for both μ- and δ-opioid receptors and are relatively potent agonists in the δ-opioid receptor system,[89] the involvement of κ-opioid receptor in any cellular regulation can only be inferred if a single concentration of dynorphin is used as the prototypic ligand.

As for μ- and δ-opioid receptors, one effect of κ-opioid receptor activation is inhibition of transmitter release. Such action of κ-opioid receptor has been reported in brain slices,[206,207] synaptosomal preparations,[208,209] spinal cord,[210,211] hippocampal slices,[212,213] striatal slices,[214] and locus coeruleus.[215,216] The κ-opioid receptor can also stimulate the release of prolactin[217] and vasopressin[202] and suppress the release of growth hormone.[217] However, unlike μ- and δ-opioid receptors, κ-opioid receptors have not been detected in any transformed clonal cell line. Tissues in which the κ-opioid receptor type is predominant include guinea pig cerebellum,[218] human placental syncytial brush border membrane,[219,220] and toad *(Bufo marinus)* brain.[221] In these preparations, κ-opioid receptor density comprises >60% of total opioid receptor density.

As with μ- and δ-opioid receptor, association of κ-opioid receptor with the G-protein family can be demonstrated by the ability of guanine nucleotides to attenuate agonist binding[222-224] and the ability of κ-opioid agonists to stimulate the low K_m GTPase activities.[225] In the initial experiments, guanine nucleotide effects on κ-opioid receptor binding were carried with ^{3}H-ethylketocyclazocine or [^{3}H]-bremazocine.[222-225] As discussed earlier, these two ligands do not exhibit high κ-opioid receptor selectivity. The guanine nucleotide effects could thus be due to the binding of the radioactive ligands to μ- or δ-opioid receptor in the preparations. But, κ-opioid agonist binding can be demonstrated to be sensitive to guanine nucleotides by the ability of GppNHp to decrease the affinity of the κ-opioid selective ligand U50,488.[226] As shown in Figure 3, the ability of U50,488 to compete for [^{3}H]-diprenorphine binding to membrane from guinea pig cerebellum was greatly attenuated by the inclusion of 10 μ*M* GppNHp. Interaction of κ-opioid receptor with a G-protein could be demonstrated further by the hydrodynamic chracterization of κ-opioid receptor in human placenta.[227] The digitonin-solubilized receptor-ligand complexes have a sedimentational coefficient in the sucrose gradients of $S_{20.w}$ = 12 S. In the presence of GppNHp, the sedimentational coefficient of the complex was decreased to 10 S. These results, together with the observations that κ-agonist can stimulate low Km GTPase activities in guinea pig cerebellum and rat brain membranes pretreated with β-funaltrexamine and *cis*-(+)-3-methylfentanyl isothiocyanate to remove μ- and δ-opioid receptors respectively,[206] clearly suggest that the κ-opioid receptor transduces its signals via the G-protein family.

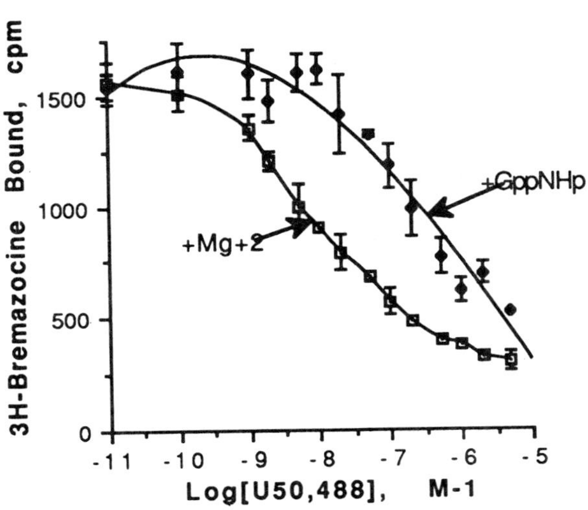

FIGURE 3. Ability of U50,488 to compete for 0.25 n*M* [^{3}H]-bremazocine binding to GP cerebellum in the presence of either 10 m*M* $MgCl_2$ or 10 μ*M* GTP.

In a recent study, Ueda et al.[78] reported that κ-opioid agonist U50,488 could attenuate the GDP/GTP exchange in guinea pig cerebellar membrane reconstituted with purified G_{i1} and G_{i2} but not G_o. In those experiments, Ueda et al. observed that U50,488 decrease the [^{3}H]-GDP/GppNHp exchange rate. It is of interest to note that the κ-opioid agonist had effects only with G_{i1} and G_{i2}, but not G_o. These interactions suggest coupling between κ-opioid receptor and some specific cellular effectors. But the fact that U50,488 inhibited the GDP/GTP exchange suggests a mechanism contrary to what has been proposed for G-proteins involved in signal transduction (see Section II.A above). Although the same authors have reported κ-opioid agonist inhibition of low K_m GTPase activity,[78] one must be cautious in the interpretation of these data. Without the prior complete elimination of the endogenous G-proteins activity, the receptor's effect on the reconstituted G-proteins could not be distinguished from the G-proteins which are present already.

Regulation of the adenylate cyclase activity by κ-opioid receptors has also been reported in membranes isolated from rat spinal cord neurons[228,229] and guinea pig cerebellum.[230] In guinea pig cerebellar membrane preparations, dynorphin analogs as well as U50,488 could inhibit the adenylate cyclase activity, but not μ- or δ-opioid agonists.[230] Similar findings were obtained with opioid inhibition of adenylate cyclase activity in the spinal cord.[228] The κ-opioid-agonist inhibition of brain adenylate cyclase activity was GTP-dependent and could be attenuated by pretreatment of the membranes with PTX.[229,230] Hence, these data suggest that the κ-opioid inhibition of adenylate cyclase activity is mediated via the G-protein family. The data also complement those of agonist affinity attenuation by GppNHp and agonist stimulation of low K_m GTPase activity.

Not only has κ-opioid agonist been found to inhibit adenylate cyclase activity, it has been reported also to alter synaptic excitability in the dorsal root ganglion (DRG) in culture[210,211,231-237] and the cell somata of guinea pig myenteric neurons,[216] by a mechanism quite different from those of μ- and δ-opioid receptors. Instead of hyperpolarizing the neuronal membrane by opening of K^+ channels, κ-opioid inhibited transmitter release in these systems by attenuation of Ca^{2+} currents. κ-Opioid agonist did not affect the resting membrane potential or conductance of the myenteric neurons, but decreased the inward Ca^{2+} current underlying the action potentials of the neurons.[216] It was demonstrated with the voltage clamp technique in DRG neurons that dynorphin A and its analogue selectively reduced a large transient (N-type) calcium current.[211] These effects of κ-opioid agonist could be blocked by naloxone or pretreatment with PTX.[216] Which member of the G-protein family mediated these κ-opioid receptor effects remains to be elucidated.

There appears to be a biphasic response to the κ-opioid agonist in DRG neurons. At a high concentration (1 μ*M*) of agonist, κ-opioid ligand reduced the Ca^{2+} component of the action potential duration (APD).[210] However, at the low concentrations (1 to 10 n*M*) of agonist, κ-opioid prolonged the APD by decreasing the K^+-conductance.[210,238,239] Although the inhibitory effect of the opioid agonists could be blocked by PTX, the prolongation effect was not affected by PTX. In contrast, treatment of the DRG neurons with cholera toxin (CTX) blocked the opioid excitatory effects,[240] thus suggesting the involvement of G_s. However, because the CTX B-subunit can block this opioid effect in a high concentration than the catalytic A-subunit, Shen and Crain[241] suggested that ganglioside (GM1) may regulate G_s-opioid receptor functions. How exactly the ganglioside may interfere with the κ-opioid effect on the Ca^{2+} current remains to be investigated.

There are some κ-opioid receptor actions that have been reported to be insensitive to PTX. In addition to the action of the κ-opioid agonist in DRG neurons as discussed earlier, the κ-opioid agonist regulation of catecholamine release from rat and guinea pig brain slices was not sensitive to PTX. Intracerebroventricular injections of PTX abolished the ability μ- and δ-opioid agonist to inhibit stimulated norepinephrine release in guinea pig brain hippocampus slices.[213,214] κ-Opioid agonist inhibition of norepinephrine release was not affected by PTX treatment.[207] Similar observations were made with PTX modulation of κ-opioid agonist inhibition of dopamine release in the striatal and cortical slices. Thus, apparently, κ-opioid receptor may be coupled to G-proteins other than the PTX substrates.

IV. CONCLUSION

The general consensus is that opioid agonist action includes the inhibition of electrical excitability and reduction in transmitter release in opioid-sensitive neurons. However, the exact molecular mechanism underlying the transduction of these opioid receptor signals has not been elucidated. The earlier hypothesis that changes in the intracellular cAMP level could be the messenger of opioid responses has been demonstrated to be most likely incorrect. There is no doubt that opioid agonists can regulate intracellular cyclic nucleotide levels, and opioid agonist has been reported to modulate the neuronal membrane phosphorylation.[242,243] Phosphorylation of neuronal membrane proteins by kinases, notably protein such as synapsins,[244] could alter the excitability of the membranes. However, the use of phosphodiesterase inhibitors, dibutyrl cAMP, forskolin, or cholera toxin treatment did not alter opioid responses in the guinea pig myenteric plexus. These findings and others do not support the role of cAMP in the principal action of opioid agonist. It is now accepted that the inhibitory effect of opioid agonist could result from hyperpolarization of the membrane by direct opening of the K^+ channels or by inhibition of Ca^{2+} currents.

Regardless of the second messenger that expresses opioid receptor functions, it is clear from all reported studies that opioid receptors transduce signals via the G-protein family. It would be a simple matter if the interaction between opioid receptor and the G-proteins were specific — δ-opioid receptor coupled only to $G_{i\alpha 2}$s, and κ-opioid receptor coupled only to the $G_{i\alpha 1}$, and so on. However, from studies of δ-opioid receptor functions in neuroblastoma × glioma NG108-15 hybrid cells, it could be demonstrated that opioid receptors activate multiple cellular effectors. Not only could δ-opioid receptor in NG108-15 hybrid cells regulate the intracellular cAMP levels, it could regulate the Ca^{2+}-channels, the Na^+/H^+ antiporter, intracellular cGMP levels, and some other effectors that have not been identified. This is concluded from the ability of δ-opioid receptor to interact with multiple G-proteins in this clonal cell line (see Figure 1). Sorting out which of these effectors are the ones primarily responsible for the principal action of opioid agonist could be difficult. However, one must consider the fact that all these cellular effectors may modulate each other. An alteration in one parameter, such as intracellular pH, could alter the activities of many enzymes, e.g., the interrelationship between the Na^+/H^+ exchanger and the intracellular cGMP in the pinealocytes.[157] The activation of Ca^{2+} channels and hence the increase of the intracellular free Ca^{2+} concentration could increase calmodulin-dependent protein kinases, adenylate cyclase, and subsequently cAMP-dependent protein kinases. These enzymes have been documented to modulate G-protein activity and Ca^{2+}-channel activity. Thus, the simultaneous activation of multiple effector systems could result in a concerted amplification of opioid agonist signals. It would be presumptuous to suggest that activation of a specific effector by a specific opioid receptor type is the mechanism of an opioid agonist action. As does δ-opioid receptor, μ- and κ-opioid receptors probably activate multiple G-proteins and subsequently elicit multiple cellular responses. Because the G-protein family is the transducer of opioid receptor functions, identification of the particular G-proteins involved in the functions of the opioid receptor in question would facilitate the identification of the cellular effectors involved.

In conclusion, from the reported studies, a picture of second messengers involved in opioid agonist actions is slowly evolving. There appear to be multiple messengers that opioid receptors could activate simultaneously. It is through the activation of these multiple messenger systems, which subsequently can modulate one another, that the final cellular response to receptor activation is observed.

ACKNOWLEDGMENT

This work is supported in parts by NIDA grants #DA5945 and #DA00564. The authors also wish to thank Dr. Andrew Smith for reading the manuscript.

REFERENCES

1. **Pert, C. B. and Snyder, S. H.,** *Science,* 179, 1011, 1973.
2. **Simon, E. J., Hiller, J. M., and Edelman, J.,** *Proc. Natl. Acad. Sci. U.S.A.,* 70, 1947, 1973.
3. **Terenius, L.,** *Acta Pharmacol. Toxicol.,* 32, 317, 1973.
4. **Chang, K. J.,** *Receptor,* 1, 1, 1984.
5. **Snyder, S. H. and Goodman, R. R.,** *J. Neurochem.,* 35, 5, 1980.
6. **Iwamoto, E. T. and Martin, W. R.,** *Med. Res. Rev.,* 1, 411, 1981.
7. **Nishimoto, I., Murayama, Y., Katada, T., Ui, M., and Ogata, E.,** *J. Biol. Chem.,* 264, 14,029, 1989.
8. **Okamoto, T., Katada, T., Murayama, Y., Ui, M., Ogata, E., and Nishimoto, I.,** *Cell,* 62, 709, 1990.
9. **Strader, C. D., Sigal, I. S., and Dixon, R. A. F.,** *FASEB J.,* 3, 1825, 1989.
10. **Lameh, J., Cone, R. I., Maeda, S., Philip, M., Corbani, M., Nadasdi, L., Ramachandran, J. J., Smith, G. M., and Sadee, W.,** *Pharmacol. Res.,* 7, 1213, 1990.
11. **Findlay, J. and Eliopoulos, E.,** *TIPS Rev.,* 11, 492, 1990.
12. **Dixon, R. A. F., Sigal, I. S., and Strader, C. D.,** Cold Spring Harbor Laboratory Symposia on Quantitative Biology LIII, Cold Spring Harbor, NY, 487, 1988.
13. **Schofield, P. R., McFarland, K. C., Hayflick, J. S., Wilcox, J. N., Cho, T. M., Roy, S., Lee, N. M., Loh, H. H., and Seeburg, P. H.,** *EMBO J.,* 8, 489, 1989.
14. **Birnbaumer, L., Abramowitz, J., and Brown, A. M.,** *Biochem. Biophys. Acta,* 1031, 163, 1990.
15. **Johnson, G. L. and Dhanasekaran, N.,** *Endocrinol. Rev.,* 10, 317, 1990.
16. **Sternweis, P. C. and Pang, E. H.,** *TINS,* 13, 122, 1990.
17. **Allende, J.,** *FASEB J.,* 2, 2356, 1988.
18. **Bourne, H. R., Sanders, D. A., and McCormick, F.,** *Nature,* 348, 125, 1990.
19. **Codina, J., Stengel, D., Woo, S. L. C., and Birnbaumer, L.,** *FEBS Lett.,* 207, 187, 1986.
20. **Fong, H. K. W., Amatruda, T. T., Birren, B. W., and Simon, M. I.,** *Proc. Natl. Acad. Sci. U.S.A.,* 84, 3792, 1987.

21. **Gao, B., Gilman, A. G., and Robishaw, J. D.,** *Proc. Natl. Acad. Sci. U.S.A.,* 84, 6122, 1987.
22. **Gao, B., Mumby, S., and Gilman, A.,** *J. Biol. Chem.,* 262, 17,254, 1987.
23. **Fong, H. K. W., Hurley, J. B., Hopkins, R. S., Miake-Lye, R., Johnson, M. S., Doolittle, R. F., and Simon, M. I.,** *Proc. Natl. Acad. Sci. U.S.A.,* 83, 2162, 1986.
24. **Levine, M. A., Smallwood, P. M., Moen, P. T., Helman, L. J., and Ahn, T. G.,** *Proc. Natl. Acad. Sci. U.S.A.,* 87, 2329, 1990.
25. **Hildebrandt, J. D., Codina, J., Rosenthal, W., Birnbaumer, L., Neer, E. J., Yamazaki, A., and Bitensky, M. W.,** *J. Biol. Chem.,* 260, 14,867, 1985.
26. **Evans, T., Fawzi, A., Fraser, E. D., Brown, M. L., and Northrup, J. K.,** *J. Biol. Chem.,* 262, 1766, 1987.
27. **Mumby, S., Pang, I. H., Gilman, A. G., and Sternweis, P. C.,** *J. Biol. Chem.,* 263, 2020, 1988.
28. **Kozasa, T., Itoh, H., Tsukamoto, T., and Kaziro, Y.,** *Proc. Natl. Acad. Sci. U.S.A.,* 85, 2082, 1988.
29. **May, D. C., Ross, E. M., Gilman, A. G., and Smigel, M. D.,** *J. Biol. Chem.,* 260, 15,829, 1985.
30. **Yatani, A., Imoto, Y., Codina, J., Hamilton, S. L., Brown, A. M., and Birnbaumer, L.,** *J. Biol. Chem.,* 263, 9887, 1988.
31. **Mattera, R., Graziano, M. P., Yatani, A., Zhou, Z., Graf, R., Codina, J., Birnbaumer, L., Gilman, A. G., and Brown, A. M.,** *Science,* 243, 804, 1989.
32. **Van Dop, C., Tsubokawa, M., Bourne, H. R., and Ramachandran, J.,** *J. Biol. Chem.,* 259, 696, 1984.
33. **Bray, P., Carter, A., Simons, C., Guo, V., Puckett, C., Kamholz, J., Spiegel, A., and Nirenberg, M.,** *Proc. Natl. Acad. Sci. U.S.A.,* 83, 8893, 1986.
34. **Suki, W. M., Abramowitz, J., Mattera, R., Codina, J., and Birnbaumer, L.,** *FEBS Lett.,* 220, 187, 1987.
35. **Jones, D. T. and Reed, R. R.,** *J. Biol. Chem.,* 262, 14,241, 1987.
36. **Yatani, A., Codina, J., Brown, A. M., and Birnbaumer, L.,** *Science,* 238, 1288, 1987.
37. **Codina, J., Olate, J., Abramowitz, J., Mattera, R., Cook, R. G., and Birnbaumer, L.,** *J. Biol. Chem.,* 263, 6746, 1988.
38. **Danilenko, M. P., Panchenko, M. P., Yesirev, O. V., and Tkachuk, V. A.,** *Eur. J. Biochem.,* 194, 155, 1990.
39. **Heschler, J., Rosenthal, W., Trautwein, W., and Schultz, G.,** *Nature,* 325, 445, 1987.
40. **Sterweis, P. C. and Robishaw, J. D.,** *J. Biol. Chem.,* 259, 13,806, 1984.
41. **Goldsmith, P., Backlund, P. S., Rossiter, K., Carker, A., Milligan, G., Unson, C. G., and Spiegel, A.,** *Biochemistry,* 227, 7085, 1988.
42. **Ewald, D. A., Sternweis, P. C., and Miller, R. J.,** *Proc. Natl. Acad. Sci. U.S.A.,* 85, 3633, 1988.
43. **Moriarty, T. M., Padrell, E., Carty, D. J., Omri, G., Landau, E. M., and Iyengar, R.,** *Nature,* 344, 836, 1990.
44. **Strathmann, M., Wilkie, T. M., and Simon, M. I.,** *Proc. Natl. Acad. Sci. U.S.A.,* 87, 6477, 1990.
45. **Jones, D. T. and Reed, R. R.,** *Science,* 244, 790, 1989.
46. **Wheeler, G. L. and Bitensky, M. W.,** *Proc. Natl. Acad. Sci. U.S.A.,* 74, 4238, 1977.
47. **Berger, S. J., DeVries, G. W., Carter, J. G., Schulz, D. W., Passonneau, P. N., Lowry, O. H., and Ferrendelli, J. A.,** *J. Biol. Chem.,* 255, 3128, 1980.
48. **Smreka, A. V., Hepler, J. R., Brown, K. O., and Sternweis, P. C.,** *Science,* 251, 804, 1991.
49. **Fong, H. K., Yoshimoto, K. K., Eversole-Cire, P., and Simon, M. I.,** *Proc. Natl. Acad. Sci. U.S.A.,* 85, 3066, 1988.

50. **Matsuoka, M., Itoh, H., Kozasa, T., and Kazior, Y.,** *Proc. Natl. Acad. Sci. U.S.A.*, 85, 5384, 1988.
51. **Holliday, K. R.,** *J. Cycl. Nucleotide Protein Phosphorylation Res.*, 9, 435, 1983—4.
52. **Kanaho, Y., Tsai, S. C., Adamik, R., Hewlett, E. L., Moss, J., Vaughn, M.,** *J. Biol. Chem.*, 259, 7378, 1984.
53. **Florio, V. A. and Sternweis, P. C.,** *J. Biol. Chem.*, 264, 3909, 1989.
54. **Logothetis, D. E., Kurachi, Y., Galper, J., Neer, E. J., and Clapham, D. E.,** *Nature,* 325, 321, 1987.
55. **Logothetis, D. E., Kim, D., Northyp, J. K., Neer, E. J., and Clapham, D. E.,** *Proc. Natl. Acad. Sci. U.S.A.*, 85, 5814, 1988.
56. **Gilman, A.,** *Cell,* 36, 577, 1984.
57. **Mumby, S. M., Casey, P. J., Gilman, A. G., Gutowski, S., Sternweis, P. C.,** *Proc. Natl. Acad. Sci. U.S.A.*, 87, 5873, 1990.
58. **Jones, T. L. Z., Somonds, W. F., Merendin, J. J., Brann, M. R., and Spiegel, A. M.,** *Proc. Natl. Acad. Sci. U.S.A.*, 87, 568, 1990.
59. **Law, P. Y., Wu, J., Koehler, J. E., and Loh, H. H.,** *J. Neurochem.*, 36, 1834, 1981.
60. **Cooper, D. M. F., Londos, C., Gill, D. L., and Rodbell, M.,** *J. Neurochem.*, 38, 1164, 1982.
61. **Blume, A. J., Lichtshtein, L., and Boone, G.,** *Proc. Natl. Acad. Sci. U.S.A.*, 76, 5626, 1979.
62. **Koski, G. and Klee, W. A.,** *Proc. Natl. Acad. Sci. U.S.A.*, 78, 4185, 1981.
63. **Vachon, L., Costa, T., and Herz, A.,** *Mol. Pharmacol.*, 31, 159, 1986.
64. **Law, P. Y., Hom, D. S., and Loh, J. J.,** *J. Pharmacol. Exp. Ther.*, 256, 710, 1991.
65. **Blume, A. J.,** *Life Sci.*, 22, 1843, 1978.
66. **Chang, K. J., Hazum, E., Killan, A., and Cuatrecasas, P.,** *Mol. Pharmacol.*, 20, 1, 1981.
67. **Law, P. Y., Hom, D. S., and Loh, H. H.,** *J. Biol. Chem.*, 260, 3561, 1985.
68. **Costa, T., Wurster, M., Gramsch, C., and Herz, A.,** *Mol. Pharmacol.*, 28, 146, 1985.
69. **Law, P. Y., Louie, A. K., and Loh, H. H.,** *J. Biol. Chem.*, 260, 14,818, 1985.
70. **Wurster, M., Costa, T., Aktories, K., and Jakobs, K. H.,** *Biochem. Biophys. Res. Commun.*, 123, 1107, 1984.
71. **Hsia, J. A., Moss, J., Hewlett, E. L., and Vaughan, M.,** *J. Biol. Chem.*, 259, 1086, 1984.
72. **Childers, S. R. and Snyder, S. H.,** *J. Neurochem.*, 34, 583, 1980.
73. **Francis, B., Moisand, C., and Meunier, J. C.,** *Eur. J. Pharmacol.*, 117, 223, 1985.
74. **Mollereau, C., Pascaud, A., Baillat, G., Mazarguil, H., Puget, A., and Meunier, J. C.,** *J. Biol. Chem.*, 263, 18,003, 1988.
75. **Fedynyshyn, J. P. and Lee, N. M.,** *Brain Res.*, 476, 102, 1989.
76. **Clark, M. J., Levenson, S. D., and Medzihradsky, F.,** *Life Sci.*, 39, 1721, 1986.
77. **Ueda, H., Harada, H., Nozaki, M., Katada, T., Ui, M., Satoh, M., Takagi, H.,** *Proc. Natl. Acad. Sci. U.S.A.*, 85, 7013, 1988.
78. **Ueda, H., Uno, S., Harada, J., Kobayashi, I., Katada, T., Ui, M., and Satoh, M.,** *FEBS Lett.*, 266, 178, 1990.
79. **Yu, V. C., Richards, M. L., and Sadee, W.,** *J. Biol. Chem.*, 261, 1065, 1986.
80. **Kazmi, S. M. I. and Mishra, R. K.,** *Mol. Pharmacol.*, 32, 109, 1987.
81. **Puttfarcken, P. S., Werling, L. L., and Cox, B. M.,** *Mol. Pharmacol.*, 33, 520, 1988.
82. **Mack, K. J., Lee, M. F., and Wehenmeyer, J. A.,** *Brain Res. Bull.*, 14, 301, 1985.
83. **Aghajanian, G. K. and Wang, Y. Y.,** *Brain Res.*, 371, 390, 1986.
84. **Crain, S. M., Crain, B., and Makman, M. H.,** *Brain Res.*, 370, 61, 1986.
85. **Yatani, A., Mattera, R., Codina, J., Graf, R., Okabe, K., Padrell, E., Iyengar, R., Brown, A. M., and Birnbaumer, L.,** *Nature,* 336, 680, 1988.
86. **McFadzean, I., Mullaney, I., Brown, D. A., and Milligan, G.,** *Neuron,* 3, 177, 1989.

87. **McKenzie, F. R. and Milligan, M.,** *Biochem. J.,* 267, 391, 1990.
88. **Murayama, T. and Ui, M.,** *J. Biol. Chem.,* 259, 761, 1984.
89. **Goldstein, A. and Naidu, A.,** *Mol. Pharmacol.,* 36, 265, 1989.
90. **McLawhon, R. W., Ellory, J. C., and Dawson, G.,** *J. Biol. Chem.,* 258, 2101, 1983.
91. **Ott, S., Costa, T., Wuster, M., Hietel, B., and Herz, A.,** *Eur. J. Biochem.,* 155, 621, 1986.
92. **Ott, S., Costa, T., and Herz, A.,** *J. Biol. Chem.,* 263, 10,524, 1988.
93. **Costa, T., Klinz, F. J., Vachon, L., and Herz, A.,** *Mol. Pharmacol.,* 34, 744, 1988.
94. **Costa, T. and Herz, A.,** *Proc. Natl. Acad. Sci. U.S.A.,* 86, 7321, 1989.
95. **Collier, H. O. J. and Roy, A. C.,** *Nature,* 248, 24, 1974.
96. **Katz, J. B. and Catravas, G. N.,** *Brain Res.,* 120, 263, 1977.
97. **Tell, G. P., Pastanak, G. W., and Cuatrecasas, P.,** *FEBS Lett.,* 51, 242, 1975.
98. **Walczak, S. A., Wikening, D., and Makman, M. H.,** *Brain Res.,* 160, 105, 1979.
99. **Sharma, S. K., Nirenberg, M., and Klee, W. A.,** *Proc. Natl. Acad. Sci. U.S.A.,* 72, 590, 1975.
100. **Traber, J., Fischer, K., Latzin, S., and Hamprecht, B.,** *Nature,* 253, 120, 1975.
101. **Duman, R. S., Tallman, J. F., and Nestler, E. J.,** *J. Pharmacol. Exp. Ther.,* 246, 1033, 1988.
102. **Childers, S. R.,** *J. Neurochem.,* 50, 543, 1988.
103. **Law, P. Y., Koehler, J. E., and Loh, H. H.,** *Mol. Pharmacol.,* 21, 483, 1982.
104. **Gilbert, J. A., Knodel, E. L., Stenstrom, S. D., and Richelson, E.,** *J. Biol. Chem.,* 257, 1274, 1982.
105. **Makman, M. H., Dvorkin, B., and Crain, S. M.,** *Brain Res.,* 445, 303, 1988.
106. **Fantozzi, R., Mullikin-Kilpatrick, D., and Blume, A. J.,** *Mol. Pharmacol.,* 20, 8, 1981.
107. **Childers, S. R. and Jackson, J. L.,** *J. Neurochem.,* 43, 1163, 1984.
108. **Lambert, S. M. and Childers, S. R.,** *J. Neurosci.,* 4, 2755, 1984.
109. **Childers, S. R., Lambert, S. M., and Lariviere, G.,** *Life Sci.,* 33 (Suppl. 1), 215, 1983.
110. **Childers, S. R. and Lariviere, G.,** *J. Neurosci.,* 4, 2764, 1984.
111. **Rasenick, M. M. and Childers, S. R.,** *J. Neurochem.,* 53, 219, 1989.
112. **Griffin, M. T., Law, P. Y., and Loh, H. H.,** *J. Neurochem.,* 45, 1585, 1985.
113. **Girardot, J. M., Kempf, J., and Cooper, D. M. F.,** *J. Neurochem.,* 41, 848, 1983.
114. **Baram, D. and Simantov, R.,** *J. Neurochem.,* 40, 55, 1983.
115. **Nehmand, R., Nadler, H., and Simantov, R.,** *Mol. Pharmacol.,* 22, 389, 1982.
116. **Katada, T., Kusakabe, K., Oinuma, M., and Ui, M.,** *J. Biol. Chem.,* 262, 11,897, 1987.
117. **Clark, M. J., Nordby, G. L., and Medzihradsky, F.,** *J. Neurochem.,* 52, 1162, 1989.
118. **Koski, G. and Klee, W. A.,** *Proc. Natl. Acad. Sci. U.S.A.,* 78, 4185, 1981.
119. **Smith, M. M. and Harden, T. K.,** *J. Pharmacol. Exp. Ther.,* 228, 425, 1981.
120. **Peralta, E. G., Ashkenazi, A., Winslow, J. W., Smith, D. H., Ramachandran, J., and Capon, D. J.,** *EMBO J.,* 6, 3923, 1988.
121. **Ashkenzi, A., Winslow, J. W., Peralta, E. G., Peterson, G. L., Schimerlik, M. I., Capon, D. J., and Ramachandran, J.,** *Science,* 238, 672, 1987.
122. **Ashkenzi, A., Peralta, E. G., Winslow, J. W., Ramachandran, J., and Capon, D. J.,** Cold Spring Harbor Symposia on Quantitative Biology V. LIII, Cold Spring Harbor, NY, 263, 1988.
123. **Vachon, L., Costa, T., and Herz, A.,** *Biochem. Pharmacol.,* 36, 2889, 1987.
124. **Mullaney, I. and Milligan, G.,** *J. Neurochem.,* 55, 1890, 1990.
125. **Milligan, G. and McKenzie, F. R.,** *Biochem. J.,* 252, 369, 1988.
126. **Klinz, F. J. and Costa, T.,** *Biochem. Biophys. Res. Commun.,* 165, 554, 1989.
127. **Roeriz, S., Law, P. Y., and Loh, H. H.,** *J. Biol. Chem.,* submitted 1991.
128. **Offermanns, S., Schultz, G., and Rosenthal, W.,** *J. Biol. Chem.,* 266, 3365, 1991.

129. **Tsunoo, A., Yoshii, M., and Narahashi, T.,** *Proc. Natl. Acad. Sci. U.S.A.,* 83, 9832, 1986.
130. **Noronha-Blob, L., Richard, C., and U'Prichard, D. C.,** *J. Neurochem.,* 50, 1381, 1988.
131. **Eckert, E., Hescheler, J., Krautwurst, D., Schultz, G., and Trautwein, W.,** *Pfleugers Arch.,* 417, 329, 1990.
132. **Higashida, H., Streaty, R. A., Klee, W., and Nirenberg, M.,** *Proc. Natl. Acad. Sci. U.S.A.,* 83, 942, 1986.
133. **Katada, T., Oinuma, M., Kusakabe, K., and Ui, M.,** *FEBS Lett.,* 213, 1987.
134. **Chalazonitis, A. and Crain, S. M.,** *Neuroscience,* 17, 1181, 1986.
135. **Crain, S. M., Crain, B., and Peterson, E. R.,** *Brain Res.,* 370, 61, 1986.
136. **Crain, S. M., Brain, B., and Makman, M. H.,** *Brain Res.,* 400, 185, 1987.
137. **Mihara, S. and North, R. A.,** *Br. J. Pharmacol.,* 88, 315, 1986.
138. **North, R. A., Williams, J. T., Supernant, A., and Christie, M. J.,** *Proc. Natl. Acad. Sci. U.S.A.,* 84, 5487, 1987.
139. **Williams, J. T., North, R. A., and Tokimasa, T.,** *J. Neurochem.,* 8, 4299, 1988.
140. **Williams, J. T., Egan, T. M., and North, R. A.,** *Nature,* 299, 74, 1988.
141. **North, R. A. and Williams, J. T.,** *Br. J. Pharmacol.,* 80, 225, 1983.
142. **North, R. A. and Williams, J. T.,** *J. Physiol.,* 364, 265, 1985.
143. **Zieglgansberger, W., French, E. D., Siggins, G. R., and Bloom, F.,** *Science,* 205, 415, 1979.
144. **Nicoll, R. A., Alger, B. E., and Jahr, C. E.,** *Nature,* 287, 22, 1980.
145. **Nicoll, R. A.,** *Science,* 241, 545, 1988.
146. **Masukawa, L. M. and Prince, D. A.,** *Brain Res.,* 249, 271, 1982.
147. **Siggins, G. R. and Zieglgansberger, W.,** *Proc. Natl. Acad. Sci. U.S.A.,* 78, 5235, 1981.
148. **Lee, H. K., Dunwiddie, T., and Hoffer, B.,** *Brain Res.,* 184, 331, 1980.
149. **Madison, D. V. and Nicoll, R. A.,** *J. Physiol.,* 398, 123, 1988.
150. **North, R. A., Katayama, Y., and Williams, J. T.,** *Brain Res.,* 165, 67, 1979.
151. **Johnson, S. M., Pillai, N. P., and Costa, M.,** *Adv. Biosci.,* 75, 169, 1989.
152. **Louis, A. K., Bass, E. S., Zhan, J., Law, P. Y., and Loh, H. H.,** *J. Pharmacol. Exp. Ther.,* 253, 401, 1990.
153. **Katada, T., Gilman, A. G., Watanabe, Y., Bauer, S., and Jakobs, K. H.,** *Eur. J. Biochem.,* 151, 431, 1985.
154. **Kurose, H. and Ui, M.,** *Arch. Biochem. Biophys.,* 238, 424, 1985.
155. **Gullis, R., Traber, J., and Hamprecht, B.,** *Nature,* 256, 57, 1975.
156. **Isom, L. L., Cragoe, E. J., and Limbird, L. E.,** *J. Biol. Chem.,* 262, 17,504, 1987.
157. **Ho, A. K., Chik, C. L., Weller, J. L., Cragoe, E. J., and Klein, D. C.,** *J. Biol. Chem.,* 264, 12,983, 1989.
158. **Minneman, K. P. and Iversen, L. I.,** *Nature,* 262, 313, 1976.
159. **Martin, W. R., Eades, C. G., Thompson, J. A., Huppler, R. E., and Gilbert, P. E.,** *Pharmacol. Exp. Ther.,* 197, 517, 1976.
160. **Chang, K. J., Hazum, E., and Cautercasas, P.,** *Proc. Natl. Acad. Sci. U.S.A.,* 77, 4469, 1980.
161. **Chang, K. J., Hazum, E., Killian, A., and Cautrecasas, P.,** *Mol. Pharmacol.,* 20, 1, 1981.
162. **Szucs, M., Spain, J. W., Oetting, G. M., Moudy, A. M., and Coscia, C. J.,** *J. Neurochem.,* 54, 841, 1990.
163. **Ueda, H., Misawa, H., Katada, T., Ui, M., Takagi, H., and Satoh, M.,** *J. Neurochem.,* 54, 841, 1990.
164. **Heyman, J. S., Mulvaney, S. A., Mosber, H. I., and Porreca, F.,** *Brain Res.,* 420, 100, 1987.
165. **Rothman, R. B. and Westfall, T. C.,** *Mol. Pharmacol.,* 21, 548, 1982.

166. **Hochhaus, G., Yu, V. C., and Sadee, W.,** *Brain Res.*, 382, 327, 1986.
167. **Yu, V. C., Richards, M. L., and Sadee, W.,** *J. Biol. Chem.*, 261, 1065, 1986.
168. **Kazmi, S. M. I. and Mishra, R. K.,** *Biochem. Biophys. Res. Commun.*, 137, 813, 1986.
169. **Toll, L.,** *Eur. J. Pharmacol.*, 176, 213, 1990.
170. **Frey, E. A. and Kebabian, J. W.,** *Endocrinologia*, 115, 1797, 1984.
171. **Puttfarccken, P. S., Weling, L. L., and Cox, B. M.,** *Mol. Pharmacol.*, 33, 520, 1988.
172. **Fedynyshyn, J. P. and Lee, N. M.,** *FEBS Lett.*, 253, 23, 1989.
173. **Schoffelmeer, A. N. M., Hogenboom, F., and Mulder, A. H.,** *Naunyn-Schmiedeberg's Arch. Pharmacol.*, 335, 278, 1987.
174. **Schoffelmeer, A. N. M., Rice, K. C., Heijna, M. H., Hogenboom, F., and Mulder, A. H.,** *Eur. J. Pharmacol.*, 149, 179, 1988.
175. **Schoffelmeer, A. N. M., Hansen, H. A., Stoof, J. C., and Mulder, A. H.,** *J. Neurosci.*, 2235, 1986.
176. **Nestler, E. J. and Tallman, J. F.,** *J. Pharmacol. Exp. Ther.*, 33, 127, 1988.
177. **Nestler, E. J., Erdos, J. J., Terwilliger, R., Duman, R. S., Tallman, J. F.,** *Brain Res.*, 476, 230, 1989.
178. **Barchfeld, C. C. and Medzihradsky, F.,** *Biochem. Biophys. Res. Commun.*, 121, 641, 1984.
179. **Franklin, P. H. and Hoss, W.,** *J. Neurochem.*, 43, 1132, 1984.
180. **Kelly, E. and Nahorski, S. R.,** *J. Neurochem.*, 47, 1512, 1986.
181. **Schoffelmeer, A. N. M., Hogenboom, F., and Mulder, A. H.,** *Naunyn-Schmiedeberg's Arch. Pharmacol.*, 338, 548, 1988.
182. **Atweh, S. F. and Kuhar, M. J.,** *Brain Res.*, 29, 1, 1977.
183. **Fedynyshyn, J. P., Kwiat, G., and Lee, N. M.,** *Eur. J. Pharmacol.*, 159, 83, 1989.
184. **Fedynyshyn, J. P. and Lee, N. M.,** *Brain Res.*, 476, 102, 1989.
185. **Morita, K. and North, R. A.,** *Brain Res.*, 242, 145, 1982.
186. **Suprenant, A. and North, R. A.,** *Neuroscience*, 16, 425, 1985.
187. **Williams, J. T. and North, R. A.,** *Mol. Pharmacol.*, 26, 489, 1984.
188. **Satao, T., Takayanagi, I., and Takagi, K.,** *Jpn. J. Pharmacol.*, 23, 665, 1973.
189. **Dingledine, R. and Goldstein, A.,** *Life Science*, 17, 62, 1975.
190. **North, R. A. and Williams, J. T.,** *Eur. J. Pharmacol.*, 45, 23, 1977.
191. **Karras, P. J. and North, R. A.,** *J. Pharmacol. Exp. Ther.*, 217, 70, 1981.
192. **Dingledine, R. and Goldstein, A.,** *J. Pharmacol. Exp. Ther.*, 196, 97, 1976.
193. **North, R. A. and Tonini, M.,** *Br. J. Pharmacol.*, 61, 541, 1977.
194. **Miyake, M., Christie, M. J., and North, R. A.,** *Proc. Natl. Acad. Sci. U.S.A.*, 86, 3419, 1989.
195. **Chapman, D. B. and Way, E. L.,** *Rev. Pharmacol. Toxicol.*, 20, 553, 1980.
196. **Huidobro-Toro, J. P., Hu, J., and Way, E. L.,** *J. Pharmacol. Exp. Ther.*, 218, 84, 1981.
197. **Sawynok, J. and Jhamandas, K. H.,** *J. Pharmacol. Exp. Ther.*, 197, 379, 1976.
198. **Opmeer, F. A. and Van Ree, J. M.,** *J. Pharmacol. Exp. Ther.*, 213, 188, 1980.
199. **Dougall, I. G. and Leff, P.,** *Br. J. Pharmacol.*, 92, 723, 1987.
200. **Remmers, A. E. and Medzihradsky, F.,** *Proc. Natl. Acad. Sci. U.S.A.*, 88, 2171, 1991.
201. **Lahti, R. A., VonVoightlander, P. F., and Barsuhn, C.,** *Life Sci.*, 31, 2257, 1982.
202. **Leander, J. D., Zerbe, R. L., and Hart, J.,** *J. Pharmacol. Exp. Ther.*, 234, 463, 1985.
203. **Chavkin, C. and Goldstein, A.,** *Proc. Natl. Acad. Sci. U.S.A.*, 78, 6543, 1981.
204. **Schulz, R., Wurster, M., and Herz, A.,** *Peptides*, 3, 973, 1982.
205. **Chavkin, C. and Goldstein, A.,** *Science*, 215, 413, 1982.
206. **Schoffelmeer, A. N. M., Rice, K. C., Jacobson, A. E., Gelderen, A. G. V., Hogenboom, F., Heijna, M. H., and Mulder, A. H.,** *Eur. J. Pharmacol.*, 154, 169, 1988.

207. **Werling, L. L., McMahon, P. N., and Cox, B. M.,** *Naunyn-Schmiedelberg's Arch. Pharmacol.*, 339, 509, 1989.
208. **Allgaier, C., Daschmann, B., Sieverling, J., and Hertting, G.,** *J. Neurochem.*, 51, 542, 1988.
209. **Adamson, P., Brammer, M. J., and Campbell, I. C.,** *J. Neurochem.*, 51, 542, 1988.
210. **Crain, S. M. and Shen, K. F.,** *Trends Pharmacol. Sci.*, 11, 77, 1990.
211. **Gross, R. A. and MacDonald, R. L.,** *Proc. Natl. Acad. Sci. U.S.A.*, 84, 5469, 1987.
212. **Bradley, P. B. and Brookes, A.,** *Br. J. Pharmacol.*, 83, 763, 1984.
213. **Micevych, P. E., Yaksh, T. L., and Go, V. L. W.,** *Brain Res.*, 290, 87, 1984.
214. **Mulder, A. H., Wardeh, G., Hogenboom, F., and Frankhyyzen, A. L.,** *Nature*, 308, 278, 1984.
215. **McFadzean, I., Lacey, M. G., Hill, R. G., and Henderson, G.,** *Neuroscience*, 20, 231, 1987.
216. **Cherubini, E. and North, R. A.,** *Proc. Natl. Acad. Sci. U.S.A.*, 82, 1860, 1985.
217. **Krulich, L., Koenig, J. I., Conway, S., McCann, S. M., and Mayfield, M. A.,** *Neuroendocrinology*, 42, 82, 1986.
218. **Itzhak, Y., Hiller, J. M., and Simon, E. J.,** *Proc. Natl. Acad. Sci. U.S.A.*, 81, 4217, 1984.
219. **Porthe, G., Valette, A., and Cros, J.,** *Biochem. Biophys. Res. Commun.*, 101, 1, 1981.
220. **Parthe, G., Valette, A., Moisand, A., Tafani, M., and Cros, J.,** *Life Sci.*, 31, 2647, 1982.
221. **Simon, E. J., Hiller, J. M., Groth, J., Itzhak, Y., Holland, M. J., and Beck, S. G.,** *Life Sci.*, 31, 1367, 1982.
222. **Mack, K. J., Lee, M. F., and Weyhenmeyer, J. A.,** *Brain Res. Bull.*, 14, 301, 1985.
223. **Pfeiffer, A., Sadee, W., and Herz, A.,** *J. Neurosci.*, 2, 912, 1982.
224. **Ishizuka, U. and Oka, R.,** *Jpn. J. Pharmacol.*, 36, 397, 1984.
225. **Clark, M. J. and Medzihradsky, F.,** *Neuropharmacology*, 26, 1763, 1987.
226. **Frances, B., Moisand, C., and Meunier, J. C.,** *Eur. J. Pharmacol.*, 117, 223, 1985.
227. **Porthe, G., Frances, B., Verrier, J., Cros, J., and Meunier, J. C.,** *Life Sci.*, 43, 559, 1988.
228. **Attali, B., Saya, D., and Vogel, Z.,** *J. Neurochem.*, 52, 360, 1989.
229. **Makman, M. H., Dvorkin, B., and Crain, S. M.,** *Brain Res.*, 445, 303, 1988.
230. **Konkoy, C. and Childers, S. R.,** *Mol. Pharmacol.*, 36, 627, 1989.
231. **Werz, M. A. and MacDonald, R. L.,** *Brain Res.*, 239, 315, 1982.
232. **Werz, M. A. and MacDonald, R. L.,** *J. Pharmacol. Exp. Ther.*, 394, 1983.
233. **Werz, M. A. and MacDonald, R. L.,** *Neurosci. Lett.*, 46, 185, 1984.
234. **Werz, M. A. and MacDonald, R. L.,** *Neuropeptides*, 5, 253, 1984.
235. **Werz, M. A. and MacDonald, R. L.,** *J. Pharmacol. Exp. Ther.*, 234, 49, 1985.
236. **Werz, M. A., Grega, D. S., and MacDonald, R. L.,** *J. Pharmacol. Exp. Ther.*, 243, 1987.
237. **MacDonald, R. L. and Werz, M. A.,** *J. Physiol.*, 398, 123, 1988.
238. **Shen, K. F. and Crain, S. M.,** *Brain Res.*, 491, 227, 1989.
239. **Shen, K. F. and Crain, S. M.,** *Neuropharmacology*, 29, 343, 1990.
240. **Shen, K. F. and Crain, S. M.,** *Brain Res.*, 525, 225, 1990.
241. **Shen, K. F. and Crain, S. M.,** *Brain Res.*, 531, 1, 1990.
242. **Guitart, X. and Nestler, E. J.,** *J. Neurosci.*, 9, 4371, 1989.
243. **Guitart, S. and Nestler, E. J.,** *Brain Res.*, 516, 57, 1990.
244. **Nestler, E. J., Walaas, S. I., and Greengard, P.,** *Science*, 225, 1357, 1984.
245. **Thayer, S.,** personal communication.

Chapter 4

THE NEUROBIOLOGY OF FENTANYL AND ITS DERIVATIVES

Paul V. Polishuk and Henry C. Nipper

TABLE OF CONTENTS

I. INTRODUCTION

Fentanyl and its derivatives have earned notoriety for their great potential for abuse. The tremendous opioid potency of these drugs, with a minimum of side effects, is an important contributor to that potential. The following pages will review what is currently known about the neurobiological activity of the fentanyls.

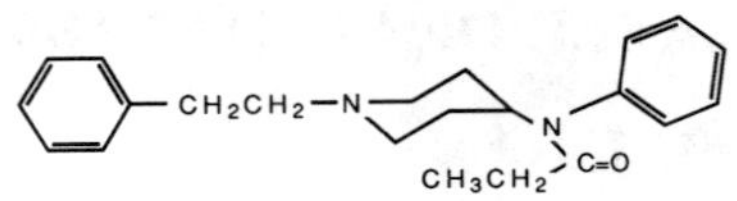

Fentanyl (**1**)

Fentanyl is the common name that was given to this molecule by Janssen Pharmaceutica of Belgium, who first synthesized it.[1] The trade name for fentanyl is Sublimaze®, which is also marketed by Janssen. The actual chemical name for this molecule is *N*-(1-phenethyl-4-piperidyl)propionanilide. Structural modifications of this molecule will be designated by conventional chemical nomenclature (i.e., α-methylfentanyl). To date, there have been a limited number of derivatives having the fentanyl structure that have appeared as street drugs. To simplify references to this class of compounds, the term fentanyls will be used as a general term to specify this whole group of related compounds.

Fentanyl is a very potent opioid narcotic estimated to be 80 to 300 times more potent than morphine.[2,3] It is short-acting and, therefore, has been found to be effective as an anesthetic during surgery.[4] Many of the derivatives of fentanyl have been shown to be even more powerful narcotics than fentanyl. Carfentanil has been reported to be over 7.8×10^3 times as potent as morphine.[5] Since 1979, fentanyl derivatives have been linked to over 110 overdose deaths,[6] which is a conservative estimate due to the difficulty of its detection in physiological fluids, and its use with other narcotics. Table 1 shows the thirteen fentanyl derivatives which have been encountered in forensic casework, and some of the other major neuroactive fentanyls.

Fentanyl and its derivatives are known as opioids; they have a similar mode of action to the naturally occurring opium-based narcotics, such as morphine. The opioids act mainly on the central nervous system (CNS) and the gastrointestinal (GI) tract. They are able to eliminate pain without initiating loss of consciousness. These compounds produce analgesia, drowsiness, changes in mood, euphoria, respiratory depression, decreased gastrointestinal motility, nausea, vomiting (except fentanyl), muscular rigidity, and changes in the endocrine and autonomic nervous system (mostly centrally).[2]

The opioids work through various types of opioid receptors, which are on surfaces of neurons and other target cells. The opioids actually mimic natural peptides found in our bodies which are known as endorphins, enkephalins, and dynorphins. When opioids bind to these receptors, they stimulate a cellular response. The potency of a substance refers, in general, to how strongly it binds to a receptor. For the more potent substances, less drug is needed to produce the same response. The opioid receptors have been categorized into subtypes: primarily, mu, kappa, sigma, and delta. These subtypes: mediate different pharmacological responses of the opioids. It has been determined that the mu subtype is specifically involved with supraspinal analgesia and addiction potential. Fentanyls have been found to be particularly potent at this receptor subtype.[7]

The opioids are known to be toxic substances because they depress respiration. This depression of respiration is due to the reaction of the opioids in the respiratory center in the brainstem which senses carbon dioxide levels in the blood. The depression of respiration is the likely cause of death in opioid overdose. Respiratory depression causes hypoxia, the lack of oxygen, which, in turn, may cause a drop in blood pressure, or hypotension. Most opioids, but not fentanyl, also stimulate the release of histamine into the circulatory system. The release of histamine also increases hypotension, and therefore may contribute to cardiovascular collapse.

Fentanyl and its analogs have the same respiratory depressant effects found with other opioids, however, its duration of action is much shorter. Fentanyl, unlike other opioids, does not promote the release of histamine, thus there is a greater margin of safety against cardiovascular collapse.

The margin of safety is measured by dividing the lethal dose of a drug for 50% of the population (LD_{50}) by the lowest effective dose to 50% of the population (ED_{50}). In general, the higher the margin of safety, the safer the drug is to use. Fentanyl's high margin of safety is valuable in medicine. The margin of safety for morphine is 71, while for fentanyl it is 277. Another fentanyl derivative, the highly potent sufentanil, which is being tested in clinical trials, has a safety margin of 26,716.[8] The lethal dose of sufentanil, however, is less than one tenth that of morphine. From recent research, it appears that the fentanyl derivatives have a mode of action very similar if not identical to fentanyl itself. It is believed that they act upon the same opioid receptors. The great potency of these analogs, as well as their large differences in potency, is of great interest to researchers. Using classical medicinal chemistry theories, one can predict potential fentanyl derivatives that are both active and inactive.

II. FENTANYL ANALGESIA

Fentanyl differs from general anesthetics in that fentanyl acts on specific nerve receptor sites, primarily in the central nervous system. General anesthetics depress all sensation, rendering the patient unconscious. Opioid sites are primarily related to pain sensation, control of respiration, physical de-

TABLE 1
Fentanyl and Its Major Derivatives

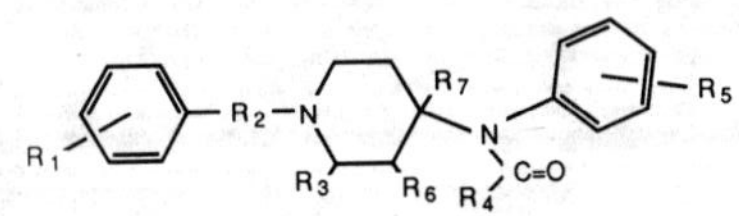

Compound Number	Common Name	R_1	R_2	R_3	R_4	R_5	R_6	R_7
1[a]	Fentanyl		CH_2CH_2		CH_2CH_3			
2			CH_2		CH_3			
3	Benzylfentanyl		CH_2		CH_2CH_3			
4			CH_2CH_2		CH_3			
5			$CH_2CH_2CH_2$		CH_3			
6			$CH_2CH_2CH_2$		CH_2CH_3			
7[a]	α-Methylfentanyl		$CH_2CH(CH_3)$		CH_2CH_3			
8			CH_3		CH_3	*o*-CH_3		
9			CH_2CH_2		CH_3	*m*-CH_3		
10			CH_2CH_2		CH_3	*p*-CH_3		
11			CH_2CH_2		CH_2CH_3	*o*-CH_3		
12			CH_2CH_2		CH_2CH_3	*m*-CH_3		
13			CH_2CH_2		CH_2CH_3	*p*-CH_3		
14		*o*-CH_3	CH_2CH_2		CH_3			
15		*m*-CH_3	CH_2CH_2		CH_3			
16		*p*-CH_3	CH_2CH_2		CH_3			
17		*o*-CH_3	CH_2CH_2		CH_2CH_3			
18		*m*-CH_3	CH_2CH_2		CH_2CH_3			
19		*p*-CH_3	CH_2CH_2		CH_2CH_3			
20			CH_2CH_2	CH_3	CH_2CH_3			
21			$CH_2CH(CH_3)$		CH_3			
22			$CH_2CH(CH_3)$		CH_2CH_3			

23[a]	*p*-Fluorofentanyl		CH_2CH_2		CH_2CH_3	p-F		
24			CH_2CH_2		CH_2CH_3	o-F		
25			CH_2CH_2		CH_3	m-F		
26			CH_2CH_2		CH_2CH_3	m-F		
27[a]	3-Methylfentanyl		CH_2CH_2		CH_2CH_3		*cis*-CH_3	
27b[a]			CH_2CH_2		CH_2CH_3		*trans*-CH_3	
28			CH_2CH_2		CH_3		*trans*-CH_3	
29[a]	Acetyl-α-methylfentanyl		$CH_2CH(CH_3)$	CH_3				
30[a]	α-Methylacrylfentanyl		$CH_2CH(CH_3)$		$CH=CH_2$			
31[a]	α-Methylthiofentanyl	Th	$CH_2CH(CH_3)$		CH_2CH_3			
32[a]	β-Hydroxyfentanyl		$CHOHCH_2$		CH_2CH_3			
33[a]	β-Hydroxy-3-methylfentanyl		$CHOHCH_2$		CH_2CH_3		CH_3	
34[a]	3-Methylthiofentanyl	Th	CH_2CH_2		CH_2CH_3		CH_3	
35[a]	Thiofentanyl	Th	CH_2CH_2		CH_2CH_3			
36[a]	Thenylfentanyl	Th	CH_2		CH_2CH_3			
37	Sufentanil	Th	CH_2CH_2		CH_2CH_3			CH_2OCH_3
38	Carfentanil		CH_2CH_2		CH_2CH_3			CO_2CH_3
39	Lofentanil		CH_2CH_2		CH_2CH_3		CH_3	CO_2CH_3
40	Alfentanil	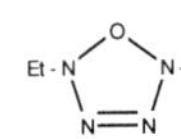	CH_2CH_2		CH_2CH_3			CH_2OCH_3

[a] Compounds encountered in forensic casework. 2-Thienyl (Th) replaces phenyl where indicated.

pendence, and miosis. It has been determined that these sites are also the sites for binding the body's opioid-like peptides, the enkephalins, endorphins, and dynorphins.

Extensive research efforts are currently underway into determining the mode of action of the synthetic opioids, and also into finding derivatives that offer advantages over existing anesthetics and analgesics. Investigations are evaluating the effectiveness of opioids in terms of: effectiveness in pain-relieving properties, or blocking of noxious stimuli (also known as antinociceptive potency); minimal behavioral effects, such as side effects; and the reversal of action by opioid antagonists. To gauge the potency of analgesic activity, a number of animal tests have been used. Most of these tests rely on the response to a noxious stimulus. Dose-response plots can be made, and an effective analgesic dose can be determined. It has been found that the pain-inhibiting action, in laboratory animals, is a general guide to their efficacy in humans. A standard drug, such as morphine, can be used as a standard for comparing their relative activity. The method of administration is important in determining potency, which can vary depending on whether given orally (po), subcutaneously (sc), intramuscularly (im), or intravenously (iv).

The techniques used most commonly to assess analgesics are the following: the rat tail-flick test (RTF) which involves the application of heat to the tip of the tail, which causes it to flick; the mouse hot-plate test (MHP), which involves the placement of the animal on a hot plate, and determining how long it takes to escape from this situation; the rat tail withdrawal (RTW) involves putting the tail directly into a warm liquid, and determining when it tries to escape. There are other tests used which employ pressure or chemical irritants, but the above mentioned tests are the most common. A more current method being used is the *in vitro* electric stimulation of smooth muscle to cause muscle contractions. The opioids will inhibit these contractions. The muscles normally used are guinea pig ileum (GPI), or mouse vas deferens (MVD). These tests can measure the potency of both opioid agonists and antagonists.

Reducing unfavorable side effects is a major concern in the development of safer analgesics and anesthetics. Most opioids such as morphine have exhibited side effects. Fentanyl has an advantage over morphine in that it has a much milder effect on the emetic trigger zone of the medulla, which causes nausea and vomiting.

In the future, fentanyl derivatives are expected to be of even greater medical importance than fentanyl. In addition to offering effective analgesia, they are already being used in combination with other drugs for anesthesia. The term neuroleptanalgesia has been coined for the use of a combination of a butyrophenone with an opioid to achieve anesthesia. By using this procedure, anesthesia is often safer. The patient can be induced quickly, and then awakened shortly after the procedure is completed. In addition, the patient is

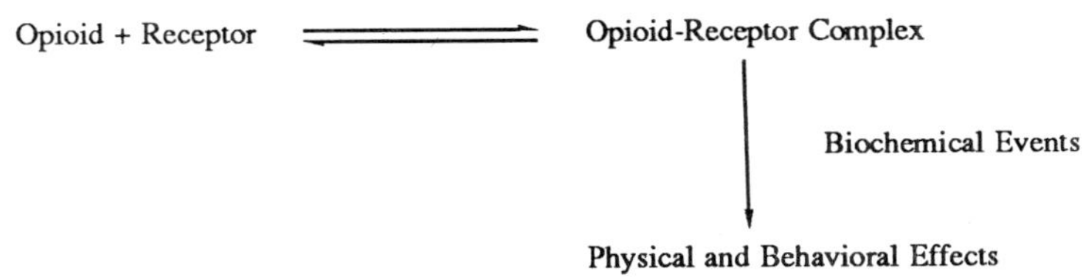

FIGURE 1. How opioids produce their effects.

stabilized easily, and the anesthesiologist has greater control in the event of medical complications. Newer fentanyls are expected to offer even greater safety.

Through structural alterations, side effects may be minimized or eliminated without compromising the therapeutic effects. Improved understanding of the natural and synthetic opioids, and opioid receptors may point the way to agents with analgesic effects but without the unwanted side effects.

III. MODE OF ACTION OF THE FENTANYLS

The exact mechanism by which the nerve cell responds to the opioid interaction, at the receptor, is unknown. The mechanism(s) by which opioids interact at the nerve cell receptors to effect cellular change is only partly known. This process is depicted in Figure 1. Evidence suggests that opioid receptors are coupled to guanine nucleotide-binding regulatory proteins, known as G-proteins. Thus opioids probably regulate transmembrane signaling systems characteristically initiated by this type of receptor (i.e., adenyl cyclase, ion channels, phospholipases).[2] Examples of this mechanism include the following steps. Stimulation of the mu receptor in the locus ceruleus can activate inward rectifying potassium channels causing membrane hyperpolarization. Mu agonists can also inhibit adenylate cyclase in many regions of the brain.[2] These are examples of signal transduction effects mediated by mu agonists, such as fentanyl, that may be involved in the inhibition of pain.

The analgesia associated with the opioids is mediated through CNS actions, both spinal and supraspinal. The mu agonists selectively inhibit nociceptive circuits resulting in potent analgesia. It has been found that the mu opioid stimulation of the ventral tegmentum activates dopaminergic neurons going to the nucleus accumbens. This may be part of the reinforcing features of opioids, and thus their euphorogenic effects on the user.[2]

The process by which the opioids exert their analgesic actions is a complex sequence of neurological events involving many parts of the nervous system. The exact routes of activation are not known, but areas of the brain which contain both opioid-binding receptors and opioid-containing neurons include the hypothalamus, midbrain gray, brainstem reticular formation, and dorsal spinal gray and sensory trigeminal nuclei.[9] Opioid receptors on the terminals of primary afferent nerves mediate the inhibition of release of substance P,[2] which is known to be involved in the mediation of pain. Mu agonists also antagonize the effects of administered substance P by postsynaptic inhibitory actions on interneurons and on output neurons of the spinothalamic tract that convey nocicoceptive signals to higher centers of the brain.[2] It seems that most areas of the brain and their circuits are affected to some degree by the action of the opioids.

A method to determine the location of the specific opioid receptors utilizes radiolabeled opioids as tracer compounds. Unfortunately, these radiolabeled opioids bind nonspecifically to many more membrane sites than just to the opioid receptors themselves. Nonetheless, specific opioid binding has been detected in the regions of the CNS listed above.

The use of *in vitro* testing techniques, such as using guinea pig ilium (GPI) and mouse vas deferens (MVD), and the use of radiolabeled opioids has made it clear that there are naturally occurring opioidlike peptides in humans. In 1975, Hughes et al.[10] characterized two pentapeptides from pig brain tissue which acted as agonists at the opioid receptor sites. These were termed enkephalins and were found to be a mixture of two peptides. One of the peptides, found in a greater concentration, is generally referred to as Met-enkephalin, and the other peptide is referred to as Leu-enkephalin. It was also astutely observed that the amino acid sequence of Met-enkephalin was present in the structure of β-lipotropin.[10]

Tyr-Gly-Gly-Phe-Met(OH)	Tyr-Gly-Gly-Phe-Leu(OH)
Met-enkephalin	Leu-enkephalin

More recently, two other classes of natural opioidlike compounds, which are composed of longer polypeptides, the endorphins and dynorphins, have been discovered. It is now generally accepted that the synthetic opioids interact with the same receptor sites as these endogenous compounds. The presence of subtypes of opioid receptors was demonstrated in 1976.[11,12] Since then, additional types of receptors have been identified, and subtypes have been further delineated.

IV. STRUCTURE-ACTIVITY RELATIONSHIPS

Fentanyl is a member of the 4-anilinopiperidine group of opioid analgesics. Meperidine, another synthetic opioid, is structurally similar to fentanyl in that it contains a piperidine ring with a tertiary amine, and substitution at the four position. Meperidine also has a phenyl group three carbons away from the amine nitrogen, as does morphine. Fentanyl, however, has its phenyl group four atoms away from the basic nitrogen. A compound structurally similar to fentanyl is diampromide, with a phenyl group also separated by four atoms from a tertiary nitrogen. Like fentanyl, it has a high narcotic potency and short duration of action.

It is informative to examine the structural features which make the fentanyls effective analgesics. By doing so, we can make inferences about the nature of the mu opioid receptor. The structure-activity relationships to be discussed are considered to reflect the properties and characteristics of the opioid receptor. A knowledge of this receptor allows us to predict from a molecular structure whether the molecule is likely to be an effective analgesic. Opioid analgesics actually inhibit processes that normally lead to the perception of pain. They have been called agonists by convention, because receptor occupation leads to a coupled intracellular response.

Those features that are known to be common to all opioid analgesics will be discussed. All of these compounds contain a basic nitrogen group. In most cases, this base is a tertiary amine, but in some cases it can be a secondary amine. The pK_a for these compounds is generally in the range of 7.8 to 9.6.[13] This pK_a range allows those bases with higher pK_a values to be almost completely protonated at the receptor surface at physiological pH values. It is most likely that the main mechanism of drug-receptor association is the formation of an electrostatic bond between the cationic center of the drug, and a center of high electron density on the receptor, usually termed the anionic site. On the basis of molecular orbital calculations, the positive charge may be distributed over the atoms directly attached to the nitrogen atom. The magnitude of this electrostatic bond has been estimated to be approximately 5 kcal/mol. This type of interaction is well known for cholinergic, adrenergic, and histaminic neurotransmitter molecules and receptors.[14]

All of the active compounds possess at least one aromatic ring. The most common aromatic structure is the phenyl group. A weak attractive force, known as a van der Waals force is believed to be formed when this flat ring is brought into close contact with the receptor surface. A close relationship is necessary, because the van der Waals force varies inversely as the seventh power of the distance between the groups. It is possible that an aromatic amino acid residue of the protein making up the structure of the receptor may provide a complementary attractive site for this group.

In morphine and the more potent opioid analgesics with rigid skeletal structures, a phenolic hydroxy group is present. In opioids, which do not have a rigid skeleton, such as fentanyl, a phenolic group is not necessary for high potency. The phenolic group is probably involved at the receptor site through hydrogen bond formation.

It has become increasingly apparent in the study of receptor binding that a specific molecular geometry is necessary in order to elicit an analgesic response. It is often observed that one enantiomer of an enantiomeric pair is an analgesic while the other may be inactive, or weakly active. Stereospecificity is an important consideration in the description of the opioid receptor. It indicates that the receptor site has a high degree of selectivity. In most cases, the required three-point attachment can only be attained with one stereoisomer. The attachment is not required to be by three groups of a chiral center, but may represent a particular structural arrangement defined by a stereochemical representation.

Casy[13] has demonstrated a model of a morphine receptor based on the four structural characteristics of this opioid. It is obvious that this rigid molecule can fit into the proposed receptor according to only one stereochemical representation. In this case, there is a cavity proposed which will accommodate two projecting carbon atoms from the piperidine ring. This correlates well with the experimental data, and strengthens the evidence for the existence of such a receptor. Additional studies into the structure-activity relationships of the fentanyl group should enable us to define a receptor model which will explain the tremendous potency of this class of compounds.

The structure-activity relationship study allows us to draw inferences about the conformation of the receptor site and the types of interactions involved. By systematically modifying molecular structure, known as the congeneric series, a particular effect can be studied, although there are limitations to this technique, as will be seen from the following discussion. This method does not allow us to positively identify and isolate the receptor itself. This can only be done by a thorough chemical analysis in which the tertiary structure of the receptor site is deduced. Because of the complexity of this problem, the true structure of the opioid receptor has not yet been determined.

The ability of the fentanyls to act as potent mu agonists makes them a good class of drugs to study this particular receptor. There has been considerable activity in the study of this problem because of the medical importance of fentanyl. The following discussion will concentrate on the potency of fentanyl analogs. It is important to note that we are studying only one action of the fentanyls, the analgesic potency, and not the side effects associated with these drugs.

Structure-activity data have been determined for many structural analogs and homologs of fentanyl. In some cases, the potency of fentanyl is enhanced, but in most cases the potency is decreased. With structural alterations, there may be other unknown effects or side effects, so potency is not the only consideration when new derivatives of fentanyl are introduced into the street drug scene, or into medicine.

The ethyl *N*-acyl group seems to be the most potent single functionality of fentanyl. Relative potencies, based on assigning the value of 1 to meperidine are: NCOEt (parent fentanyl), 700; NCOMe, 50; and NCOPr, 90.[5,15] The *N*-phenethyl group is the most effective length for an unbranched N group. The potency ranking for this series is: *N*-phenethyl > *N*-benzyl > *N*-methyl (almost inactive).[5,16] The relative potencies, again assigning a value of 1 to meperidine are: NCH_2CH_2Ph (fentanyl), 700; NCH_2Ph, 0.7; $NCH_2CH_2CH_2Ph$, 14.[5,15]

The insertion of a methyl group into the phenethyl group at a location alpha to the piperidine nitrogen also enhances potency. Insertion of a hydroxyl group beta to the piperidine nitrogen decreases potency, but a combination of α-methyl and β-hydroxyl groups is the most potent of this series: Ph-

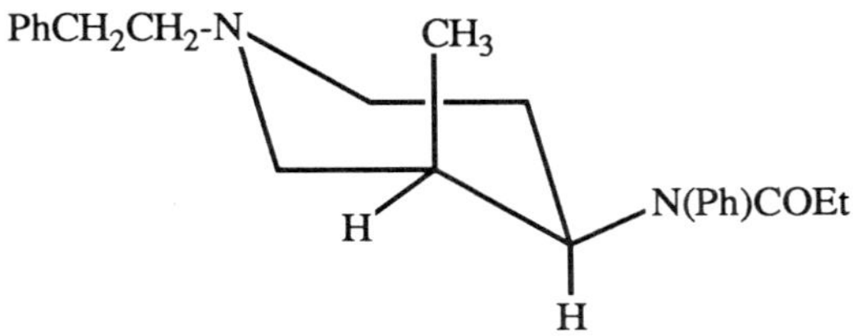

FIGURE 2. Conformation of *cis*-(+)-3-methylfentanyl.

CH_2CH_2– (fentanyl), 700 ± 250; $PhCH_2CH(Me)$–, 1150 ± 150; PHCH(OH)CH_2–, 150; PhCH(OH)CH(Me)–, 1300; PHCH(OH)CH(Et)–, 900.[5,15] α-Methylfentanyl was the first illicitly manufactured fentanyl derivative to be identified, and it was responsible for many overdose deaths. β-Hydroxyfentanyl has been found on the street and has been made illegal, but the potent combination of α-methyl together with β-hydroxy moieties has not been seen in forensic casework.

The piperidine ring substituents of fentanyl have a great influence on the reactivity of these derivatives. The most potent derivative so far found on the street is 3-methylfentanyl. The *cis*-(+) enantiomer, shown in Figure 2, was found to be sixteen times more potent than fentanyl, and up to 6684 times more potent than morphine, while the *cis*-(−) enantiomer was found to be inactive.[3] This indicates a specific drug-receptor association based on both absolute geometry and shape. These potency values were based on pure enantiomers, and the compounds that appeared on the street were most likely a combination of enantiomers with a corresponding intermediate net potency.

The most substantial enhancement in potency by a substituent on the fentanyl backbone reported so far has been for a carbomethoxy group at the 4-position of the piperidine ring. This compound is known as carfentanil, and has a potency of 7862 times that of morphine. Substitution with methoxymethyl and methylacyl has also been shown to be effective (potencies 4038 and 4291 times that of morphine, respectively).[5] This group of compounds represents some of the most versatile fentanyl derivatives studied so far. Substitution of the phenethyl phenyl group tends to give similar potencies with a shorter duration of action. Sufentanil (compound 37), composed of thiofentanyl with a 4-methoxymethyl group, is expected to have considerable promise in its application to medicine because of its high potency, short duration of action, and high margin of safety of 26,716. A variant of sufentanil, known as alfentanil, with 4,5-dihydro-5-oxotetrazolethyl as the *N*-substituent (Figure 3), is a highly potent drug with an even shorter action than sufentanil.

Within this group of fentanyl derivatives, it seems that the additional polar group at the 4-position enhances the potency of these derivatives. From our knowledge of the geometry of the molecule, this group is probably facing away from the receptor site when binding takes place. How this factor influ-

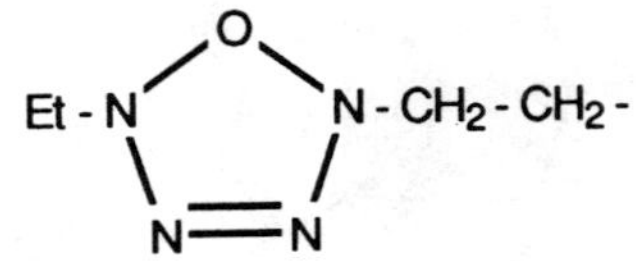

FIGURE 3. 4,5-Dihydro-5-oxotetrazolethyl group.

ences potency is unclear. Perhaps a three-dimensional model for the opioid receptor, which could accommodate such a polar group, would be a fruitful area of future research. To date, no illicit fentanyl that has been substituted at the 4-position of the piperidine ring has been detected. Because one of the unfavorable characteristics mentioned about fentanyl from drug users is the short duration of action compared to heroin, it is likely that in the future clandestine chemists will try to develop derivatives such as lofentanil, with substitutions at the 3- and 4-positions that have longer durations of action. This class of fentanyls represents an even greater threat to society than the previous derivatives because of their extreme potencies and length of action.

V. FACTORS AFFECTING THE DISTRIBUTION OF FENTANYLS

There are many other factors that contribute to the potency of fentanyls. The characteristic feature which distributes fentanyl so quickly in the body is its great solubility in lipids. In comparison to morphine, which is quite polar, fentanyl has a lipophilic partition coefficient of 4.05, compared to 1.03 for morphine.[17] Fentanyl readily crosses the brain barrier, while morphine is transported with greater difficulty. The relative potential for entering the CNS is 155 times greater for fentanyl than morphine.[18] Using labeled fentanyl, the brief duration of action of fentanyl in humans was shown to be due to its redistribution from the brain to the body rather than its rapid metabolism and excretion. It has been determined that the analgesic effect of fentanyl is brought about by the unaltered drug because the metabolites identified were inactive.[19]

In studies of relative receptor binding, fentanyl was shown to have a tenfold greater association constant than morphine. However, in a measure of the binding times, morphine had duration-distribution coefficients that were four times greater than of fentanyl.[17] These data show that fentanyl binds tightly, but not for as long as morphine due to rapid redistribution.

VI. THE EFFECT OF FENTANYLS ON THE DRUG USER

There is some confusing information in the literature concerning the effects of fentanyls on the user. It has been pointed out that most users cannot tell the difference between equipotent doses of heroin and fentanyls.[20] The euphoric effects of fentanyl have been verified in medical studies.[21,22] Fentanyls produce both tolerance and physical dependence for the user. Thus, they have a strong addiction potential. The long-range effects of the various designer fentanyls is unknown, but with their greater potency, we should expect a greater decrease in respiration, heart rate, and blood pressure.

There has been evidence to suggest that the effects of fentanyl can be distinguished from heroin and other opiates. There are also variations, however, based upon dose and method of administration. Fentanyl effects have been described as being more intense, and the initial effects, or "rush," overpowering.[20] Others, however, have found that the initial effects of fentanyl are more subtle than heroin.[23] Some have reported a burning sensation after injection of fentanyl. In a survey of opioid drug users at a treatment clinic in Contra Costa County, California, 89% of the respondents felt the effects of fentanyl were different from heroin and 88% indicated that they preferred heroin, even though they found both experiences similar. They reasoned that fentanyl is possibly a more powerful analgesic, while heroin seems to be a more powerful euphoriant.[23]

The normal mode for using fentanyls, like heroin, is by iv injection. It has been found that because of the high lipid solubility, fentanyls can be used by insufflation. Overdoses of fentanyls have been found with no injection marks, or with no hypodermic needles and syringes present, pointing to this mode of usage.[23] This may be an appealing factor in taking this drug by those who are reluctant to use needles.

Novice users have been involved in many of the overdose deaths related to the fentanyls. The victims have been described "white, upper-middle class, educated, but dumb about drugs".[20] Novice or infrequent users have not built up the tolerance and experience that addicted users have. Addicts often will "shoot-up" together, and have a supply of cocaine or methamphetamine nearby as an "antidote" to overdose.

Addiction is the major curse of these drugs. The withdrawal symptoms are believed to be due to the body's failure to produce endorphins, enkephalins, and dynorphins while under the influence. Symptoms include running nose, tearing, loss of appetite, sneezing, abdominal cramps, irritability, sweating, insomnia, nausea, tremors, and an increase in heart rate and blood pressure. These are probably due to the actions of the neurotransmitter, norepinephrine, that are not being regulated by the endorphins and enkephalins. Withdrawal is the body's way of asking for an artificial replacement for the diminished actions of its opioids. Studies have shown that the neurotransmitter, dopamine, is also responsible for euphoria and pleasure. Heroin and cocaine appear to activate the same reward system in the brain that involves dopamine. Cocaine blocks dopamine re-uptake at the synapse, and heroin causes the initial release of dopamine. As described earlier, fentanyl has also demonstrated a reinforcing feature in laboratory animals. Addiction is often a vicious cycle due to the tolerance that develops to the drug.

VII. THE FUTURE OF THE FENTANYLS

The fentanyls thus far have shown two surges in use. The first was due to α-methylfentanyl which created a problem because it took over a year to identify what was killing the drug users. The second wave was even more pronounced because the drug, 3-methylfentanyl, possessed such great potency and accounted for many more deaths. No new fentanyl derivative has been reported since the late 1980s. Due to the fact that new fentanyls have been discovered which are many times more potent than 3-methylfentanyl, it is possible and even likely that clandestine chemists will also try to manufacture these deadly drugs. This would surely usher in a third surge in use that would be even more disasterous than the previous two. It is also possible that a clever clandestine chemist will design fentanyls, such as lofentanil, which have a longer duration of action. These would be more acceptable to the heroin user, who expects a long, euphoric experience. It is also very likely that a clandestine chemist will use the present knowledge of fentanyls as a springboard to the manufacture of other novel compounds and again make them available to unsuspecting drug users.

Much research is needed in the investigation of effects of fentanyl derivatives, especially their neurologic effects. Because these drugs may become a major source of addiction and overdose deaths, we need to know how to treat those addicted to these substances. This requires a more in-depth knowledge of the physiological responses to the fentanyls. Perhaps the fentanyls can become a replacement for methadone addiction therapy. Without the same level of euphoric activity of opiates, and thus less craving, perhaps a regime of tapering of dosage fentanyls could become a more successful therapy.

ACKNOWLEDGMENTS

The authors would like to thank Dr. Frank Dowd for his critical review of this manuscript.

REFERENCES

1. **Janssen, Gardocki,** U.S. Pattent 3,141,823, 1964.
2. **Jaffe, J. H. and Martin, W. R.,** Opioid analgesics and antagonists, in *The Pharmacological Basis of Therapeutics,* 8th ed., Gilman, A. G., Rall, T. W., Nies, A. S., and Taylor, P., Eds., Pergamon Press, New York, 1990, 485.
3. **Van Bever, W. F. M., Niemegeers, C. J. E., and Janssen, P. A. J.,** Synthetic analgesics, *J. Med. Chem.,* 17, 1047, 1974.
4. **Baun, R.,** New variety of street drugs poses growing problem, *Chem. Eng. News,* Sept. 9, p. 7, 1985.
5. **Casy, A. F.,** Fentanyl and the 4-anilidopiperidine group of analgesics, in *Opioid Analgesics,* Plenum Press, New York, 1986, chap. 6.

6. **Henderson, G. L.,** Fentanyl-related deaths: demographics, circumstances, and toxicology of 112 cases, *J. Forensic Sci.,* 36, 2, 1991.
7. **Brown, R. M., Clouet, D. H., and Freeman, D. P.,** Opiate receptor subtypes and brain function, *NIDA Monograph 71,* National Institute on Drug Abuse, Washington, DC, ix, 1986.
8. **Janssen, P. A. J.,** The development of new synthetic narcotics, in *Opioids in Anesthesia,* Eastafanous, F. G., Ed., Butterworths, Reading, MA, 1984, 40.
9. **Greene, N. M. and Hug, C. J.,** in *Narcotic Analgesics in Anesthesiology,* Kitahata, L. M. and Collins, J. G., Eds., Williams & Wilkins, Baltimore, MD, 1982, 46.
10. **Hughes, J., Smith, T. W., and Kosterlitz, et al.,** Identification of two related pentapeptides from the brain with potent opiate agonist activity, *Nature (London),* 258, 577, 1975.
11. **Martin, W. R., Eades, C. G., and Thompson, J. A., et al.,** The effects of morphine- and nalorphine-like drugs in the nondependent and morphine-dependent chronic spinal dog, *J. Pharmacol. Exp. Ther.,* 197, 517, 1976.
12. **Gilbert, P. E. and Martin, W. R.,** The effects of morphine- and nalorphine-like drugs in the nondependent, morphine-dependent and cyclazine-dependent chronic spinal dog, *J. Pharmacol. Exp. Ther.,* 198, 66, 1976.
13. **Casy, A. F.,** Analgesic receptors, in *A Guide to Molecular Pharmacology-Toxicology,* Marcel Dekker, New York, 1973, 217.
14. **Goldstein, A., Aronow, L., and Kalman, S. M.,** Binding forces in the drug-receptor interaction, in *Principles of Drug Action, The Basis of Pharmacology,* 2nd ed., John Wiley & Sons, New York, 1974, 14.
15. **Janssen, P. A. J. and van der Eycken, C. A. M.,** The chemical anatomy of potent morphine-like analgesics, in *Drugs Affecting the CNS,* Burger, A., Ed., Marcel Dekker, New York, 1968, 25.
16. **Casy, A. F., Hassan, M. M., Simmonds, A. B., and Staniforth, D.,** Structure-activity relations in analgesics based on 4-anilinopiperidine, *J. Pharm. Pharmacol.,* 21, 434, 1969.
17. **Rugheimer, E. and Zindler, M., Eds.,** *Anaesthesiology, Proceedings of the 7th World Congress of Anaesthesiologists,* Excerpta Medica, Princeton, NJ, 1981, 1003.
18. **Hug, J. C.,** Pharmacokinetics of new synthetic narcotic analgesics, in *Opioids in Anesthesia,* Butterworths, Reading, MA, 1984, 155.
19. **Lehman, K. A., Weski, C., Hunger, L., Heinrich, C., and Daub, D.,** Biotransformations of fentanyl, *Anaesthesist,* 31, 221, 1982.
20. **Kirsch, M. M.,** China white, *Designer Drugs,* CompuCare, Minneapolis, MN, 1986, chap. 1.
21. **Hoehe, M., Duka, T., and Doenicke, A.,** Human studies on the mu opiate receptor agonist fentanyl: neuroendocrine and behavioral responses, *Psychoneuroendocrinology,* 13(5), 397, 1988.
22. **Gorodetzky, C. W. and Martin, W. R.,** A comparison of fentanyl, droperidol and morphine, *Clin. Pharmacol. Ther.,* 6(6), 73, 1965.
23. **LaBarbera, M. and Wolfe, T. J.,** Characteristics, attitudes and implications of fentanyl use based on reports from self-identified fentanyl users, *J. Psychoactive Drugs,* 15(4), 293, 1983.

Chapter 5

NEUROCHEMISTRY OF NICOTINE

Cecilia J. Hillard

TABLE OF CONTENTS

I. INTRODUCTION

Although it does not receive the "press" of other drugs of abuse, cigarette smoking continues to be the most popular and persistent method of recreational drug use in the U.S. Many compounds have been identified in cigarette smoke, several of which may contribute to the overall pharmacological and toxicological effects of smoking.[1] However, the evidence is convincing that smoking behavior is maintained by the pharmacological effects of the tobacco alkaloid nicotine.[2] Smokers will not willingly smoke cigarettes that lack nicotine, and, within certain limits, smokers will alter their smoking behavior to titrate nicotine dose.[3] The role of nicotine in tobacco dependence is supported by the success of nicotine-substitution therapy as an adjunct to behavioral therapy in smoking cessation programs.[4] Studies demonstrating that administration of the nicotinic cholinergic receptor blocker mecamylamine to smokers caused increased cigarette consumption also provide evidence that the reinforcing properties of cigarette smoking result from effects of nicotine.[5-7]

The mechanism or mechanisms by which nicotine maintains smoking behavior are complex and probably vary among individuals.[8] Nicotine exerts a wide range of psychopharmacological and physiological effects that may be perceived as pleasurable by the smoker. These include increased arousal, muscular relaxation, reduced anxiety, and decreased appetite, as well as enhanced attention to and memory of certain tasks.[9] However, as tolerance develops to these effects of nicotine during a day of smoking, smoking behavior may be maintained more by a desire to avoid the unpleasant symptoms of nicotine abstinence than by pleasure seeking.[10] In most individuals, smoking behavior is maintained by a complex interaction between nicotine dependence and a host of secondary, conditioned behaviors.[11]

Behavioral studies have shown that the effects of nicotine in animals parallel its effects in humans to a greater or lesser extent, depending upon the particular behavioral characteristic.[12] Nicotine has both stimulant and depressant activity in humans[13] and produces both increases and decreases in activity levels of animals, depending on the dose, time after administration, and previous exposure of the animal to nicotine.[14,15] Nicotine is self-administered by humans and, under some circumstances, by animals as well. Cigarette-deprived smokers will self-administer nicotine as cigarettes[16] or as nicotine chewing gum.[4] While nicotine self-administration in animals is difficult to demonstrate, with long treatment protocols, rats have been shown to administer nicotine preferentially over saline and to increase their response rates when the nicotine dose is reduced.[17] Squirrel monkeys self-administered nicotine when the nicotine injection was paired to a secondary reinforcer, such as a visual stimulus.[18] Nicotine withdrawal has been demonstrated both in humans[19,20] and in animals.[21] Therefore, studies of the neurochemical effects of nicotine in animals should provide information concerning how and where nicotine exerts its effects in humans.

The primary focus of this review is the neurochemistry of nicotine in the mammalian central nervous system. Nicotine is an agonist of nicotinic cholinergic receptor (nAChR) in the central nervous system, and it is widely accepted that it exerts its pharmacological effects as a result of interactions with these receptors. Studies of the effects of mecamylamine on cigarette consumption provide evidence that the reinforcing properties of nicotine result from stimulation of nAChR.[5-7] There have been suggestions that nicotine may produce effects that are not mediated by nAChR, suggestions based primarily on the lack of blockade of nicotine effects by peripheral nicotinic receptor antagonists.[22,23] However, it is becoming clear that the CNS nicotinic receptor is heterogeneous and that different subtypes exhibit different structure-activity profiles, particularly for the various antagonists of peripheral nAChRs (see References 24-26 and below). The preponderance of evidence supports nAChR as the primary site of nicotine action in the CNS.

II. STRUCTURAL ASPECTS OF CENTRAL NICOTINIC CHOLINERGIC RECEPTORS

The nicotinic cholinergic receptors from electric organ and muscle have been extensively studied and are known to be pentamers composed of four different but highly homologous subunits that may have evolved from repeated duplication of a single ancestoral gene.[27] The nAChRs of the CNS are heteromeric proteins consisting of at least two types of subunits, α and β.[28] Four distinct α subunits have been identified in brain tissue and have been sequenced, α2, α3, and α4[29-31] and, recently, α5.[32] These subunits all contain acetylcholine (ACh) binding sites and are structurally similar to the α subunit found in muscle, particularly in the M1 membrane spanning helix region and in the putative binding site of the N-terminal extracellular region.[25] At least three non-ACh binding subunits have also been sequenced: β2;[33] nonα, or β3;[34] and β4.[35] The subunit stoichiometry of the nAChRs is not known; however, microsequencing of the receptor[36] and the molecular mass of the isolated receptor[28] predict that the nAChR is composed of several copies each of one α- and one β-subunit.

The evidence that the identified subunits are constituents of functional neuronal nAChRs comes from several studies. First, mRNA for the four α-subunits are expressed in brain in a regionally specific manner.[31,32] Second, reconstitution studies that employ *Xenopus* oocytes have demonstrated that functional neuronal nAChR are expressed when mRNA encoding a β-subunit and mRNA encoding an α-subunit are co-injected.[31,37-39] The resulting oocytes express functional nicotinic receptor ionophores with pharmacological profiles that are similar to neuronal type receptors. Functional receptors are not formed if only one of the two subunits is injected.[38] Recent studies have shown that each subunit combination has a different pattern of sensitivity to a series of nicotinic agonists, particularly when the potency of ACh is compared with that of DMPP.[39] These studies also suggest that the β-subunit contributes to the ultimate agonist sensitivity of the receptor. Finally, the functional relevance of these subunits is supported by a study in which monoclonal antibodies raised against nAChR isolated from chick brain were shown to cross-react with a functional nAChR in PC12 cells.[36]

III. CNS DISTRIBUTION OF MESSAGE FOR nAChR

The distribution of mRNA for the identified subunits of central nervous system nAChR has been determined.[29,30,32,33,40] Overall, the receptor genes are widely expressed in the CNS, attesting to the importance of the cholinergic system in the brain.[40] The four α-subunits are expressed in distinct patterns with some regions of overlap. The mRNA for the α4-subunit is the most abundant and ubiquitous, while the mRNA for the α2, α3, and α5 is localized to smaller regions of the CNS. There are several regions, including the thalamus, that exhibit significant expression of mRNA for multiple α subunits, which raises the possibility that some receptors may be composed of more than one type of α-subunit. The two nonbinding subunits, nonα and β2, are expressed in most regions of the CNS. In general, there is a correspondence between regions that show hybridization to an α probe and the presence of hybridization to either the nonα or β2 probes,[25] which agrees with the results of the expression studies that a functional receptor requires both α- and β-subunits. The α3 subunit is also expressed in cells of neural crest origin that receive cholinergic input, including rat adrenal glands,[29] chick ciliary and sympathetic ganglia,[41] and chick dorsal root ganglion cells.[42]

IV. AGONIST BINDING SITE IDENTIFICATION

An important tool in the study of central nAChR has been the radioligand binding assay. Specific agonist binding to the nAChR of membrane or synaptosomal preparations from rat and mouse brain has been demonstrated using

[^{3}H]-nicotine[43-52] and [^{3}H]ACh.[43,51,53-56] The binding of both ligands is saturable,[45,50,52,53] reversible,[43,48] and stereoselective.[45,53,57]

In spite of the fact that structural studies suggest that there are different subtypes of nicotinic receptor in the CNS, multiple classes of binding sites for nicotine and acetylcholine have not been convincingly demonstrated. In brain membrane preparations, [^{3}H]-(-)-nicotine binding is characterized by a single affinity site with a K_d in the range of 2 to 10 n*M* under experimental conditions in which the assay temperature is 25°C or greater and bound and free ligands are separated using rapid filtration through glass fiber filters.[50-52] [^{3}H]-Nicotine binding to a second, low affinity site has been reported by some investigators.[45,47,49] However, it is likely that low affinity binding of [^{3}H]-nicotine is an artifact resulting from one of the following: receptor changes as a result of proteolysis during the membrane isolation;[50] impurities in the ligand, including use of the racemic mixture of [^{3}H]-nicotine;[50] the use of very high concentrations of unlabeled nicotine, resulting in an underestimate of the contribution of nonspecific binding to the total; or low temperature effects.[48] With the exception of one study,[55] radioligand binding assays have shown that [^{3}H]-ACh binds to a single site (in the presence of atropine to block muscarinic receptors) with a K_d in the range of 10 to 20 n*M*.[43,51,53,54,56]

There is biochemical and anatomical evidence that [^{3}H]-nicotine and [^{3}H]-ACh bind to the same nicotinic cholinergic recognition site in rat and mouse brain. Autoradiographic comparison of the binding of the two ligands in rat brain demonstrates a high degree of concordance in both areas rich in binding and areas devoid of binding.[59] The density of binding sites labeled by the two ligands is similar, and the binding of both radioligands is decreased to the same extent by dithiothreitol treatment of the membranes to destroy disulfide bonds, by heating the membranes to temperatures above 60° and by membrane protease treatment.[51,56] Furthermore, the binding of both ligands is increased in membranes from rats[51] and mice[56] exposed to chronic nicotine *in vivo* and decreased in membranes from rats exposed to the irreversible cholinesterase inhibitor, soman.[51]

Although the affinity of nicotine for the binding site is higher than that of ACh, nicotinic cholinergic agonists compete for binding with the two ligands with similar affinities.[51,56] Agonists that bind with high affinity include agonists of peripheral nAChR such as cytisine, 1,1-dimethyl-4-phenylpiperazinium (DMPP), carbachol, lobeline, and anabasine[45,46] and anatoxin-a, an exotoxin isolated from a blue-green algae.[60] In general, antagonists of peripheral nAChRs, including mecamylamine, hexamethonium, and α-bungarotoxin (α-BuTx) are poor competitors of [^{3}H]-nicotine and [^{3}H]-ACh binding in the CNS.[45,56] Antagonists that have been shown to compete for binding with nicotine and ACh do so with relatively low affinity; these include dihydro-β-erythroidine (K_i = 100 n*M*),[51] d-tubocurarine (K_i = 30 to 50 μ*M*),[51,56] and neosurugatoxin (K_i = 50 to 100 n*M*).[61] Recently, however, it has been suggested that a contaminant of commercially available preparations of α-BuTx — referred to as bungarotoxin 3.1, toxin F, or κ-bungarotoxin by various investigators — may be an antagonist of nAChR in the nervous system.[62,63] In fact, two peptides of amino acid sequences identical to segments located within the N-terminal extracellular domain of the α3-subunit of the nAChR of PC12 cells have been shown to bind κ-bungarotoxin with K_ds of 300 to 500 n*M*.[64]

Two other agonists of the nAChR have been used to study binding to the receptor in the CNS: [^{3}H]*N*-methyl-carbamylcholine [^{3}H]-MCC)[65-68] and [^{3}H]-cytisine.[69] Both ligands appear to bind to the same recognition site as nicotine and ACh. Nicotine and ACh compete with high affinity for [^{3}H]-MCC labeled sites[65,68] and [^{3}H]-cytisine labeled sites[69], and the distribution of the binding sites is very similar to the distribution of sites labeled by [^{3}H]-nicotine and [^{3}H]-ACh.[66,69,70] The advantage [^{3}H]-MCC has over [^{3}H]-nicotine is better chemical stability, and its selectivity for nicotinic cholinergic receptors gives [^{3}H]-MCC some advantage over [^{3}H]-ACh.[66] [^{3}H]-Cytisine binds with very high affinity (K_d = 1 n*M*) and has low nonspecific binding that may be an advantage in some circumstances.[69]

V. IS THE AGONIST BINDING SITE PART OF A FUNCTIONAL nAChR?

One important question concerning the nicotinic binding site labeled by [^{3}H]-nicotine and [^{3}H]-ACh is whether it identifies a functional nicotinic receptor. It is, of course, impossible to identify a receptor by binding measurements alone: binding studies must be correlated with biochemical, electrophysiological, or behavioral studies of the response to receptor activation, and these will be discussed further below. However, some basic information about the relationship of the [^{3}H]-nicotine/[^{3}H]-ACh binding site to the nicotinic receptor is available from binding studies. A minimal requirement for binding to be associated with a receptor is that the binding sites be found in brain regions where the endogenous transmitter is found. Autoradiographic studies have shown that the regional distribution of nicotinic agonist binding is not uniform throughout the brain. Areas having high concentrations of agonist binding sites include the interpeduncular nucleus, thalamus, medial habenula, superior colliclus, and cerebral cortex.[66,70-73] All these areas have been shown to be major terminal regions of the cholinergic cells of the laterodorsal tegmental nucleus or the basal nuclear complex (reviewed in Reference 74), and these studies have been complemented by and agree with the results of studies of the localization of nAChR in the CNS using immunohistochemical methods.[75,76] The relative distribution of nicotinic agonist binding differs from the distribution of muscarinic binding sites,[77] and from the pattern of α-BGT binding.[59] When subcellular fractions of rat brain were prepared using discontinuous sucrose gradients, binding of [^{3}H]-nicotine and [^{3}H]-ACh was confined to the synaptosomal fraction; no binding was found in mitochondrial or myelin fractions.[44,53,57] One of the best described CNS effects of nicotine is the modulation of neurotransmitter release (see below), and therefore, a high density of [^{3}H]-nicotine and [^{3}H]-ACh binding sites in synaptosomal membranes supports their functional relevance.

Although studies of the structure-activity relationships and localization of agonist binding sites in the CNS support the argument that binding sites identified by agonist binding studies correspond to nAChR, it is likely that the receptor population that is detected in binding studies are in a desensitized state. In their original paper, Romano and Goldstein[45] suggested that prolonged exposure of the receptor to agonist during the radioligand binding assay produced a desensitization of the receptor that was manifested as a shift to a high affinity, agonist-prefering state. This hypothesis is supported by studies of nAChR of Torpedo membranes and mammalian neuromuscular junction, which demonstrate that agonist activation of the receptor is followed by a rapid desensitization during which the receptor becomes functionally deactivated and has an increased affinity for agonists.[24,78,79] The active or functional form of the CNS nAChR may, like the peripheral receptor,[24] have a K_d for nicotine in the μM range. In fact, functional studies of nAChR activation in the CNS almost always require micromolar concentrations of nicotine to elicit a pharmacological response (see below). Unfortunately, one limitation to the use of radioligand binding studies for the identification of receptors is that low affinity sites (i.e., K_ds in the micromolar range) are essentially impossible to detect. Therefore, until an agonist with a high affinity for the nondesensitized form of the receptor is identified, it will be impossible to characterize completely the agonist binding site of the CNS nAChR using radioligand binding assays.

VI. STUDIES OF ANTAGONIST BINDING TO nAChR IN CNS

In the periphery, nicotinic receptors fall into two classes, those found on the muscle cell membranes that are blocked by d-tubocurarine, α-bungarotoxin, decamethonium, and dihydro-β-erythroidine and those found in the peripheral autonomic ganglia that are blocked by hexamethonium and mecamylamine.[24,80] In spite of the fact that several of these antagonists have been shown to inhibit various electrophysiological,[81] behavioral,[82] and neurochemical[83] effects of nicotine, with the exception of dihydro-β-erythroidine, these antagonists are not effective competitors of agonist binding to putative nAChR in the CNS.[45,46] In attempts to understand the relationships among the CNS recognition sites for the nicotinic agonists and various antagonists, radioligand binding studies have been carried out using [^{3}H]-dihydro-β-erythroidine (DHβE),[84] [^{125}I]-α-BuTx, [^{3}H]d-tubocurarine,[49] and [^{3}H]-mecamylamine.[85]

The binding site of rat brain membranes labeled with [^{3}H]-DHβE strongly resembles the agonist site labeled by nicotine and ACh.[84] Cytisine, nicotine, and ACh compete for [^{3}H]-DHβE binding sites with nanomolar K_i values while the antagonists mecamylamine and hexamethonium have no affinity for the site. This agrees with the finding that DHβE competes with [^{3}H]-nicotine and [^{3}H]-ACh for binding.[51,52] Furthermore, identical to [^{3}H]-nicotine and [^{3}H]-ACh binding, the highest concentration of [^{3}H]-DHβE binding sites was found in the thalamus, and preferential enrichment was seen in the synaptosomal subfraction. DHβE has been shown to inhibit several neurochemical effects of nicotine, including increased release of dopamine from striatal synaptosomes.[83] These findings suggest that DHβE is a competitive antagonist at the agonist binding site of CNS nAChR and that, like agonists, it has a fairly high affinity for the desensitized form of the binding site.

The snake neurotoxin α-BuTx binds specifically and nearly irreversibly to the ACh binding site of nAChR of the neuromuscular junction.[86] When high affinity binding of [^{125}I]-α-BuTx was also detected in neuronal tissues, both in autonomic ganglia[87] and in the CNS,[88,89] it was thought to label binding sites on functional neuronal nAChR.[90] However, localization studies, functional studies, studies involving binding site purification and studies of binding site regulation all largely support the thesis that the α-BuTx binding protein and the nAChR involved in synaptic events in the mammalian CNS are distinct entities (see References 91 and 92 for reviews). The α-BuTx binding site in the mammalian CNS appears, however, to be a member of the same ligand-gated ion channel gene superfamily that also includes the muscle endplate nAChR and GABA and glycine receptors.[93,94] In addition, although α-BuTx has no affinity at the binding site labeled by [^{3}H]-nicotine,[56] nicotine competes for [^{125}I]-α-BuTx binding with a K_i in the range of 1 *nM*,[48] suggesting that the α-BuTx binding site could mediate other physiological effects of nicotine.

VII. IS THERE A ROLE FOR NICOTINE IN DEVELOPMENT OF THE CNS?

A hypothesis concerning a role for α-BuTx binding proteins in the development, maturation, and/or maintenance of synaptic contacts in the mammalian CNS has been put forth.[90,92,95] Developmental studies have shown that β-BuTx binding site on neurons appears earlier than evidence of choline acetyltransferase activity, suggesting that α-BuTx binding sites may be involved in the initiation of synaptic contacts.[93] Furthermore, neonatal mouse brain has approximately twice the number of α-BuTx binding sites as brains from adult mice, leading to the suggestion that they play a role in directing CNS development.[97] In another study of the regulation of synapse formation, α-BuTx induced sprouting in retinal terminals in the toad tectum.[98] However, nicotinic antagonists without affinity for α-BuTx binding sites, such as d-tubocurare and mecamylamine, have also been shown to enhance neuronal process outgrowth, which calls into question the specific role of α-BuTx sites in this effect.[99] In converse experiments, ACh has been shown to inhibit neurite or dendrite outgrowth in several nervous tissues.[100] The possibility that nicotine, via effects on α-BuTx binding sites, may play a role in the formation and maintenance of synapses in the CNS has much importance, particularly in light of the well-documented effects of maternal tobacco use on neonatal brain development.[101]

Thymopoeitin is a polypeptide isolated from the thymus gland that is involved in a variety of hematopoietic processes including the induction of T cell and B cell differentiation.[102] A study demonstrating that patients with myasthenia gravis showed clinical improvement following thymectomy led to the suggestion that thymopoeitin might have effects on nicotinic receptors at the neuromuscular junction.[103] In myotubes and nicotinic receptors isolated from Torpedo membranes, thymopoietin accelerates agonist-induced desensitization of nAChR, supporting a physiological role for thymopoeitin as a regulator of the ion channel of nAChR at the neuromuscular junction.[104,105] Recently, the thymic peptide thymopoietin has been shown to compete with high affinity for binding sites labeled by [^{125}I]-α-BuTx on brain membranes,[106] on neuroblastoma cells,[107] on PC12 pheochromocytoma cells,[108] and on C2 muscle cells.[109] Treatment of PC12 cells with either thymopoietin or α-BuTx induces process formation, leading to the suggestion that these substances are neuronotrophic to these normally endocrinoid cells.[110] Thymopoietin has been found in the brain[102] and in serum[103] leading to the suggestion that it is an endogenous ligand for the α-BuTx binding site of muscle and nervous tissue.[109] Future studies will give insight into the relationship between the apparent trophic role of α-BuTx binding sites, synaptic regulation by thymopoietin, and nicotinic agonists.

VIII. LIGANDS OF THE NICOTINIC CHOLINERGIC RECEPTOR ION CHANNEL

In spite of the fact that mecamylamine does not compete for the agonist binding site of brain nAChR,[45,56] it clearly blocks many of the neurochemical,[83,111,112] physiological,[113] and behavioral[12] effects of nicotine in the CNS. Electrophysical studies have clarified the antagonist action of mecamylamine, suggesting that it is a noncompetitive antagonist that blocks the ionic channels associated with the nAChR.[114,115] A recent study reported that [^{3}H]-mecamylamine ([^{3}H]-MEC) binds to rat brain membranes in a specific, saturable manner.[85] Other nicotinic channel blockers, such as phencyclidine,[116] compete for [^{3}H]-MEC binding with K_i values in the range of 0.1 to 10 μM. The presence of nicotine in the incubation mixture did not enhance [^{3}H]-MEC binding, and in fact nicotine decreased binding, which argues that the channel of the nAChR need not be open for MEC to reach its binding site. Unfortunately, [^{3}H]-MEC binding sites were not enriched in the synaptosomal fraction of brain homogenate compared to myelin and mitochondria, suggesting that nonspecific binding may be a problem.

A second channel-blocking antagonist of nAChR that may be useful for the study of the ion channel in the CNS is [^{3}H]-perhydrohistrionicotoxin ([^{3}H]-H_{12}-HTX). [^{3}H]-H_{12}-HTX has been shown to bind to the ionic channel of the nAChR in membranes from the electric organ of Torpedo.[117] The initial rate of [^{3}H]-H_{12}-HTX binding was enhanced several hundred fold in the presence of agonists while pre-exposure of the receptor to agonist decreased the binding of [^{3}H]-H_{12}-HTX, suggesting that its binding is sensitive to the conformational state of the receptor (i.e., whether the channel is open, closed, or desensitized). In one study, perhydrohistrionicotoxin blocked the effects of nicotine on transmitter release in both striatum and hippocampus,[118] which leads to the exciting possibility that the ligand may provide a tool with which to study the functional status of nAChR in the CNS. To the author's knowledge, however, no studies of [^{3}H]-H_{12}-HTX binding studies in brain have been reported.

IX. RECEPTOR REGULATION—ACUTE CHANGES IN SENSITIVITY

Rapid desensitization following activation of nAChR by agonists is an intrinsic property of muscle and electric organ nAChR. When the muscle-type nAChR is in a desensitized state, it is functionally inactive and has increased affinity for nicotinic agonists.[24] Electrophysiological studies of the effects of ACh and nicotine on rat locus coeruleus neurons demonstrate that this population of CNS nAChR also exhibits desensitization.[119] Studies of catecholamine release in isolated adrenal chromaffin cells have also demonstrated that desensitization of neural-type nAChR occurs following exposure to nicotinic agonists.[120,121] Neuronal nAChR also have the capacity to increase sensitivity to nicotinic agonists under certain conditions. For example, in PC12 cells, nerve growth factor enhances the release of dopamine in response to nicotine;[122] substance P protects against agonist-induced desensitization of catecholamine release from isolated adrenal medullary cells;[123] and rat superior cervical ganglion cells exhibit long-term potentiation in response to tetanic preganglionic stimulation, owing, in part, to an increased nicotinic response.[124]

Several investigations have provided insight into possible mechanisms by which the sensitivity of neuronal nAChR are regulated. Studies of peripheral-type nAChR from muscle and electric organ have provided evidence that phosphorylation of the nAChR occurs and is involved in receptor desensitization.[125,126] Studies of neuronal nAChR on chick ciliary ganglion neurons have demonstrated that the sensitivity of the receptors to ACh is enhanced via a cyclic adenosine 3′:5′-monophosphate (cAMP)-dependent mechanism.[127] The increased sensitivity is most consistent with an increase in the number of functional receptors in the membrane as a result of a conversion of pre-existing nonfunctional receptors either in the membrane or from an internal pool.[128] The sensitivity of nAChRs of bovine adrenal chromaffin cells to agonist stimulation is also enhanced via a cAMP-dependent process.[129] Neither *de novo* synthesis nor the transport of internal receptors to the membrane is required for increased sensitivity, suggesting that phosphorylation of pre-existing, nonfunctional receptors in the membrane occurs via cAMP-dependent protein kinase. Studies of long term potentiation in superior cervical ganglion cells also provide some evidence that β-adrenergic agonists and increased intracellular cAMP can enhance nicotinic synaptic transmission, but it is unclear whether or not the enhancement is a result of presynaptic or postsynaptic events.[124] Several of these studies have used forskolin to activate adenylate cyclase and increase intracellular cAMP levels; however, forskolin has been shown to be a direct inhibitor of ion flux through the nAChR ionophore in PC12 cells in a manner that is not related to adenylate cyclase inhibition.[130,131] To the author's knowledge, the role of cAMP-mediated processes in the control of nAChR sensitivity in the CNS has not been explored.

X. RECEPTOR REGULATION IN RESPONSE TO AGONIST EXPOSURE *IN VIVO*

Several laboratories have demonstrated that prolonged exposure of mice and rats to nicotine and other nicotinic agonists produces significant increases in the number of agonist binding sites in several brain regions, including cortex, striatum, thalamus, hippocampus, and hypothalamus (see References 132-135, and for a review see Reference 136). This effect of nicotine occurs as soon as 4 d after first nicotine exposure and is reversible following cessation of treatment.[137] Rats are more sensitive to this effect of nicotine than mice: rats exhibit significant increases in agonist binding after nicotine doses of less than 1 mg/kg/d, whereas mice require 3 mg/kg/h.[132,134] There is also evidence from human postmortem studies that cigarette smokers have more [^{3}H]-nicotine binding sites than nonsmokers.[138] It has been postulated by several investigators that the increase in agonist binding sites represents an increase in the number of receptors occurring as a homeostatic response to a chronic state of receptor desensitization.[134,136] As is pointed out in Reference 136, however, if this hypothesis is correct, chronic treatment with nicotinic antagonists would be expected to produce the same effect. In fact, studies to date have not demonstrated up-regulation in response to nicotinic antagonist treatment.[134] Another hypothesis concerning nAChR up-regulation in response to nicotinic agonist treatment is based upon the suggestion discussed above that ligand binding studies of nAChR detect only a population of receptors that has a high affinity for nicotine and that this population of receptors is desensitized. The alternative hypothesis, then, is that agonist treatment results in a decrease in the proportion of nAChR that are in active, low affinity states compared to inactive, high affinity states. One study has reported that nAChR of cerebral cortex of rats treated with nicotine exhibited a shift in the proportion of high- and low-affinity [^{3}H]-nicotine binding sites in favor of increased high-affinity sites.[139] As discussed above, however, other investigators do not detect multiple binding sites with radioligand binding assays of nAChR; so these results have not been repeated. The two proposed models differ in their prediction of the effect of the change in binding site number on the overall effect on the target neuron: in the first model, one would expect an increased response to nicotine owing to an increased number of receptors, and in the second model, one would expect decreased responsiveness to nicotine because the number of low-affinity functional receptors is decreased. Studies of the effects of chronic nicotine exposure on assays of nAChR function will be useful in determining whether either of these models is correct.

Repeated administration of nicotine to rats or mice for several days produces both increases in agonist binding sites in the CNS and tolerance to the effects of nicotine on locomotor activity.[140] It is not clear whether the changes in nicotine binding are causally related to the development of tolerance. Collins and co-workers have carried out an interesting series of studies in which they have measured the effects of repeated administration of nicotine both on the number of agonist binding sites and on physiological behavioral responses to nicotine in similarly treated animals under various experimental conditions in order to determine the degree to which increases in binding parallel the development of tolerance.[113,137,141-144] Overall, the results of these studies lead to the conclusion that tolerance to nicotine cannot be completely explained by changes in agonist binding. The strongest evidence for this conclusion comes from a comparison of the effects of nicotine in four genetically different mouse strains. All four strains showed an increase in the number of agonist binding sites following repeated nicotine dosing; yet the strains exhibited different patterns of tolerance to nicotine.[113,141] Recently, this group has also demonstrated in rats that the time courses for development and loss of tolerance to the effect of nicotine on locomotor activity and for increase in agonist binding in brain were dissimilar: the effect of nicotine treatment on agonist binding lasted longer than the state of tolerance.[144] Not surprisingly, then, these studies lead to the conclusion that tolerance to nicotine is complex and involves more than changes in the number of agonist binding sites.

XI. NICOTINE AND CEREBRAL METABOLISM

Techniques that estimate changes in local cerebral glucose utilization (LCGU) have been used to trace alterations in brain activity following *in vivo* drug administration or other manipulations that produce changes in behavior. The technique is based on the assumption that local rates of energy metabolism and, therefore, glucose uptake and utilization in the brain are correlated with local functional activity under normal conditions.[145] Cerebral metabolism has been mapped using 2-deoxy-D[1-^{14}C]glucose after acute administration of nicotine; nicotine produced increases in LCGU in anatomical regions that paralleled the distribution of [^{3}H]-nicotine and [^{3}H]-ACh binding sites in the brain.[73,146] In specific, the regions exhibiting the greatest increases in LCGU were the thalamic nuclei, cerebellum, geniculate nuclei, medial habenula, and superior colliculus. Interestingly, in both studies, no effect was seen either in the cerebral cortex or in any limbic structures, although these areas have significant numbers of agonist binding sites.

Changes in LCGU after chronic administration of nicotine have also been studied. After the administration of nicotine via osmotic minipump for 14 days, LCGU was increased in the lateral geniculate body, superior colliculus, and interpeduncular nucleus, all regions that were similarly affected by acute nicotine.[147,148] However, LCGU was also increased in the globus pallidus and in the septal nucleus, regions that were not affected by acute administration. In a study designed to determine whether tolerance to the effects of nicotine on LCGU occurred, rats were treated with nicotine or saline for 10 d and then both groups were challenged with an acute dose of nicotine.[149] Tolerance to the nicotine challenge was seen in the thalamic nuclei, in the relay structures of the visual system, in the cerebellum, and in the vestibular nuclei. In this study, in contrast to the study mentioned above, no regions of increased sensitivity to nicotine after chronic exposure were found. Whether changes in regional brain activity play a role in the development of tolerance to nicotine remains an important question to be answered.

XII. NICOTINE AND ION FLUX THROUGH THE nAChR IONOPHORE

The function of the nAChR is to bind ACh and to convert the binding into an electrical signal as a result of an increase in cation conductance through an ionophore that is part of the receptor complex.[24,150] The ionophore of the muscle and electrical organ has been extensively studied, and results of studies of the mechanisms of signal transduction and regulation of the receptor are now being applied to studies of neuronal nAChR ionophores. One significant impediment to the study of ion flux through nAChR ionophores in the CNS has been the lack of a preparation sufficiently enriched in nAChR to reliably detect changes using biochemical methods. As a result, most of the direct studies of nicotine effects on ion flux have been carried out using electrophysiological methods in the CNS or have been inferred from studies of peripheral nAChR.

Results of electrophysiological studies of the effects of nicotine on CNS neurons suggest that nicotine produces multiple effects depending upon the cell type and the location. For example, nicotine produces depolarization of Renshaw cells,[151] neurons of the cerebral cortex and ventrobasal thalamus;[152] of nigrostriatal dopamine neurons,[153] in the rat dorsal motor nucleus of the vagus;[154] and of optic nerve terminal.[155] The electrophysiological effects of nicotine in the cerebellum have been found to be both excitatory and inhibitory, depending upon the cell type.[156] Interestingly, the two effects are also heterogeneous in their sensitivity to blockade by antagonists, the excitatory effects of nicotine being blocked by curare[156] and α-bungarotoxin[157] while the inhibitory effects were sensitive to hexamethonium[156] and mecamylamine.[81]

Characteristics of the nAChR ionophore have also been investigated using single-channel recording with patch clamp. Neurons from the hippocampus and brain stem of fetal rats have been cultured, and the channels that were activated in the presence of the nicotinic agonists ACh and anatoxin-a have been characterized.[158] The single channel conductance was small (20 pS) and similar to that seen in embryonic and denervated muscle nAChR, and the kinetics of opening and closing were faster than seen in muscle. Rat retinal ganglion cells in culture have also been studied:[159] the agonist-induced currents were due to cations and had a unitary conductance of 48 pS; the effects of nicotine and ACh were blocked by hexamethonium, mecamylamine, curare, DHβE, and neuronal bungarotoxin, but not by α-bungarotoxin. Recently, studies of neurons acutely isolated from rat medial habenular nucleus have been shown to have differences from previously studied CNS nAChR.[160] The channels activated by nicotinic agonists were nonselective for cations, were insensitive to either α- or neuronal bungarotoxin, and had a unitary conductance of 26 pS. A fundamental conclusion that must be drawn from the electrophysiological evidence is that the nAChR in the CNS is functionally heterogeneous and that, in spite of the evidence from binding studies, there are clearly several subtypes of nicotinic receptor in the CNS that differ in their agonist/antagonist profiles.

Influx of ^{22}Na through nAChR channels has been demonstrated in cultured adrenal chromaffin cells.[161] The effect of nicotinic agonist stimulation was very rapid and was blocked by both curare and hexamethonium, but not by tetrodotoxin. To the author's knowledge, measurement of nAChR ionophore conductance in the CNS with ^{22}Na or other isotopic cations has proven to be difficult, largely because of very small signals. Another approach that has been used to assay nAChR function in neurons employs agmatine, a guanidine derivative that can apparently enter neurons via opened nicotinic receptor channels.[162] In the original experiment, [^{3}H]-agmatine was incubated with frog sympathetic ganglia, and autoradiographic evidence was presented demonstrating that ganglia that were incubated with 1 m*M* nicotine, carbachol, or ACh preferentially accumulated the marker. These studies have been extended to rat sympathetic ganglia, where 50 μ*M* nicotine produced a significant increase in accumulation that was inhibited by curare and neuronal bungarotoxin.[163] Nevertheless, the usefulness of agmatine as a probe for nicotinic receptor function may be limited by the finding that at high concentrations (1 m*M*), it antagonizes the depolarization of rat superior cervical ganglion by DMPP.[164]

We have recently used the membrane permeant cation, [^{3}H]-tetraphenylphosphonium ([^{3}H]-TPP$^+$) to study the effects of nicotinic receptor activation on synaptosomal membrane potential.[165] [^{3}H]-TPP$^+$ has been shown to distribute across synaptosomal[166,167] and cell membranes[168] following its Nernst potential, and, thereby, its concentration gradient provides an estimate of the membrane potential difference. Since the physiological consequence of increased membrane cation conductance through nicotinic receptor ion channels is membrane depolarization, the effects of nicotine on synaptosomal [^{3}H]-TPP$^+$ accumulation should reflect activation of the nicotinic receptor ionophore. We have demonstrated that concentrations of nicotine between 10 and 100 μM decrease synaptosomal accumulation of [^{3}H]-TPP$^+$ and that this effect is mimicked by other nicotinic agonists and attenuated by hexamethonium and mecamylamine. Although further characterization must be done, this method has the advantages of being technically simple, sensitive, and useful in synaptosomal preparations, and it may provide a useful tool with which to assess nAChR function in a test tube assay.

XIII. EFFECTS OF NICOTINE ON SELECTED BRAIN NEUROTRANSMITTER SYSTEMS

Nicotinic receptors have been found on both presynaptic and postsynaptic sites in the CNS.[72] Electrophysiological studies provide evidence that the activation of postsynaptic nAChR by acetylcholine likely represents a classical neurotransmitter signaling system, resulting in an epsp in the target neuron in a manner analogous to cholinergic transmission at the neuromuscular junction (see above). However, there is increasing evidence that nicotine also acts as a neuromodular via activation of a presynaptic population of nAChR which play a role in the regulation of neurotransmitter release. This nicotinic effect is analogous to the effect of nicotine on catecholamine release in the adrenal medulla, and its mechanism is that opening of the nAChR channel results in depolarization, increased intraneuronal calcium, and activation of exocytosis. There are several comprehensive and excellent reviews of the effects of nicotine on brain neurotransmitter systems;[169-171] so only some of the recent publications will be highlighted here, particularly those concerned with *in vitro* models of release.

A. ACETYLCHOLINE

There is clear evidence that presynaptic nAChRs function as autoreceptors on cholinergic terminals where they mediate a positive feedback mechanism.[169] Presynaptic nAChRs have been shown to increase the release of ACh from motor nerve preparations, including rat phrenic nerve.[172] This population of receptors is activated by ACh and DMPP, inhibited by both hexamethonium and tubocurare, and desensitized by high agonist concentrations.[173] It is evident that released ACh itself is an endogenous ligand for these receptors since tubocurare reduces ACh release in the absence of exogenous agonist.[174] Several techniques have been used to demonstrate increased release of ACh following activation of nAChR in several regions of the CNS. Nicotine increases ACh release from guinea pig cerebral cortical slice preparations,[175] rat cortical and hippocampal slices,[67] and rat cerebellar slices.[176] Nicotine increases ACh release from [^{3}H]-choline-loaded cerebral cortical synaptosomes.[177] *In vivo* studies of release during brain dialysis have shown significant effects of nicotine on release in frontal cortex, hippocampus, and striatum.[178] However, *in vitro* studies have not shown nicotinic modulation of ACh release in striatum.[179,180] Overall, these findings demonstrate that ACh release is enhanced in the presence of agonist concentrations in the range of 10 to 1000 μM, is partially calcium-dependent,[177] and is antagonized by tubocurare,[175] hexamethonium,[177] and DHβE.[67]

The effect of chronic nicotine treatment on autoreceptor regulation of ACh release from slices of rat frontal cortex has been reported.[180] Nicotine was administered to rats for 10 d at a dose of 1.2 mg base/kg/d, and, at this dose, [^{3}H]-MCC binding site density in the cerebral cortex was increased by nearly 100%. The same treatment regimen attenuated nicotinic-agonist-induced ACh release from cortical slices. These results support the hypothesis that the increased agonist binding sites in the cerebral cortex following multiple exposure to nicotine *in vivo* are functionally inactive, possibly as a result of chronic agonist stimulation.

B. NOREPINEPHRINE

Although it is clear that stimulation of nAChR on adrenal medullary chromaffin cells induces the release of norepinephrine,[181] the effects of nicotine on norepinephrine release in the CNS have not been clearly demonstrated.[169] However, evidence has been reported recently that nicotine produces excitation of the noradrenergic neurons of the locus coeruleus as a result of peripheral activation of primary sensory fibers.[182] The pathway by which nicotine activates these neurons involves ganglionic-type nAChR on primary sensory neurons and an excitatory projection from the nucleus paragigantocellularis that releases excitatory amino acids that increase the activity of the locus coeruleus neurons.[183,184] It is possible that this mechanism is involved in the effects of nicotine on increased alertness and attention.[183]

C. DOPAMINE

It is clear from behavioral and biochemical studies that several of the physiological effects of nicotine are mediated via increased dopaminergic activity in the CNS, and this evidence has been recently reviewed.[171] Biochemical studies have demonstrated that nicotine enhances the release of dopamine from striatal slices[185] and from perfused synaptosomes.[83,186] Studies have also demonstrated activation of dopaminergic systems by nicotine in the mesolimbic dopaminergic system.[187,188] Enhancement of dopamine release from striatal synaptosomes occurs at concentrations of 1 to 100 μM,[186] is dependent on calcium,[189] and has a pharmacological profile that supports action via nAChR.[83] In addition to their effects on release, high concentrations of nicotine have been shown to inhibit dopamine accumulation into synaptic storage vesicles,[190] and low concentrations may indirectly inhibit dopamine uptake into striatal minces.[191] It is unclear, however, whether these effects on dopamine *in vitro* are operative *in vivo*.

The effects of chronic exposure to nicotine on locomotor activity, an effect that is thought to be mediated by effects of nicotine on striatal dopaminergic activity, are complex, and it is unclear whether the changes induced by chronic exposure are owing to tolerance to depressant effects of nicotine or sensitization to stimulant effects.[171] After continuous nicotine administration via osmotic pumps, Fung and co-workers found that agonist binding-site density in the striatum was increased and locomotor response to nicotine was enhanced.[192] Exposure of rats to the nicotinic agonist anatoxin for 7 d via osmotic minipumps also increased the number of agonist binding sites in the striatum and significantly enhanced nicotine-induced dopamine release from isolated striatal synaptosomes.[193] These studies support the hypothesis that the increase in nicotinic agonist binding in brain regions containing dopaminergic terminal fields corresponds to an increase in functional nAChR. Several studies have also demonstrated that nicotine exposure *in vivo* enhances the stimulant activity of nicotine *in vitro* in regions containing nerve terminals of the mesolimbic dopaminergic system,[194,195] although other studies indicate that there may be differences in the response of the two dopaminergic systems to nicotine.[196] A recent study also suggests that environmental temperature may alter the pattern of changes in striatal dopamine metabolism produced by nicotine exposure in mice;[197] so there may be several confounding factors to consider.

D. PROLACTIN

Nicotine induces the release of prolactin in rats.[198,199] Nicotine produces this effect following intravenous administration of fairly low doses, and its pharmacological characteristics are consistent with nicotine action at central nAChR.[200] The fact that the effect of nicotine on prolactin release can be assessed *in vivo* makes it an attractive model with which to study the regulation of CNS nAChR after chronic nicotine exposure. Several investigators have reported that chronic exposure to nicotine abolishes the prolactin response to an acute nicotine challenge[201,202] and that the lack of response correlates both with the onset and with the loss of increased agonist binding site density in the hypothalamus.[202] The authors suggest that these results are consistent with the hypothesis that repeated exposure to nicotine inactivates nAChR in the hypothalamus and that this prolonged inactivation leads to increased number. It is also possible that the exposure to nicotine produces a functional down-regulation due to a shift in the existing receptors to a desensitized, high-affinity state.

XIV. NICOTINE AND OTHER SECOND MESSENGERS

One consequence of the membrane depolarization that results from nAChR activation in cultured adrenal medullary cells is increased intracellular calcium concentration. The increase in intracellular calcium is recognized to be the trigger for exocytosis of catecholamines in adrenal chromaffin cells and for the exocytosis of neurotransmitters in neurons. There is experimental evidence for nicotinic receptor activation resulting in calcium entry through open nAChR channels,[203] increased influx of calcium through voltage-dependent calcium channels,[161,204,205] and increased release of intracellular calcium.[206] Recent evidence also suggests that, in bovine adrenal chromaffin cells, nAChR activation may also lead to a decrease in calcium influx through voltage-dependent channels, an effect that may be involved in negative feedback control of catecholamine secretion.[207] In studies using rat cerebral cortical synaptosomes, my laboratory has recently reported that nicotine (30 to 300 μM) increases intrasynaptosomal calcium concentrations as measured using the fluorescent calcium indicator FURA-2.[208]

Calcium may not be the only second-messenger signal generated by nicotinic receptor activation in adrenal medullary cells. Manipulations, such as high extracellular potassium levels and calcium ionophores that raise intracellular calcium levels to the same degree, result in a smaller degree of catecholamine secretion.[209] Nicotinic stimulation of adrenal glands results in an increase in adrenomedullary cAMP levels.[210] The increase in cAMP is calcium-dependent, but it is not accompanied by an increase in membrane adenylate cyclase activity. The mechanism of this increase is obscure, and the role of cAMP in catecholamine secretion from these cells is unclear. It has been suggested that the time-course of cAMP accumulation is too long to explain release and that it is likely that the role of cAMP is a modulatory one. In some studies, manipulations that increase cAMP, such as nerve growth factor treatment,[122] enhance nicotine-induced release; in other studies, increased cAMP attenuates nicotine-induced actin disassembly and catecholamine release.[211] As outlined above, it is possible that some effects of cAMP may stem from changes in nAChR sensitivity, although definitive evidence of causal relationships among nicotinic receptor activation, cAMP levels, and modulation of physiological responses has not been demonstrated. Whether nAChR activation in the CNS produces similar effects on cAMP levels is yet to be determined.

Nicotinic stimulation of adrenal chromaffin cells produces an increase in the concentration of inositol triphosphate ($InsP_3$)[212,213] as well as increases in $InsP_4$ and $InsP_5$[214] and enhanced translocation of protein kinase C from the cytosol to membranes.[215] This effect is calcium-dependent and is mimicked by stimulation of the cells with a depolarizing concentration of potassium, suggesting that it is not a specific effect of nicotinic receptor activation but secondary to an increase in intracellular calcium.[215] Prior exposure of cells to a protein kinase C activating phorbol ester results in an enhancement of nicotine-induced calcium accumulation and catecholamine release,[216,217] suggesting that protein kinase C mediated events may also be involved in regulating adrenal medullary responsiveness to nicotinic stimulation.

XV. SUMMARY

In spite of the large body of literature regarding the neurochemical effects of nicotine, there are many important questions that remain to be answered. While it is clear that radioligand binding studies of the agonist binding site have provided and continue to provide important information concerning the pharmacology, localization, and, possibly, regulation of nAChR, it is also clear that these studies have serious limitations. In particular, the heterogeneity of the CNS nAChR that is supported by structural studies is not seen in the results of binding studies. Furthermore, agonist binding site up-regulation correlates strongly with the development of tolerance (or sensitization) to nicotine following chronic exposure, but it is clear that functional studies and studies of the changes in receptor mRNA are needed to answer fundamental questions about the mechanisms involved. These questions point to the pivotal role studies of nAChR in the CNS continue to play in our understanding of brain neurotransmission.

ACKNOWLEDGMENT

The author was supported by USPHS grant R29 DA04800 during the preparation of this review.

REFERENCES

1. **Henningfield, J. E. and Woodson, P. P.,** Dose-related actions of nicotine on behavior and physiology: review and implications for replacement therapy for nicotine dependence, *J. Sub. Abuse,* 1, 301, 1989.
2. **Jaffe, J. H. and Kanzler, M.,** Smoking as an addictive disorder, in *Cigarette Smoking as a Dependence Process,* Krasnegor, N. A., Ed., NIDA, Rockville, MD, 1979, 4.
3. **Griffiths, R. R. and Henningfield, J. E.,** Pharmacology of cigarette smoking behavior, *Trends Pharmacol. Sci.,* 3, 260, 1982.
4. **Hughes, J. R. and Miller, S. A.,** Nicotine gum to help stop smoking, *J. Am. Med. Assoc.,* 252, 2855, 1984.
5. **Nemeth-Coslett, R., Henningfield, J. E., O'Keefe, M. K., and Griffiths, R. R.,** Effects of mecamylamine on human cigarette smoking and subjective ratings, *Psychopharmacologia,* 88, 420, 1986.
6. **Stolerman, I. P., Goldfarb, T., Fink, R., and Jarvik, M. E.,** Influencing cigarette smoking with nicotine antagonists, *Psychopharmacologia,* 28, 247, 1973.
7. **Pomerleau, C. S., Pomerleau, O. F., and Majchrzak, M. J.,** Mecamylamine pretreatment increases subsequent nicotine self-administration as indicated by changes in plasma nicotine level, *Psychopharmacologia,* 91, 391, 1987.

8. **Henningfield, J. E.,** Behavioral pharmacology of cigarette smoking, *Adv. Behav. Pharmacol.,* 4, 131, 1984.
9. **Benowitz, N. L.,** Clinical pharmacology of nicotine, *Annu. Rev. Med.,* 37, 21, 1986.
10. **Benowitz, N. L.,** Pharmacologic aspects of cigarette smoking and nicotine addiction, *N. Engl. J. Med.,* 319, 1318, 1988.
11. **Henningfield, J. E. and Nemeth-Coslett, R.,** Nicotine dependence: interface between tobacco and tobacco-related disease, *Chest,* 93, 37S, 1988.
12. **Clarke, P. B. S.,** Nicotine and smoking: a perspective from animal studies, *Psychopharmacologia,* 92, 135, 1987.
13. **Gilbert, D. G.,** Paradoxical tranquilizing and emotion-reducing effects of nicotine, *Psychol. Bull.,* 86, 643, 1979.
14. **Stolerman, I. P., Fink, R., and Jarvik, M. E.,** Acute and chronic tolerance to nicotine measured by activity in rats, *Psychopharmacologia,* 30, 329, 1973.
15. **Clarke, P. B. S., and Kumar, R.,** The effects of nicotine on locomotor activity in nontolerant and tolerant rats, *Br. J. Pharmacol.,* 78, 329, 1983.
16. **Henningfield, J. E., and Goldberg, S. R.,** Nicotine as a reinforcer in human subjects and laboratory animals, *Pharmacol. Biochem. Behav.,* 19, 989, 1983.
17. **Cox, B. M., Goldstein, A., and Nelson, W. T.,** Nicotine self-administration in rats, *Br. J. Pharmacol.,* 83, 49, 1984.
18. **Goldberg, S. R., Spealman, R. D., and Goldberg, D. M.,** Persistent high rate behavior maintained by intravenous administration of nicotine, *Science,* 214, 573, 1981.
19. **Chessick, R. D.,** The problem of tobacco habituation, *J. Am. Med. Assoc.,* 188, 208, 1964.
20. **Hughes, J. R., Hatsukami, D. K., Pickens, R. W., Krahn, D., Malins, S., and Luknic, A.,** Effect of nicotine on the tobacco withdrawal syndrome, *Psychopharmacologia,* 83, 82, 1984.
21. **Hall, G. H. and Morrison, C. F.,** New evidence for a relationship between tobacco smoking, nicotine dependence and stress, *Nature,* 243, 199, 1973.
22. **Abood, L. G., Reynolds, D. T., Booth, H., and Bidlack, J. M.,** Sites and mechanisms for nicotine's action in the brain, *Neurosci. Biobehav. Rev.,* 5, 479, 1981.
23. **Rosecrans, J. A. and Meltzer, C. T.,** Central sites and mechanisms of action of nicotine, *Neurosci. Biobehav. Rev.,* 5, 497, 1981.
24. **Colquhoun, D., Ogden, D. C., and Mathie, A.,** Nicotinic acetylcholine receptors of nerve and muscle: functional aspects, *Trends Pharmacol. Sci.,* 8, 465, 1987.
25. **Steinbach, J. H. and Ifune, C.,** How many kinds of nicotinic cholinergic receptors are there?, *Trends Neurosci.,* 12, 3, 1989.
26. **Luetje, C. W., Patrick, J., and Seguela, P.,** Nicotine receptors in the mammalian brain, *FASEB J.,* 4, 2753, 1990.
27. **Stroud, R. M. and Finer-Moore, J.,** Acetylcholine receptor structure, function, evolution, *Annu. Rev. Cell. Biol.,* 1, 317, 1985.
28. **Whiting, P. and Lindstrom, J.,** Purification and characterization of a nicotinic acetylcholine receptor from rat brain, *Proc. Natl. Acad. Sci.,* 84, 595, 1987.
29. **Boulter, J., Evans, K., Goldman, D., Martin, G., Treco, D., Heinemann, S., and Patrick, J.,** Isolation of a cDNA clone coding for a possible neural nicotinic acetylcholine receptor alpha subunit, *Nature,* 319, 368, 1986.
30. **Goldman, D., Deneris, E., Luyten, W., Kochhar, A., Patrick, J., and Heinemann, S.,** Members of a nicotinic acetylcholine receptor gene family are expressed in different regions of the mammalian nervous system, *Cell,* 48, 965, 1987.
31. **Wada, K., Ballivet, M., Boulter, J., Connolly, J., Wada, E., Deneris, E. S., Swanson, L. W., Heinemann, S., and Patrick, J.,** Functional expression of a new pharmacological subtype of brain nicotinic acetylcholine receptor, *Science,* 240, 330, 1988.

32. **Wada, E., McKinnon, D., Heinemann, S., Patrick, J., and Swanson, L. W.,** The distribution of mRNA encoded by a new member of the neuronal nicotinic acetylcholine receptor gene family (alpha5) in the rat central nervous system, *Brain Res.,* 526, 45, 1990.
33. **Deneris, E. S., Connolly, J., Boulter, J., Wada, E., Wada, K., Swanson, L. W., Patrick, J., and Heinemann, S.,** Primary structure and expression of β2: a novel subunit of neuronal nicotinic acetylcholine receptors, *Neuron,* 1, 45, 1988.
34. **Nef, P., Oneyser, C., Alliod, C., Couturier, S., and Ballivet, M.,** Genes expressed in the brain define three distinct neuronal nicotinic acetylcholine receptors, *EMBO J.,* 7, 595, 1988.
35. **Duvoisin, R. M., Deneris, E. S., Patrick, J., and Heinemann, S.,** The functional diversity of the neuronal nicotinic acetylcholine receptors is increased by a novel subunit, β4, *Neuron,* 3, 487, 1989.
36. **Whiting, P. J., Schoepfer, R., Swanson, L. W., Simmons, D. M., and Lindstrom, J. M.,** Functional acetylcholine receptor in PC12 cells reacts with a monoclonal antibody to brain nicotinic receptors, *Nature,* 327, 515, 1987.
37. **Boulter, J., Connolly, J., Deneris, E., Goldman, D., Heinemann, S., and Patrick, J.,** Functional expression of two neuronal nicotinic acetylcholine receptors from cDNA clones identifies a gene family, *Proc. Natl. Acad. Sci.,* 84, 7763, 1987.
38. **Bertrand, D., Ballivet, M., and Rungger, D.,** Activation and blocking of neuronal nicotinic acetylcholine receptor reconstituted in Xenopus oocytes, *Proc. Natl. Acad. Sci.,* 87, 1993, 1990.
39. **Luetje, C. W. and Patrick, J.,** Both alpha and beta subunits contribute to the agonist sensitivity of neuronal nicotinic acetylcholine receptors, *J. Neurosci.,* 11, 837, 1991.
40. **Wada, E., Wada, E., Boulter, J., Deneris, E., Heinemann, S., Patrick, J., and Swanson, L. W.,** Distribution of alpha2, alpha3, alpha4, and beta2 neuronal nicotinic receptor subunit mRNAs in the central nervous system: a hybridization histochemical study in the rat, *J. Comp. Neurol.,* 284, 314, 1989.
41. **Boyd, R. T., Jacob, M. H., Couturier, S., Ballivet, M., and Berg, D. K.,** Expression and regulation of neuronal acetylcholine receptor mRNA in chick ciliary ganglia, *Neuron,* 1, 495, 1988.
42. **Boyd, R. T., Jacob, M. H., McEachern, A. E., Caron, S., and Berg, D. K.,** Nicotinic acetylcholine receptor mRNA in dorsal root ganglion neurons, *J. Neurobiol.,* 22, 1, 1990.
43. **Schleifer, L. S. and Eldefrawi, M. E.,** Identification of the nicotinic and muscarinic acetylcholine receptors in subcellular fractions of mouse brain, *Neuropharmacology,* 13, 53, 1974.
44. **Yoshida, K. and Imura, H.,** Nicotinic cholinergic receptors in brain synaptosomes, *Brain Res.,* 172, 453, 1979.
45. **Romano, C. and Goldstein, A.,** Stereospecific nicotinic receptors on rat brain membranes, *Science,* 210, 647, 1980.
46. **Abood, L. G., Reynolds, D. T., and Bidlack, J. M.,** Stereospecific [^{3}H]-nicotine binding to intact and solubilized rat brain membranes and evidence for its noncholinergic nature, *Life Sci.,* 27, 1307, 1980.
47. **Abood, L. G., Grassi, S., and Costanza, M.,** Binding of optically pure (-)-[^{3}H]-nicotine to rat brain membranes, *FEBS Lett.,* 157, 147, 1983.
48. **Marks, M. J. and Collins, A. C.,** Characterization of nicotine binding in mouse brain and comparison with the binding of alpha-bungarotoxin and quinuclidinyl benzilate, *Mol. Pharmacol.,* 22, 554, 1982.
49. **Larsson, C. and Nordberg, A.,** Comparative analysis of nicotine-like receptor-ligand interactions in rodent brain homogenate, *J. Neurochem.,* 45, 24, 1985.
50. **Lippiello, P. M. and Fernandes, K. G.,** The binding of L-[^{3}H]-nicotine to a single class of high affinity sites in rat brain membranes, *Mol. Pharmacol.,* 29, 448, 1986.
51. **Martino-Barrows, A. M. and Kellar, K. J.,** [^{3}H]-Acetylcholine and [^{3}H](-)nicotine label the same recognition site in rat brain, *Mol. Pharmacol.,* 31, 169, 1987.

52. **Reavill, C., Jenner, P., and Stolerman, I. P.,** High affinity binding of [^{3}H](-)-nicotine to rat brain membranes and its inhibition by analogues of nicotine, *Neuropharmacology,* 27, 235, 1988.
53. **Schwartz, R. D., McGee, R., Jr., and Kellar, K. J.,** Nicotinic cholinergic receptors labeled by [^{3}H]-acetylcholine in rat brain, *Mol. Pharmacol.,* 22, 56, 1982.
54. **Schwartz, R. D. and Kellar, K. J.,** [^{3}H]-Acetylcholine binding sites in brain, *Mol. Pharmacol.,* 24, 387, 1983.
55. **Reulecke, M. and Hucho, F.,** High and low affinity binding of [^{3}H]-acetylcholine at nicotinic cholinergic receptors in rat brain, *Neurosci. Lett.,* 59, 271, 1985.
56. **Marks, M. J., Stitzel, J. A., Romm, E., Wehner, J. M., and Collins, A. C.,** Nicotinic binding sites in rat and mouse brain: comparison of acetylcholine, nicotine and alpha-bungarotoxin, *Mol. Pharmacol.,* 30, 427, 1986.
57. **Wonnacott, S.,** Stereospecificity of nicotine binding to rat brain, *Biochem. Soc. Trans.,* 14, 597, 1986.
58. **Sloan, J. W., Martin, W. R., Hernandez, J., and Hook, R.,** Binding characteristics of (-)- and (+)-nicotine to the rat brain P2 fraction, *Pharmacol. Biochem. Behav.,* 23, 987, 1985.
59. **Clarke, P. B. S., Schwartz, R. D., Payl, S. M., Pert, C. B., and Pert, A.,** Nicotinic binding in rat brain: autoradiographic comparison of [^{3}H]-acetylcholine, [^{3}H]-nicotine and [^{125}I]-alpha-bungarotoxin, *J. Neurosci.,* 5, 1307, 1985.
60. **Zhang, X., Stjernlof, P., Adem, A., and Nordberg, A.,** Anatoxin-a a potent ligand for nicotinic cholinergic receptors in rat brain, *Eur. J. Pharmacol.,* 135, 457, 1987.
61. **Hayashi, E., Isogai, M., Kagawa, Y., Takayanagi, N., and Yamada, S.,** Neosurugatoxin, a specific antagonist of nicotinic acetylcholine receptors, *J. Neurochem.,* 42, 1491, 1984.
62. **Chiappinelli, V. A.,** Kappa-bungarotoxin: a probe for the neuronal nicotinic acetylcholine receptor, *Trends Pharmacol. Sci.,* 5, 425, 1984.
63. **Loring, R. H., Schulz, D. W., and Zigmond, R. E.,** Characterization of neuronal nicotinic receptors using neuronal bungarotoxin, *Prog. Brain Res.,* 79, 109, 1989.
64. **McLane, K. E., Tang, F., and Conti-Tronconi, B. M.,** Localization of sequence segments forming a kappa-bungarotoxin-binding site on the alpha3 neuronal nicotinic receptor, *J. Biol. Chem.,* 265, 1537, 1990.
65. **Abood, L. G. and Grassi, S.,** [^{3}H]-Methylcarbamylcholine, a new radioligand for studying brain nicotinic receptors, *Biochem. Pharmacol.,* 35, 4199, 1986.
66. **Boksa, P. and Quirion, R.,** [^{3}H]-N-methyl-carbamylcholine, a new radioligand specific for nicotinic receptors in brain, *Eur. J. Pharmacol.,* 139, 323, 1987.
67. **Araujo, D. M., Lapchak, P. A., Collier, B., and Quiron, R.,** Characterization of *N*-[^{3}H]-methylcarbamylcholine binding sites and effect of *N*-methylcarbamylcholine on acetylcholine release in rat brain, *J. Neurochem.,* 51, 292, 1988.
68. **Takayama, H., Majewska, M. D., and London, E. D.,** Interactions of noncompetitive inhibitors with nicotinic receptors in the rat brain, *J. Pharmacol. Exp. Ther.,* 253, 1083, 1989.
69. **Pabreza, L. A., Dhawan, S., and Kellar, K. J.,** [^{3}H]-Cytisine binding to nicotinic cholinergic receptors in brain, *Mol. Pharmacol.,* 39, 9, 1991.
70. **Yamada, S., Gehlert, D. R., Hawkins, K. N., Nakayama, K., Roeske, W. R., and Yamamura, H. I.,** Autoradiographic localization of nicotinic receptor binding in rat brain using [^{3}H]-methylcarbamylacholine, a novel radioligand, *Life Sci.,* 41, 2851, 1987.
71. **Rainbow, T. C., Schwartz, R. D., Parsons, B., and Kellar, K. J.,** Quantitative autoradiography of nicotinic [^{3}H]-acetylcholine binding sites in rat brain, *Neurosci. Lett.,* 50, 193, 1984.
72. **Clarke, P. B. S., Pert, C. B., and Pert, A.,** Autoradiographic distribution of nicotine receptors in rat brain, *Brain Res.,* 323, 390, 1984.
73. **London, E. D., Waller, S. B., and Wamsley, J. K.,** Autoradiographic localization of [^{3}H]-nicotine binding sites in the rat brain, *Neurosci. Lett.,* 53, 179, 1985.

74. **Semba, K. and Fibiger, H. C.,** Organization of central cholinergic systems, *Prog. Brain Res.*, 79, 37, 1989.
75. **Whiting, P. and Lindstrom, J.,** Pharmacological properties of immuno-isolated neuronal nicotinic receptors, *J. Neurosci.*, 6, 3061, 1986.
76. **Deutch, A. Y., Holliday, J., Roth, R. H., Chun, L. L. Y., and Hawrot, E.,** Immunohistochemical localization of a neuronal nicotinic acetylcholine receptor in mammalian brain, *Proc. Natl. Acad. Sci.*, 84, 8697, 1987.
77. **Schwartz, R. D.,** Autoradiographic distribution of high affinist muscarinic and nicotinic cholinergic receptors labeled with [^{3}H]-acetylcholine in rat brain, *Life Sci.*, 38, 2111, 1986.
78. **Boyd, N. D. and Cohen, J. B.,** Kinetics of binding of [^{3}H]-acetylcholine and [^{3}H]-carbamylcholine to torpedo postsynaptic membranes; slow conformational transitions of the cholinergic receptor, *Biochem.*, 19, 5344, 1980.
79. **Boyd, N. D. and Cohen, J. B.,** Desensitization of membrane-bound Torpedo acetylcholine receptor by amine noncompetitive antagonists and aliphatic alcohols: studies of [^{3}H]-acetylcholine binding and $^{22}Na^+$ ion fluxes, *Biochem.*, 23, 4023, 1984.
80. **Paton, W. D. M. and Zaimis, E. J.,** *Br. J. Pharmacol.*, 6, 155, 1951.
81. **Garza, R. de la, Freedman, R., and Hoffer, B. J.,** Nicotine-induced inhibition of cerebellar purkinje neurons: specific actions of nicotine and selective blockade by mecamylamine, *Neuropharmacology*, 28, 495, 1989.
82. **Collins, A. C., Evans, C. B., Miner, L. L., and Marks, M. J.,** Mecamylamine blockade of nicotine responses: evidence for two brain nicotinic receptors, *Pharmacol. Biochem. Behav.*, 24, 1767, 1986.
83. **Rapier, C., Lunt, G. G., and Wonnacott, S.,** Nicotinic modulation of [^{3}H]-dopamine release from striatal synaptosomes: pharmacological characterisation, *J. Neurochem.*, 54, 937, 1990.
84. **Williams, M. and Robinson, J. L.,** Binding of the nicotinic cholinergic antagonist, dihydro-beta-erythroidine, to rat brain tissue, *J. Neurosci.*, 4, 2906, 1984.
85. **Banerjee, S., Punzi, J. S., Kreilick, K., and Aboood, L. G.,** [^{3}H]-Mecamylamine binding to rat brain membranes: studies with mecamylamine and nicotine analogs, *Biochem. Pharmacol.*, 40, 2105, 1990.
86. **McCarthy, M. P., Earnest, J. P., Young, E. F., Choe, S., and Stroud, R. M.,** The molecular neurobiology of the acetylcholine receptor, *Annu. Rev. Neurosci.*, 9, 383, 1986.
87. **Patrick, J. and Stallcup, B.,** Alpha-bungarotoxin binding and cholinergic receptor function ion a rat sympathetic nerve line, *J. Biol. Chem.*, 252, 8629, 1977.
88. **McQuarrie, C., Salvaterra, P. M., De Blas, A., Routes, J., and Mahler, H. R.,** Studies on nicotinic acetylcholine receptors in mammalian brain, *J. Biol. Chem.*, 251, 6335, 1976.
89. **Schmidt, J.,** Drug binding properties of an alpha-bungarotoxin binding component from rat brain, *Mol. Pharmacol.*, 13, 283, 1977.
90. **Oswald, R. E. and Freeman, J. A.,** Alpha-bungarotoxin binding and central nervous system nicotinic acetylcholine receptors, *Neuroscience*, 6, 1, 1981.
91. **Schmidt, J.,** Biochemistry of nicotinic acetylcholine receptors in the vertebrate brain, *Int. Rev. Neurobiol.*, 30, 1, 1988.
92. **Quik, M. and Geertsen, S.,** Neuronal nicotine alpha-bungarotoxin sites, *Can. J. Physiol. Pharmacol.*, 66, 971, 1988.
93. **Conti-Tronconi, B. M., Dunn, S. M. J., Barnard, E. A., Dolly, J. O., Lai, F. A., Ray, N., and Raftery, M. A.,** Brain and muscle nicotinic acetylcholine receptors are different homologous proteins, *Proc. Natl. Acad. Sci.*, 82, 5208, 1985.
94. **Schoepfer, R., Conroy, W. G., and Whiting, P., Gore, M., and Lindstrom, J.,** Brain alpha-bungarotoxin binding protein cDNAs and MAbs reveal subtypes of this branch of the ligand-gated ion channel gene superfamily, *Neuron*, 5, 35, 1990.

95. **Freeman, J. A.,** Possible regulatory function of acetylcholine receptor in maintenance of retinotectal synapses, *Nature,* 269, 218, 1977.
96. **Chiappinelli, V. A. and Giacobini, E.,** Time course of appearance of alpha-bungarotoxin sites during development of chick ciliary ganglion and iris, *Neurochem. Res.,* 3, 465, 1978.
97. **Fiedler, E. P., Marks, M. J., and Collins, A. C.,** Postnatal development of cholinergic enzymes and receptors in mouse brain, *J. Neurochem.,* 49, 983, 1987.
98. **Freeman, J. A.,** Possible regulatory function of acetylcholine receptor in maintenance of retinotectal synapses, *Nature,* 269, 218, 1977.
99. **Lipton, S. A., Frosch, M. P., Phillips, M. D., Tauck, D. L., and Aizenman, E.,** Nicotinic antagonists enhance process outgrowth by rat retinal ganglion cells in culture, *Science,* 239, 1293, 1988.
100. **Lipton, S. A. and Kater, S. B.,** Neurotransmitter regulation of neuronal outgrowth, plasticity and survival, *Trends Neurosci.,* 12, 265, 1989.
101. **Lichtensteiger, W., Ribary, U., Schlumpf, M., Odermatt, B., and Widmer, H. R.,** Prenatal adverse effects of nicotine on the developing brain, *Prog. Brain Res.,* 73, 137, 1988.
102. **Brown, R. H., Schweitzer, T., Audhya, T., Goldstein, G., and Dichter, M. A.,** Immunoreactive thymopoietin in the mouse central nervous system, *Brain Res.,* 381, 237, 1986.
103. **Twomey, J. J., Lewis, V. M., Pattn, B. M., Goldstein, G., and Good, R. A.,** Myasthenia gravis, thymectomy and serum thymic hormone activity, *Am. J. Med.,* 66, 639, 1979.
104. **Revah, F., Mulle, C., Pinset, C., Audhya, T., Goldstein, G., and Changeux, J. P.,** Calcium-dependent effect of the thymic polypeptide thymopoietin on the desensitization of the nicotinic acetylcholine receptor, *Proc. Natl. Acad. Sci.,* 84, 3477, 1987.
105. **Ochoa, E. L. M., Medrano, S., deCarlin, M. C. L., and Dilonardo, A. M.,** Arg-Lys-Asp-Val-Tyr (Thymopentin) accelerates the cholinergic-induced inactivation (desensitization) of reconstituted nicotinic receptor, *Cell. Mol. Biol.,* 8, 325, 1988.
106. **Quik, M., Afar, R., Audhya, T., and Goldstein, G.,** Thymopoietin, a thymic polypeptide, specifically interacts at neuronal nicotinic alpha-bungarotoxin receptors, *J. Neurochem.,* 53, 1320, 1989.
107. **Lukas, R. J., Audhya, T., Goldstein, G., and Lucero, L.,** Interactions of the thymic polypeptide hormone thymopoietin with neuronal nicotinic alpha-bungarotoxin binding sites and with muscle-type, but not ganglia-type, nicotinic acetylcholine receptor, *Mol. Pharmacol.,* 38, 887, 1990.
108. **Quik, M., Cohen, R., Audhya, T., and Goldstein, G.,** Thymopoietin interacts at the alpha-bungarotoxin site of and induces process formation in PC12 pheochromocytoma cells, *Neuroscience,* 39, 139, 1990.
109. **Quik, M., El-Bizri, H., Audhya, T., and Goldstein, G.,** Thymopoietin, a potent antagonist at nicotinic receptors in C2 muscle cell cultures, *Mol. Pharmacol.,* 39, 324, 1991.
110. **Quik, M., Afar, R., Geertsen, S., Audhya, T., Goldstein, G., and Trifaro, J. M.,** Thymopoietin, a thymic polypeptide, regulates nicotinic alpha-bungarotoxin sites in chromaffin cells in culture, *Mol. Pharmacol.,* 37, 90, 1990.
111. **Andersson, K.,** Mecamylamine pretreatment counteracts cigarette smoke induced changes in hypothalamic catecholamine neuron systems and in anterior pituitary function, *Acta Physiol. Scand.,* 125, 445, 1985.
112. **Barrios, V., Rodriguez-Snachez, M. N., Colas, B., and Arilla, E.,** Effects of acute nicotine and mecamylamine administration on somatostatin concentration and binding in the rat brain, *Eur. J. Pharmacol.,* 179, 263, 1990.
113. **Marks, M. J., Romm, E., Gaffney, D. K., and Collins, A. C.,** Nicotine-induced tolerance and receptor changes in four mouse strains, *J. Pharmacol. Exp. Ther.,* 237, 809, 1986.

114. **Lingle, C.,** Blockade of cholinergic channels by chlorisondamine on a crustacean muscle, *J. Physiol.,* 339, 395, 1983.
115. **Varanda, W. A., Arcava, Y., Sherby, S. M., Van Meter, W. G., Eldefrawi, M. E., and Albuquerque, E. X.,** The acetylcholine receptor of the neuromuscular junction recognizes mecamylamine as a noncompetitive antagonist, *Mol. Pharmacol.,* 28, 128, 1985.
116. **Albuquerque, E. X., Tsai, M. C., Aronstam, R. S., Eldefrawi, A. T., and Eldefrawi, M. E.,** A possible involvement of cyclic AMP in the expression of desensitization of the nicotinic acetylcholine receptor, *Proc. Natl. Acad. Sci.,* 77, 1224, 1980.
117. **Aronstam, R. S., Eldefrawi, A. T., Pessah, I. N., Daly, J. W., Albuquerque, E. X., and Eldefrawi, M. E.,** Regulation of [^{3}H]-perhydrohistrionicotoxin binding to Torpedo ocellate electroplax by effectors of the acetylcholine receptor, *J. Biol. Chem.,* 256, 2843, 1981.
118. **Rapier, C., Wonnacott, S., Lunt, G. G., and Albuquerque, E. X.,** The neurotoxin histrionicotoxin interacts with the putative ion channel of the nicotinic acetylcholine receptors in the central nervous system, *FEBS Lett.,* 212, 292, 1987.
119. **Egan, T. M. and North, R. A.,** Actions of acetylcholine and nicotine on rat locus coeruleus neurons *in vitro, Neuroscience,* 19, 565, 1986.
120. **Douglas, W. W. and Rubin, R. P.,** The role of calcium in the secretory response of the adrenal medulla to acetylcholine, *J. Physiol.,* 159, 40, 1961.
121. **Boksa, P. and Livett, B. G.,** Desensitization to nicotinic cholinergic agonists and K^+ agents that stimulate catecholamine secretion, in isolated adrenal chromaffin cells, *J. Neurochem.,* 42, 607, 1984.
122. **Baizer, L. and Weiner, N.,** Nerve growth factor treatment enhances nicotine-stimulated dopamine release and increases in cyclic adenosine 3′:5′-monophosphate levels in PC12 cell cultures, *J. Neurosci.,* 5, 1176, 1985.
123. **Boksa, P. and Livett, B. G.,** Substance P protects against desensitization of the nicotinic response in isolated adrenal chromaffin cells, *J. Neurochem.,* 42, 618, 1984.
124. **Briggs, C. A., McAfee, D. A., and McCaman, R. E.,** Long-term regulation of synaptic acetylcholine release and nicotinic transmission: the role of cyclic AMP, *Br. J. Pharmacol.,* 93, 399, 1988.
125. **Albuquerque, E. X., Deshpande, S. S., Aracava, Y., Alkondon, M., and Daly, J. W.,** A possible involvement of cyclic AMP in the expression of desensitization of the nicotinic acetylcholine receptor, *FEBS Lett.,* 199, 113, 1986.
126. **Huganir, R. L., Delcour, A. H., Greengard, P., and Hess, G. P.,** Phosphorylation of the nicotinic acetylcholine receptor regulates its rate of desensitization, *Nature,* 321, 774, 1986.
127. **Margiotta, J. F., Berg, D. K., and Dionne, V. E.,** Cyclic AMP regulates the proportion of functional acetylcholine receptors on chicken ciliary ganglion neurons, *Proc. Natl. Acad. Sci.,* 84, 8155, 1987.
128. **Stollberg, J. and Berg, D. K.,** Neuronal acetylcholine receptors: fate of surface and internal pools in cell culture, *J. Neurosci.,* 7, 1809, 1987.
129. **Higgins, L. S. and Berg, D. K.,** Cyclic AMP-dependent mechanism regulates acetylcholine receptor function on bovine adrenal chromaffin cells and discriminates between old and new receptors, *J. Cell. Biol.,* 107, 1157, 1988.
130. **McHugh, E. M. and McGee, R.,** Direct anesthetic-like effects of forskolin on the nicotinic acetylcholine receptors of PC12 cells, *J. Biol. Chem.,* 261, 3103, 1986.
131. **Nishizawa, Y., Seamon, K. B., Daly, J. W., and Aronstam, R. S.,** Effects of forskolin and analogues on nicotinic receptor-mediated sodium flux, voltage-dependent calcium flux and voltage-dependent rubidium efflux in pheochromocytoma PC12 cells, *Cell. Mol. Neurobiol.,* 10, 351, 1990.
132. **Marks, M. J., Burch, J. B., and Collins, A. C.,** Effects of chronic nicotine infusion on tolerance development and nicotinic receptors, *J. Pharmacol. Exp. Ther.,* 226, 817, 1983.

133. **Schwartz, R. D. and Kellar, K. J.,** Nicotinic cholinergic receptor binding sites in the brain: regulation *in vivo*, *Science,* 220, 214, 1983.
134. **Schwartz, R. D. and Kellar, K. J.,** *In vivo* regulation of [^{3}H]-acetylcholine recognition sites in brain by nicotinic cholinergic drugs, *J. Neurochem.,* 45, 427, 1985.
135. **Bhat, R. V., Turner, S. L., Marks, M. J., and Collins, A. C.,** Selective changes in sensitivity to cholinergic agonists and receptor changes elicited by continuous physostigmine infusion, *J. Pharmacol. Exp. Ther.,* 255, 187, 1990.
136. **Wonnacott, S.,** The paradox of nicotinic acetylcholine receptor up-regulation by nicotine, *Trends Pharmacol. Sci.,* 11, 216, 1990.
137. **Marks, M. J., Stitzel, J. A., and Collins, A. C.,** Time course study of the effects of chronic nicotine infusion on drug response and brain receptors, *J. Pharmacol. Exp. Ther.,* 235, 619, 1985.
138. **Benwell, M. E. M., Balfour, D. J. K., and Anderson, J. M.,** Evidence that tobacco smoking increases the density of (−)-[^{3}H] nicotine binding sites in human brain, *J. Neurochem.,* 50, 1243, 1988.
139. **Romanelli, L., Ohman, B., Adem, A., and Nordberg, A.,** Subchronic treatment of rats with nicotine: interconversion of nicotinic receptor subtypes in brain, *Eur. J. Pharmacol.,* 148, 289, 1988.
140. **Schecter, M. D. and Rosecrans, J. A.,** Behavioral tolerance to an effect of nicotine in the rat, *Arch. Int. Pharmacodyn.,* 195, 52, 1972.
141. **Marks, M. J., Stitzel, J. A., and Collins, A. C.,** Dose-response analysis of nicotine tolerance and receptor changes in two inbred mouse strains, *J. Pharmacol. Exp. Ther.,* 239, 358, 1986.
142. **Marks, M. J., Stitzel, J. A., and Collins, A. C.,** Influence of kinetics of nicotine administration on tolerance development and receptor levels, *Pharmacol. Biochem. Behav.,* 27, 505, 1987.
143. **Collins, A. C., Romm, E., and Wehner, J. M.,** Nicotine tolerance: an analysis of the time course of its development and loss in the rat, *Psychopharmacologia,* 96, 7, 1988.
144. **Collins, A. C., Romm, E., and Wehner, J. M.,** Dissociation of the apparent relationship between nicotine tolerance and up-regulation of nicotinic receptors, *Brain Res. Bull.,* 25, 373, 1990.
145. **Sokoloff, L., Revich, M., Kennedy, C., Des Rosiers, M. H., Patlak, C. S., Pettigrew, K. D., Sakurada, O., and Shinohara, M.,** The [^{14}C]-deoxyglucose method for the measurement of local cerebral glucose utilization: theory, procedure and normal values in the conscious and anesthetized albino rat, *J. Neurochem.,* 28, 897, 1977.
146. **McNamara, D., Larson, D. M., Rapoport, S. I., and Soncrant, T. T.,** Preferential metabolic activation of subcortical brain areas by acute administration of nicotine to rats, *J. Cerebral Blood Flow Metab.,* 10, 48, 1990.
147. **Grunwald, F., Schrock, H., and Kuchinsky, W.,** The effect of acute nicotine infusion on the local cerebral glucose utilization of the awake rat, *Brain Res.,* 400, 232, 1987.
148. **Grunwald, F., Schrock, H., Theilen, H., Biber, A., and Kuchinsky, W.,** Local cerebral glucose utilization of the awake rat during chronic administration of nicotine, *Brain Res.,* 456, 350, 1988.
149. **London, E. D., Fanelli, R. J., Kimes, A. S., and Moses, R. L.,** Effects of chronic nicotine on cerebral glucose utilization in the rat, *Brain Res.,* 520, 208, 1990.
150. **Hucho, F.,** The nicotinic acetylcholine receptor and its ion channel, *Eur. J. Biochem.,* 158, 211, 1986.
151. **Curtis, D. R. and Ryall, R. W.,** The excitation of Renshaw cells by cholinomimetics, *Exp. Brain Res.,* 49, 65, 1966.
152. **McLennan, H. and Hicks, T. P.,** Pharmacological characterization of the excitatory cholinergic receptors of rat central neurones, *Neuropharmacology,* 17, 329, 1978.
153. **Lichtensteiger, W., Heifti, F., Felix, D., Uwyler, T., Melamed, E., and Schlumpf, M.,** Stimulation of nigrostriatal dopaminergic neurones by nicotine, *Neuropharmacology,* 21, 963, 1982.

154. **Ito, C., Fukuda, A., Nabekura, J., and Oomura, Y.,** Acetylcholine causes nicotinic depolarization in rat dorsal motor nucleus of the vagus, *in vitro, Brain Res.,* 503, 44, 1989.
155. **King, W. M.,** Nicotinic depolarization of optic nerve terminals augments synaptic transmission, *Brain Res.,* 527, 150, 1990.
156. **Garza, R. de la, Bickford-Wimer, P. C., Hoffer, B. J., and Freedman, R.,** Heterogeneity of nicotine actions in the rat cerebellum: an *in vivo* electrophysiological study, *J. Pharmacol. Exp. Ther.,* 240, 689, 1987.
157. **Garza, R. de la, McGuire, T. J., Freedman, R., and Hoffer, B. J.,** Selective antagonism of nicotine actions in the rat cerebellum with alpha-bungarotoxin, *Neuroscience,* 23, 887, 1987.
158. **Arcava, Y., Deshpande, S. S., Swanson, K. L., Rapoport, H., Wonnacott, S., Lunt, G., and Albuquerque, E. X.,** Nicotinic acetylcholine receptors in cultured neurons from the hippocampus and brain stem of the rat characterized by single channel recording, *FEBS Lett.,* 222, 63, 1987.
159. **Lipton, S. A., Aizenman, E., and Loring, R. H.,** Neuronal nicotinic acetylcholine responses in solitary mammalian retinal ganglion cells, *Pfluegers Arch.,* 410, 37, 1987.
160. **Mulle, C. and Changeux, J.-P.,** A novel type of nicotinic receptor in the rat central nervous system characterized by patch-clamp techniques, *J. Neurosci.,* 10, 169, 1990.
161. **Wada, A., Takara, H., Izumi, F., Kobayashi, H., and Yanagihara, N.,** Influx of ^{22}Na through acetylcholine receptor-associated Na channels: relationship between ^{22}Na influx, ^{45}Ca influx and secretion of catecholamines in cultured bovine adrenal medulla cells, *Neuroscience,* 15, 283, 1985.
162. **Yoshikami, D.,** Transmitter sensitivity of neurons assayed by autoradiography, *Science,* 212, 929, 1981.
163. **Quik, M.,** Inhibition of nicotinic receptor mediated ion fluxes in rat sympathetic ganglia by BGT II-S1 a potent phospholipase, *Brain Res.,* 325, 79, 1985.
164. **Loring, R. H.,** Agmatine acts as an antagonist of neuronal nicotinic receptors, *Br. J. Pharmacol.,* 99, 207, 1990.
165. **Hillard, C. J. and Pounds, J. J.,** Effects of nicotine *in vitro* and *in vivo* on synaptosomal membrane potential measured using [^{3}H]-tetraphenylphosphonium (3H-TPP), *Pharmacologist,* 32, 192, 1990.
166. **Ramos, S., Grollman, E. F., Lazo, P. S., Dyer, S. A., Habig, W. H., Hardergree, M. C., Kaback, H. R., and Kohn, L. D.,** Effect of tetanus toxin on the accumulation of the permeant lipophilic cation tetraphenylphosphonium by guinea pig brain synaptosomes, *Proc. Natl. Acad. Sci.,* 76, 4783, 1979.
167. **Pauwels, P. J. and Laduron, P. M.,** TPP^+ accumulation in rat brain synaptosomes as a probe for Na^+ channels, *Eur. J. Pharmacol.,* 132, 289, 1986.
168. **Lichtshtein, D., Kaback, H. R., and Blume, A. J.,** Use of a lipophilic cation for determination of membrane potential in neuroblastoma-glioma hybrid cell suspensions, *Proc. Natl. Acad. Sci.,* 76, 650, 1979.
169. **Balfour, D. J. K.,** The effects of nicotine on brain neurotransmitter systems, *Pharmacol. Ther.,* 16, 269, 1982.
170. **Pomerleau, O. F. and Rosecrans, J.,** Neuroregulatory effects of nicotine, *Psychoneuroendocrinology,* 14, 407, 1989.
171. **Clarke, P. B. S.,** Dopaminergic mechanisms in the locomotor stimulant effects of nicotine, *Biochem. Pharmacol.,* 40, 1427, 1990.
172. **Wessler, I.,** Control of transmitter release from the motor nerve by presynaptic nicotinic and muscarinic autoreceptors, *Trends Pharmacol. Sci.,* 10, 110, 1989.
173. **Wessler, I., Halank, M., Rasbach, J., and Kilbinger, H.,** Presynaptic nicotine receptors mediating a positive feed-back on transmitter release from the rat phrenic nerve, *Arch. Pharmacol.,* 334, 365, 1986.

174. **Wessler, I., Apel, C., and Hillen, U.,** Increase by tubocurare of [^{3}H]-acetylcholine release from the rat phrenic nerve after desensitization of nicotinic receptors, *Eur. J. Pharmacol.,* 151, 139, 1988.
175. **Beani, L., Bianchi, C., Nilsson, L., Nordberg, A., Romanelli, L., and Sivilotti, L.,** The effect of nicotine and cytisine on [^{3}H]-acetylcholine release from cortical slices of guinea-pig brain, *N.S. Arch. Pharmacol.,* 331, 293, 1985.
176. **Lapchack, P. A., Araujo, D. M., Quirion, R., and Collier, B.,** Presynaptic cholinergic mechanisms in the rat cerebellum: evidence for nicotinic, but not muscarinic autoreceptors, *J. Neurochem.,* 1843, 1989.
177. **Rowell, P. P. and Winkler, D. L.,** Nicotine stimulation of [^{3}H]-acetylcholine release from mouse cerebral cortical synaptosomes, *J. Neurochem.,* 43, 1593, 1984.
178. **Toide, K. and Arima, T.,** Effects of cholinergic drugs on extracellular levels of acetylcholine and choline in rat cortex, hippocampus and striatum studied by brain dialysis, *Eur. J. Pharmacol.,* 173, 133, 1989.
179. **Weiler, M. H., Misgeld, U., and Cheong, D. K.,** Presynaptic muscarinic modulation of nicotinic excitation in the rat neostriatum, *Brain Res.,* 296, 111, 1984.
180. **Lapchak, P. A., Araujo, D. M., Quirion, R., and Collier, B.,** Effect of chronic nicotine treatment on nicotinic autoreceptor function and N-[^{3}H]-methylcarbamylcholine binding sites in the rat brain, *J. Neurochem.,* 52, 483, 1989.
181. **Leszczyszyn, D. J., Jankowski, J. A., Viveros, O. H., Dilberto, E. J., Jr., Near, J. A., and Wightman, R. M.,** Nicotinic receptor-mediated catecholamine secretion from individual chromaffin cells, *J. Biol. Chem.,* 265, 14,736, 1990.
182. **Hajos, M. and Engberg, G.,** Role of primary sensory neurons in the central effects of nicotine, *Psychopharmacologia,* 94, 468, 1988.
183. **Engberg, G.,** Nicotine induced excitation of locus coeruleus neurons is mediated via release of excitatory amino acids, *Life Sci.,* 44, 1535, 1989.
184. **Tung, C. S., Ugedo, L., Grenoff, J., Engberg, G., and Svensson, T. H.,** Peripheral induction of burst firing in locus coeruleus neurons by nicotine mediated via excitatory amino acids, *Synapse,* 4, 313, 1989.
185. **Giorguieff-Chesselet, M. F., Kemel, M. L., Wandscheer, D., and Glowinski, J.,** Regulation of dopamine release by presynaptic nicotinic receptors in rat striatal slices: effect of nicotine in a low concentration, *Life Sci.,* 25, 1257, 1979.
186. **Rapier, C., Lunt, G. G., and Wonnacott, S.,** Stereoselective nicotine-induced release of dopamine from striatal synaptosomes: concentration dependence and repetitive stimulation, *J. Neurochem.,* 50, 1123, 1988.
187. **Imperato, A., Mulas, A., and DiChiara, G.,** Nicotine preferentially stimulates dopamine release in the limbic system of freely moving rats, *Eur. J. Pharmacol.,* 132, 337, 1986.
188. **Grenoff, J. and Svensson, T. H.,** Selective stimulation of limbic dopamine activity by nicotine, *Acta Physiol. Scand.,* 133, 595, 1988.
189. **Rapier, C. M., Lunt, G. G., and Wonnacott, S.,** Ion fluxes associated with transmitter release in perfused synaptosomes, *Biochem. Soc. Trans.,* 14, 598, 1986.
190. **Kramer, H. K., Sershen, H., Lajtha, A., and Reith, M. E. A.,** The effect of nicotine on catecholaminergic storage vesicles, *Brain Res.,* 503, 296, 1989.
191. **Izenwasser, S., Jacocks, H. M., Rosenberger, J. G., and Cox, B. M.,** Nicotine indirectly inhibits [^{3}H]-dopamine uptake at concentrations that do not directly promote [^{3}H]-dopamine release in rat striatum, *J. Neurochem.,* 56, 603, 1991.
192. **Fung, Y. K. and Lau, Y. S.,** Receptor mechanisms of nicotine-induced locomotor hyperactivity in chronic nicotine-treated rats, *Eur. J. Pharmacol.,* 152, 263, 1988.
193. **Rowell, P. P. and Wonnacott, S.,** Evidence for functional activity of up-regulated nicotine binding sites in rat striatal synaptosomes, *J. Neurochem.,* 55, 2105, 1990.
194. **Carr, L. A., Rowell, P. P., and Pierce, W. M., Jr.,** Effects of subchronic nicotine administration on central dopaminergic mechanisms in the rat, *Neurochem. Res.,* 14, 511, 1989.

195. **Fung, Y. K.,** Effects of chronic nicotine pretreatment on (+)-amphetamine and nicotine-induced synthesis and release of [^{3}H]-tyrosine in rat nucleus accumbens, *J. Pharm. Pharmacol.,* 41, 66, 1989.
196. **Fung, Y. K. and Lau, Y. S.,** Differential effects of chronic nicotine administration on dopaminergic receptor binding sites in rat nigrostriatal and mesolimbic regions, *Gen. Pharmacol.,* 22, 117, 1991.
197. **Leikola-Pelho, T., Heinamaki, J., Laakso, I., and Ahtee, L.,** Chronic nicotine treatment changes differentially the effects of acute nicotine on the three main dopamine metabolites in mouse striatum, *N.S. Arch. Pharmacol.,* 342, 400, 1990.
198. **Andersson, K., Siegel, R., Fuxe, K., and Eneroth, P.,** Intravenous injections of nicotine induce very rapid and discrete reductions of hypothalamic catecholamine levels associated with increases in ACTH, vasopressin, and prolactin secretion, *Acta Physiol. Scand.,* 118, 35, 1983.
199. **Sharp, B. M. and Beyer, H. S.,** Rapid desensitization of the acute stimulatory effects of nicotine on rat plasma adrenocorticotropin and prolactin, *J. Pharmacol. Exp. Ther.,* 238, 486, 1986.
200. **Hulihan-Giblin, B. A., Lumpkin, M. D., and Kellar, K. J.,** Acute effects of nicotine on prolactin release in the rat: agonist and antagonist effects of a single injection of nicotine, *J. Pharmacol. Exp. Ther.,* 252, 15, 1990.
201. **Sharp. B. M., Beyer, H. S., Levine, A. S., Moreley, J. E., and McAllen, K. M.,** Attenuation of plasma prolactin response to restraint stress after acute and chronic administration of nicotine to rats, *J. Pharmacol. Exp. Ther.,* 241, 438, 1987.
202. **Hulihan-Giblin, B. A., Lumpkin, M. D., and Kellar, K. J.,** Effects of chronic administration of nicotine on prolactin release in the rat: inactivation of prolactin response by repeated injections of nicotine, *J. Pharmacol. Exp. Ther.,* 252, 21, 1990.
203. **Miledi, R., Parker, I., and Schalow, G.,** Transmitter induced calcium entry across the post-synaptic membrane of frog end-plates measured using arsenazo III, *J. Physiol.,* 300, 197, 1980.
204. **Cena, V., Nicholasm, G. P., Sanchez-Garcia, P., Kirpekar, S. M., and Garcia, A. G.,** Pharmacological dissection of receptor-associated and voltage-sensitive ionic channels involved in catecholamine release, *Neuroscience,* 10, 1455, 1983.
205. **Boarder, M., Marriott, D., and Adams, M.,** Stimulus secretion coupling in cultured chromaffin cells: dependency on external sodium and on dihydropyridine-sensitive calcium channels, *Biochem. Pharmacol.,* 36, 163, 1987.
206. **Sasakawa, N., Ishii, K., and Kato, R.,** Nicotinic receptor-mediated intracellular calcium release in cultured bovine adrenal chromaffin cells, *Neurosci. Lett.,* 63, 275, 1986.
207. **Klepper, M., Hans, M., and Takeda, K.,** Nicotinic cholinergic modulation of voltage-dependent calcium current in bovine adrenal chromaffin cells, *J. Physiol.,* 428, 545, 1990.
208. **Hillard, C. J.,** Effects of nicotine and nicotinic agonist on calcium influx into brain synaptosomes, *NIDA Monogr. Ser.,* 81, 324, 1988.
209. **Burgoyne, R. D.,** The relationship between secretion and intracellular free calcium in bovine adrenal chromaffin cells, *Biosci. Rep.,* 4, 605, 1984.
210. **Guidotti, A. and Costa, E.,** A role for nicotinic receptor in the regulation of adenylate cyclase of adrenal medulla, *J. Pharmacol. Exp. Ther.,* 189, 665, 1974.
211. **Cheek, T. R. and Burgoyne, R. D.,** Cyclic AMP inhibits both nicotine-induced actin disassembly and catecholamine secretion from bovine adrenal chromaffin cells, *J. Biol. Chem.,* 262, 11,663, 1987.
212. **Eberhard, D. A. and Holz, R. W.,** Cholinergic stimulation of inositol monophosphate formation in bovine adrenal chromaffin cells: distinct nicotinic and muscarinic mechanisms, *J. Neurochem.,* 49, 1634, 1987.
213. **Nakaki, T., Sasakawi, N., Yamamoto, S., and Kato, R.,** Functional shift from muscarinic to nicotinic cholinergic receptors involved in inositol triphosphate and cyclic GMP accumulation during the primary culture of adrenal chromaffin cells, *Biochem. J.,* 251, 397, 1988.

214. **Sasakawa, N., Nakaki, T., and Kato, R.,** Rapid increases in inositol pentakisphosphate accumulation by nicotine in cultured adrenal chromaffin cells, *FEBS Lett.,* 261, 378, 1990.
215. **TerBush, D. R., Bittner, M. A., and Holz, R. W.,** Ca^{2+} influx causes rapid translocation of protein kinase C to membranes, *J. Biol. Chem.,* 263, 18873, 1988.
216. **Forsberg, E. J., Rojas, E., and Pollard, H. B.,** Muscarinic receptor enhancement of nicotine-induced catecholamine secretion may be mediated by phosphoinositide metabolism in bovine adrenal chromaffin cells, *J. Biol. Chem.,* 261, 4915, 1986.
217. **Wakade, A. R., Malhotra, R. K., and Wakade, T. D.,** Phorbol ester facilitates 45Ca accumulation and catecholamine secretion by nicotine and excess K^+ but not by muscarine in rat adrenal medulla, *Nature,* 321, 698, 1986.

Chapter 6

PHARMACOKINETICS AND PHARMACODYNAMICS OF NICOTINE: IMPLICATIONS FOR TOBACCO ADDICTION APPREHENSION

Hervé Claude Porchet

TABLE OF CONTENTS

I. INTRODUCTION

Tobacco abuse, especially when smoked as cigarettes, remains the main avoidable cause of mortality and morbidity in the western world. The estimated annual tobacco-related mortalities in the U.S. of 350,000 is substantially in excess of those related to all other drugs of abuse, traffic accidents, and suicides combined.[1] It was also shown that life expectancy is increased by 5 years if individuals quit smoking at age 40 and is still increased by more than 1 year if they quit smoking only at age 65.[2]

Although these numbers are not well known by smokers, who are not always conscious of the narrow relationship between tobacco use and morbidity and mortality, the great majority of them know that smoking is bad for their health and most of them have tried, at least once, to quit smoking. More than 80% of them fail on their first effort and seven attempts later, more than 60% relapse to cigarette smoking.[3] In more motivated subjects, as those who survived uncomplicated myocardial infarction, the rate of success remains poor and is rarely higher than one out of three.[4]

These facts reveal that the regular use of tobacco generates a strong dependence phenomenon. There is now compelling evidence that tobacco use results in addiction to nicotine. Dependence and its correlated drug tolerance are only partly due to the intrinsic pharmacological properties of nicotine but also highly related to its mode of administration. In this context, pharmacokinetic and pharmacodynamic properties of nicotine play a major role. A good knowledge of these characteristics is crucial to a closer comprehension of tobacco addiction and the limits of substitution therapies.

II. PHARMACOLOGICAL PROPERTIES RELATED TO DEPENDENCE

A. DETERMINANT FACTORS

The three main factors influencing the appearance of dependence are the dose, the duration of drug exposure, and mode of administration to the target organ (generally the brain). The relationship defining the threshold for the induction of physiological dependence and tolerance looks like an hyperbolic curve: the weaker the exposure (dose), the longer the time needed to induce dependence or tolerance. Conversely, the stronger the exposure, the shorter it would be.[5]

B. PHARMACOKINETIC FACTORS

The pharmacokinetic properties of psychoactive drugs associated with dependence can be classified according to the different components of drug addiction.[6] The pharmacological *positive reinforcement,* which is related to establishing and maintaining habits leading to a psychological status characterized by a subjective sensation of pleasure, requires that the addictive drug be rapidly absorbed in blood, diffused quickly in the brain, and that its psychoactivity be closely related to its cerebral concentration. The development of *tolerance* requires a cumulative exposure (dose, frequency, duration), a long half-life, or a low clearance. The *withdrawal syndrome,* corresponding to a *negative reinforcement,* is bound to physiological dependence, and appears if the drug is rapidly cleared from the brain (redistribution or elimination), has a short half-life, or a high clearance.

Some of these properties might seem to be contradictory. Thus, a drug with a long half-life, which is able to induce tolerance, should not promote a withdrawal syndrome. This is true if the elimination from the site of action is also slow. However, this elimination may be relatively rapid when it mainly depends on tissular redistribution. In this case, the onset of a withdrawal syndrome is facilitated, but the development of dependence will be possible only if the administration mode is cumulative and/or repetitive. These important points will be extensively discussed here for nicotine.

III. PHARMACOKINETICS OF NICOTINE

A. ABSORPTION

Nicotine is a weak base and, consequently, the pH of tobacco smoke or of other nicotine substrates (chewing tobacco, snuff, nicotine gum) is a major determinant of its absorption. Thus, the acidic smoke of flue-cured tobacco (blond) found in most cigarettes induces an important ionization of nicotine, which, in this state, does not easily cross organic membranes. Its mouth absorption is very limited and the smoker needs to deeply inhale in order to allow absorption through the lungs. At this level, absorption is apparently not limited by the acidic pH because of the wide exchange surface provided by the pulmonary alveoli. The smoke of air-cured tobacco (brown), found in pipe, cigar, and some European cigarettes, is alkaline and thus nicotine is essentially nonionized and is relatively well absorbed through the mouth mucosa. Nicotine gum or snuff tobacco are also buffered at alkaline pH to facilitate its mouth absorption. The average dose of nicotine delivered by a cigarette is approximately 1 mg, for a tobacco snuff maintained for 30 min in the mouth the dose is from 3 to 5 mg; 4 mg of nicotine gum delivers only around 1.5 mg after being chewed conscientiously and regularly for 30 min.[7]

B. DISTRIBUTION

The concentration-time curves of nicotine vary considerably according to its different routes of administration. Whereas the smoking of one cigarette induces an acute increase of nicotine concentrations with a peak of ~15 ng/ml reached in less than 10 min, the increment induced by snuff is much less abrupt, but its amplitude is quite similar. With nicotine gum the rise is still slower and peak concentration is two times lower.[7] With cigarettes, once maximal concentration is reached, one can observe an initial acute fall of the concentration-time curve corresponding to the tissular redistribution of nicotine. This redistribution half-life is approximately 10 min. With oral administration of nicotine (nicotine chewing gum or chewing tobacco), the tissular redistribution phase cannot be observed as it is hidden by the slower absorption rate. Nicotine is only weakly bound to proteins in the blood (~10%), which facilitates its tissular diffusion. Its volume of distribution is 2 to 3 l/kg body weight.[8]

C. ELIMINATION

Nicotine is rapidly and extensively metabolized primarily in the liver, but also to a small extent in the lungs. Its renal excretion depends on urinary pH and urine flow and accounts for 2 to 35% of total elimination.[9] Its elimination half-life averages 2 h and its primary metabolites are cotinine and nicotine-*N*-oxide, both of which are pharmacologically inactive.[8] Cotinine, because of its long half-life (16 to 20 h), is used as a marker of nicotine intake.

D. BRAIN DISTRIBUTION

Nicotine penetration in the central nervous system is outstanding and essential because it is the target organ. The pulmonary administration mode through inhalation is remarkable because it allows an extremely rapid entry of nicotine in the systemic circulation.[10] In fact, the lag time between smoking and entry of nicotine into the brain is shorter than after intravenous injection. It was shown in the mouse with autoradiographic technique that 5 min after the iv administration of nicotine, brain concentrations were high but declined very rapidly and had almost completely disappeared after 30 min.[11] Intravenously injected [^{14}C]-nicotine is almost immediately taken up in the brain of mice, reaching a maximum concentration within 1 min after injection.[12] Comparable findings were obtained in the monkey using positron emission tomography (PET) of the brain after injection of [^{11}C]-nicotine.[13]

Using the partition coefficients of nicotine between tissues and blood obtained in rabbits and organ weights and tissue blood flows obtained in man, it is possible, by means of physiological perfusion kinetic models, to predict nicotine concentration-time profiles in various organs after smoking one cigarette (Figure 1). Arterial, brain, and lung concentrations increase and then decrease very rapidly during the first 20 min, whereas nicotine is distributed much slower in other tissues, especially fat and muscles. It can be observed that nicotine concentrations are still increasing in these tissues when they are declining in the others. It is important to note that the venous concentration-time profile is clearly different from that of the brain. Thus, after rapid administration, venous concentrations do not reflect what is happening in the target organ: the brain.

The rapid absorption of nicotine through the lungs into the blood and into the brain provides the possibility for the smoker to precisely modulate levels of nicotine in the brain and, therefore, its pharmacological effects. To better

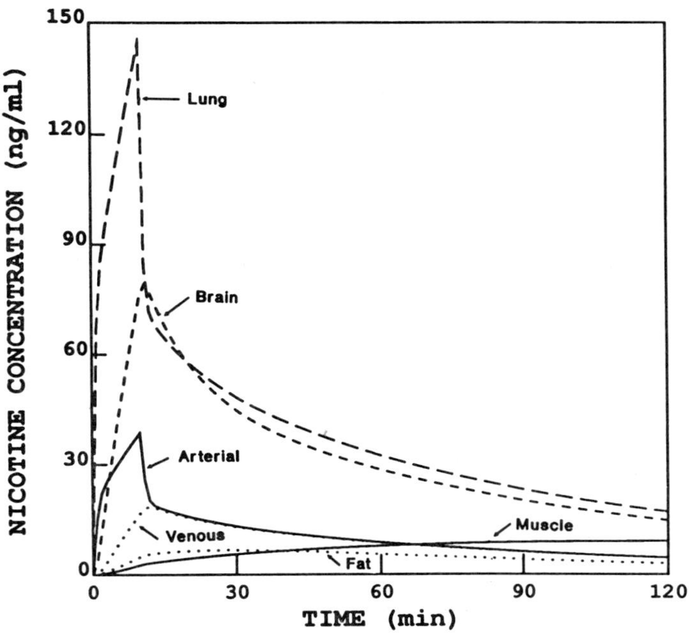

FIGURE 1. Simulation of nicotine tissular distribution after constant infusion of 1.5 mg of nicotine in 10 min in the lung compartment according to a physiological perfusion model. The dose, the route of administration, and the parameters roughly simulate the inhalation of a standard cigarette by a man of 70 kg.

understand this process, one must examine in more detail the concentration-effect relationship of nicotine, especially in the central nervous system (CNS).

IV. PHARMACODYNAMICS OF NICOTINE

The dose-response relationship of nicotine is complex. It is characterized by a biphasic pattern where low doses induce stimulation and high doses induce initial short stimulations followed by ganglionic blockade. This biphasic response pattern was also observed in the human CNS. Using contingent negative variation, a component of the evoked electroencephalographic potential, it was shown that low doses of nicotine increased while high doses decreased its amplitude.[14] Thus, it was proposed that low doses induced arousal, increased attention, and vigilance while higher doses produced relaxation and reduced stress. It is possible that the smoker by "fine tuning" his level of nicotine in the brain achieves the desired state.

The rate of nicotine absorption in the body strongly influences its effects, although it is still uncertain whether this is due to the faster rate of increase

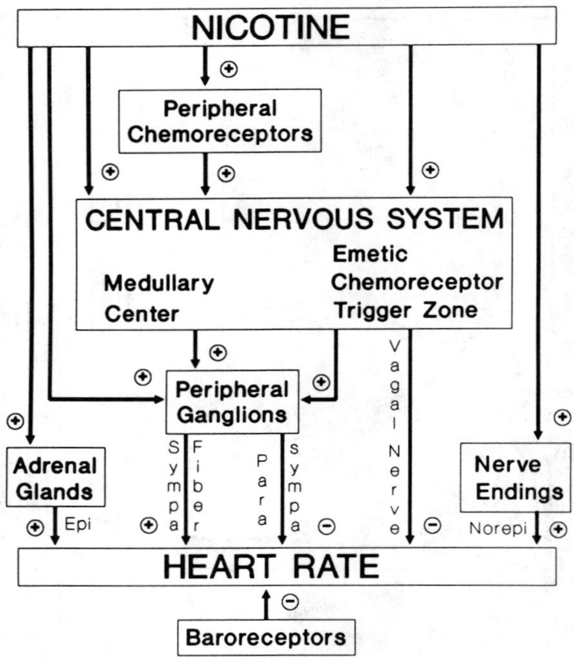

FIGURE 2. Multiple sites where nicotine may influence heart rate. Heart rate increase is indicated by "+" and decrease by "−".

of nicotine or to its higher absolute levels. Heart rate is a good index of the sympathetic neural activity and seems to be relatively well correlated to the importance of subjective effects of nicotine.[15,16] As shown in Figure 2, the acceleration of heart rate induced by nicotine is mediated via the peripheral chemoreceptor afferent pathways, direct effects in the CNS at the brainstem level, and the release of norepinephrine by nerve endings.[17-19] The net result is sympathetic neural discharge with an increase in blood pressure and heart rate. At higher doses, there may be direct effects on the peripheral nervous system, such as ganglionic stimulation, with the release of adrenal catecholamines. With extremely high doses, there may be hypotension and heart rate slowing, mediated either by peripheral vagal activation or by direct depressor effects on the brain.[20,21] Rosenberg[15] showed that with repeated intravenous administration of nicotine, which partly simulated cigarette smoking, heart rate increased and subjective effects were greater with the first series of injections. These effects decreased over time with successive injections, although nicotine plasma levels were higher. This clearly indicated the development of an acute tolerance.[15]

A. IMPORTANCE OF ADMINISTRATION RATE

Before examining the details of tolerance to nicotine, we need to take a

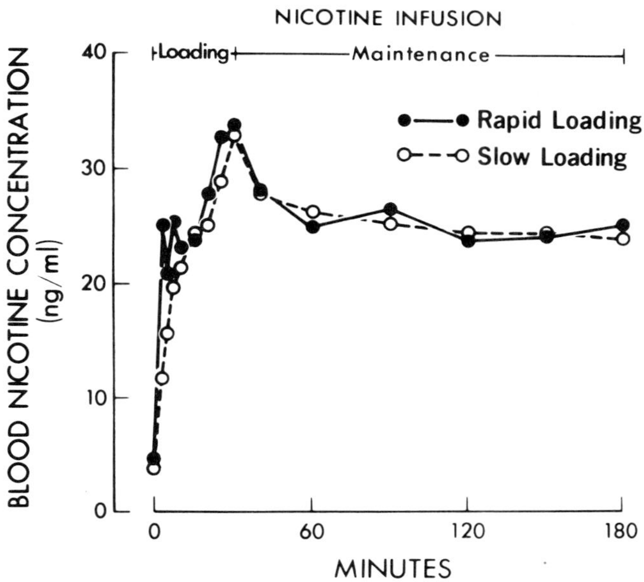

FIGURE 3. Average blood nicotine concentrations in seven volunteers. The rapid loading dose was 17.5 μg/kg/min for 1.5 min, then 1.75 μg/kg/min up to 30 min. The slow loading dose was 2.5 μg/kg/min during the first 30 min. Both total loading doses were similar (75 μg/kg). Maintenance infusions were 0.35 μg/kg/min on both cases. (Reprinted from Porchet, H. C. et al., *J. Clin. Invest.*, 80, 1466, 1987, by copyright permission of the American Society for Clinical Investigation, Inc.)

closer look at the importance of the rate of nicotine administration on heart rate acceleration. We compared the effects of a slow and rapid dosing rate of nicotine in a cross-over study in seven volunteer smokers who had abstained from smoking overnight.[22] Figure 3 shows that both average venous concentration-time profiles are almost superimposable except for the first 5 min, where the increase was greater with the rapid infusion. Peak concentrations were observed at 30 min and were almost identical for both infusions. The corresponding heart rate accelerations depicted in Figure 4 show that maximal effect is already reached in 3 min with the rapid infusion and is much greater than with the slow infusion with which maximal effect is reached only at 30 min and is synchronous with the corresponding peak venous concentration. The rapid loading dose was also associated with a marked transient dizziness and euphoria in all volunteers. These subjective effects were almost absent with the slow infusion. Using kinetic parameters obtained from subjects of this study and studies in rabbits and by mean of a physiological perfusion model (Figure 1), we estimated levels of nicotine in human brain during both infusions. Figure 5 shows that predicted "brain" concentrations of nicotine

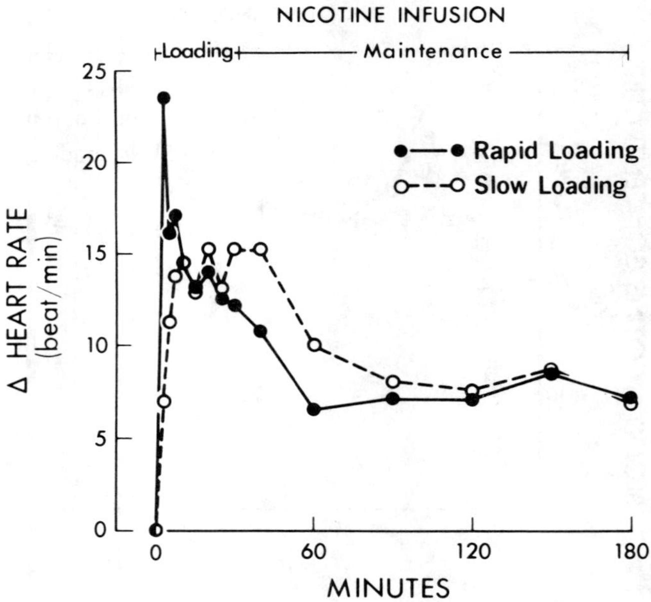

FIGURE 4. Average heart rate acceleration over time after two different nicotine infusions (see legend of Figure 3). (Reprinted from Porchet, H. C. et al., *J. Clin. Invest.*, 80, 1466, 1987, by copyright permission of the American Society for Clinical Investigation.)

rose much more rapidly and to a higher level after rapid infusion, with at this point, a concentration-time profile closer to that of heart rate acceleration. Peak brain concentrations are synchronous with maximal effect for both infusions. Brain concentration curves tended to be more or less parallel to respective effect curves. These findings indicate that this effect of nicotine is tracking predicted brain concentrations over time and illustrates that these are augmented and psychoactive effects are greater when nicotine is dosed more rapidly.

Rapid equilibration between arterial blood and the brain is a characteristic shared by other drugs of abuse, such as heroin and cocaine. This model illustrates and quantitates the concept that the level in the brain and the magnitude of the psychoactive effects of such drugs will be greater when the drug is dosed rapidly. This also indicates that the magnitude and duration of effects of a dose of many drugs of abuse is determined in part by the distribution kinetics of the drug to the brain. This may explain why smoking a cigarette may be preferred over the use of smokeless tobacco where absorption is slow.

B. TOLERANCE TO NICOTINE

The development of tolerance, which is a decreased effect with prolonged

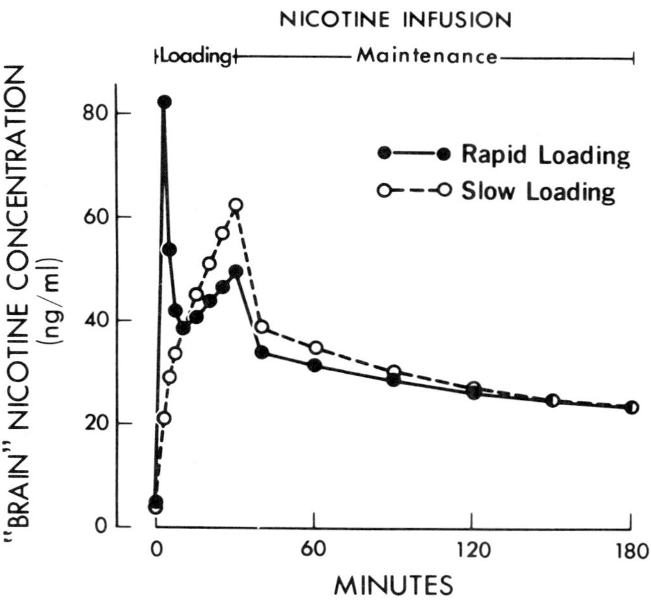

FIGURE 5. Predicted mean "brain" concentrations of nicotine over time (based on a physiological pharmacokinetic model) according to infusion rates described in Figure 3. (Reprinted from Porchet, H. C. et al., *J. Clin. Invest.*, 80, 1466, 1987, by copyright permission of the American Society for Clinical Investigation.)

exposure to a drug, represents an important issue in drug dependence. When it occurs during the time course of a single dose, it is called acute tolerance or tachyphylaxis. Studies in animals demonstrated rapid development of tolerance to many effects of nicotine.[23,24] Furthermore, smokers well know that tolerance develops to some effects of smoking. Thus, the first cigarette ever smoked is almost always associated with dizziness, nausea and/or vomiting and very rarely felt as pleasant. These effects progressively disappear with subsequent cigarettes as the smoker develops tolerance. The first cigarette of the day after overnight abstinence is often considered as the best or the most potent. This effectiveness is more pronounced if abstinence is prolonged, as shown by West and Russel,[16] probably because tolerance has more time to dissipate. Pharmacodynamic tolerance to heart rate acceleration and subjective effects can be observed within the time of a 30-min intravenous infusion of nicotine.[8] Thus tolerance to nicotine effects is gained apparently rapidly and lost to some degree after overnight abstinence.

To characterize and quantify nicotine tolerance in terms of pharmacokinetics and pharmacodynamics, we conducted a study where 8 volunteer smokers received, after overnight abstinence, paired intravenous infusions of nicotine separated by different time intervals.[25] Despite higher blood

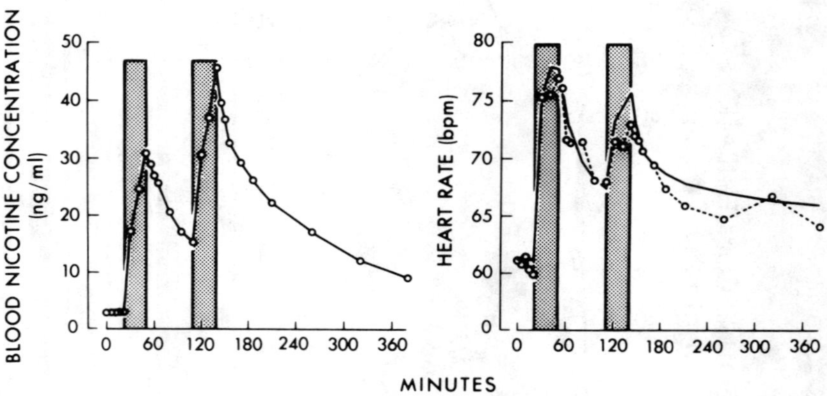

FIGURE 6. Average blood concentrations of nicotine (left panel) and average heart rate acceleration (right panel) in 8 volunteers after two 30-min intravenous infusion of 2.5 μg/kg/min of nicotine beginning 60 min apart. The shadowed areas indicate the periods during which nicotine was infused. The solid line in the right panel shows the fit of the model of Figure 8 to the effect data. (Reprinted from Porchet, H. C. et al., *J. Pharmacol. Exp. Ther.*, 244, 231, 1988, © by American Society for Pharmacology and Experimental Therapeutics.)

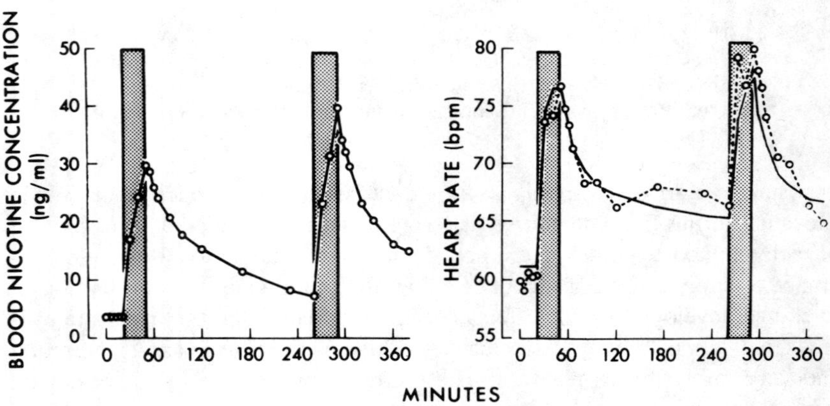

FIGURE 7. The same as Figure 6 for two 30-min infusions of 2.5 μg/kg/min of nicotine, but beginning 210 min apart. (Reprinted from Porchet, H. C. et al., *J. Pharmacol. Exp. Ther.*, 244, 231, 1988, © by American Society for Pharmacology and Experimental Therapeutics.)

concentrations, heart rate and subjective effects were much less pronounced when the second infusion was given 60 min after the end of the first infusion (Figure 6). However, the effectiveness of nicotine was partly restored when the second infusion was administered 210 min after the first one, indicating partial regression of tolerance (Figure 7).

An integrated pharmacokinetic-pharmacodynamic model (Figure 8) allowing the quantitative description of tolerance development was fitted to

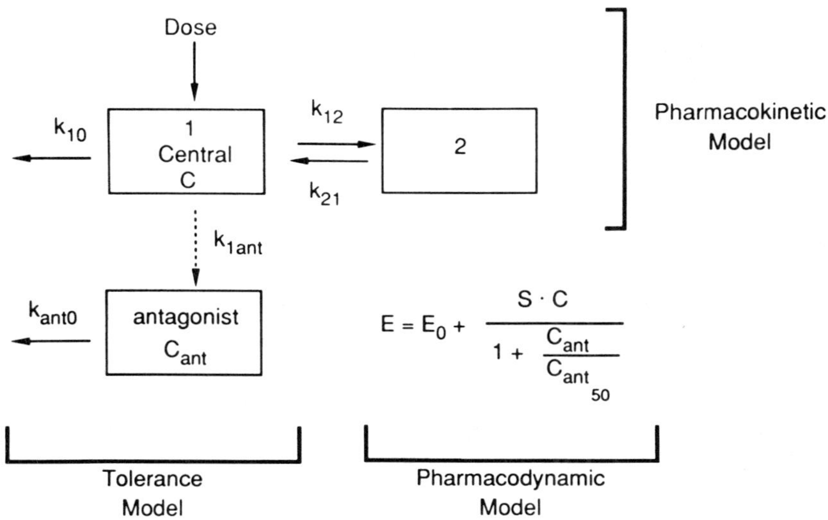

FIGURE 8. Diagrammatic representation of the integrated pharmacokinetic and pharmacodynamic model proposed for nicotine tolerance. The k_{ij} are the intercompartmental and elimination rate constants, C is the concentration of the agonist, C_{ant} is the concentration of the hypothetical antagonist, C_{ant50} quantifies the degree of tolerance attainable for a given steady-state concentration of the agonist, k_{ant0} is the elimination rate constant of the antagonist and expressed the rate of development and of disappearance of tolerance, S is the slope of the linear relationship between effect and concentration, E is the effect, and E_0 is the base-line effect. (Reprinted from Porchet, H. C. et al., *J. Pharmacol. Exp. Ther.*, 244, 231, 1988, © by American Society for Pharmacology and Experimental Therapeutics.)

these data. The lines of best fit are shown in Figures 6 and 7, which describe them relatively well.

The model postulates the existence of a *"force"* governing tolerance, which acts as a noncompetitive antagonist. The greater the intensity of past exposure to the agonist, the stronger this driving force and vice versa. There is but a single rate constant, K_{ant0}, governing the waxing and the waning of tolerance in response to exposure to the agonist. Accordingly, if agonist concentration were suddenly to rise and stay at a fixed level, the rate of appearance of tolerance would be the same as its rate of disappearance if agonist concentrations were to suddenly drop from a previous sustained fixed value to zero. The parameter C_{ant50} quantifies the degree of tolerance attainable for a given steady-state nicotine concentration. The model also postulates that the concentration of the antagonist (C_{ant}) at steady state is equal to the steady-state concentration of the agonist.

Although this model is empirical, it can be interpreted as a model of receptor downregulation. For nicotine, the idea of receptor downregulation may seem inappropriate because studies conducted *in vivo* in rats and mice have shown that the number of nicotine receptors in the brain increases after

TABLE 1
Pharmacokinetic and Pharmacodynamic Parameter Estimates of Nicotine Tolerance Model

Parameters	k_{10} (min^{-1})	k_{12} (min^{-1})	k_{21} (min^{-1})	V_c (L)	k_{ant0} (min^{-1})	S (bpm/ng/ml)	C_{ant50} (ng/ml)	E_0 (bpm)
Estimates	0.0112	0.03	0.0325	114	0.0195	1.31	7.72	61.2
C.V. (%)[a]	5.2	15.0	9.2	5.1	13.8	12.8	18.8	0.5

[a] C.V. corresponds to the estimation error on the parameter.

chronic administration of nicotine and induction of tolerance.[23,26] Further studies have shown that changes in receptor are correlated with the acquisition and loss of tolerance.[24] The different explanations put forward to explain this finding lead to the same net result: an increase in the total number of receptors but a decrease in the number of receptors that are functional. This is functionally equivalent to downregulation of active receptors, and the kinetics of tolerance by this mechanism could be described by the proposed model.

Parameter estimates of the model are presented in Table 1. This modeling indicates a half-life of development and regression of tolerance of 35 min. Thus, after four to five half-lives or approximately 3 h after a cigarette, nearly full sensitivity should have been regained. The model also predicts that, at a steady-state concentration of 30 ng/ml, the response is attenuated to 20% of the initial response. Thus, tolerance to nicotine is not complete and some persistent effect may be expected despite a substantial degree of tolerance. These findings are in agreement with a study showing that, when smokers smoke ad libitum, the increase of heart rate above first morning values is greater in the first few hours in the morning (when tolerance is not fully developed) than it is later in the day.[27] After this, the heart rate follows a circadian pattern similar to that seen during abstinence from smoking but at a slightly faster rate.

V. SIMULATION

This model allows simulation of the effect of nicotine according to different doses and routes of administration. Such a simulation is shown in Figure 9 where heart rate acceleration is stimulated after cigarette smoking (1.5 mg of nicotine in 10 min every 40 min; total dose, 36 mg) and nicotine gum (4 mg of nicotine in 30 min every 90 min; total dose, 44 mg). With smoking, maximal effect is reached at the end of the first cigarette. The following cigarettes have progressively smaller maximal effect, but each cigarette is still able to induce a well-defined peak. If change in heart rate does, in fact, mirror smoking satisfaction, we can better understand the daily smoking cycle. Thus, the first cigarette of the day, which is often reported as the

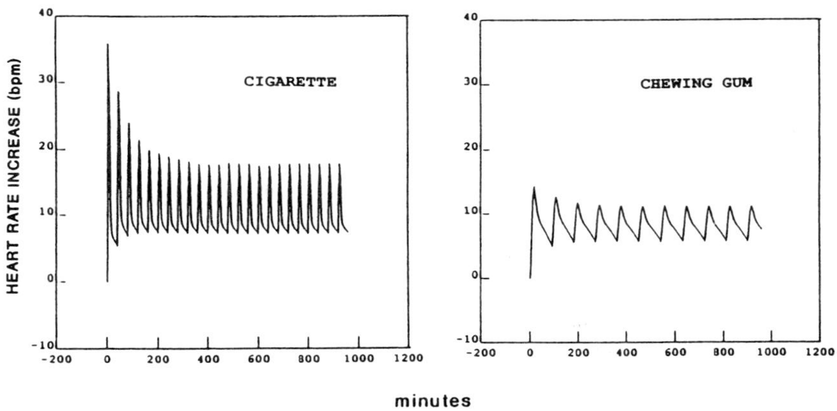

FIGURE 9. Simulation of heart rate response to usual daily intake of cigarette (left panel) and nicotine gum (right panel) versus time of day. Dosing for cigarette is: 1.5 mg of nicotine given intrapulmonary in 10 min, every 40 min, corresponding to a total dose of 36 mg. Dosing for nicotine gum is 4 mg of nicotine chewed during 30 min, every 90 min, corresponding to a total dose of 44 mg.

best one, induces a sharp effect, but tolerance has already started to develop. With the following cigarettes tolerance increases but can be at least partly overriden by the rapid and acute increase of nicotine concentration induced by each cigarette or, perhaps, each puff. The rapid fluctuations also induce some sort of ''mini-withdrawal'' between each cigarette and one is smoking mostly to avoid abstinence craving rather than for the pleasure. Finally, overnight abstinence allows considerable resensitization to the actions of nicotine. Thus, because of the dose-response characteristics, habitual smokers need to smoke at least 15 cigarettes per day to achieve the desired effects and to minimize withdrawal discomfort throughout the day.

In contrast with nicotine gum, effects are much lower than with cigarettes, but the rate and degree of increase with each intake is also less. This is not a consequence of a lower total dose of gum versus cigarette, but of the slower input rate of gum. In fact, the 4 mg nicotine gum will deliver a lower amount of nicotine than the simulation does, as its systemic bioavailability is estimated to be equal to 53%.[28] Thus tolerance already develops during input, thereby decreasing peak effect and rate of increase. It can be seen that the gum produces a state of smoking satisfaction equivalent to that experienced by a habitual smoker who is about to reach for another cigarette. This state is less uncomfortable than total abstinence, but will never rival with smoking. It is now obvious why the gum does not help very much with cigarette abstinence and is not perceived as very satisfying.

Thus, nicotine, as it is dosed by cigarette smokers, meets all kinetic properties for positive reinforcement: rapid and extensive absorption and penetration in the brain; for tolerance: cumulative exposure is obtained by repeated administrations; and for withdrawal syndrome: rapid exit from the brain through extensive redistribution. Nicotine gum is only able to meet tolerance criteria and thus is only useful to fight withdrawal syndrome but not positive reinforcement. This is confirmed by clinical studies investigating the efficacy of nicotine gum or nicotine transdermal patch in smoking cessation programs. These substitution therapies generally show some efficacy during the first month of treatment, but then relapse rates are parallel.[29-32] Thus, these substitution therapies help to fight physical withdrawal, which is maximal at the beginning of abstinence, and then progressively disappears over 2 to 3 weeks. As they are not effective in fighting a positive reinforcement, which lasts much longer and is responsible for most relapses, they should normally not be used for more than 1 month.

VI. CONCLUSION

The relationship between rate of administration and the development of tolerance helps, at least partly, to understand why so many drugs capable of inducing toxicomania are self-administered by intravenous route or by inhalation and, conversely, why both these routes bestow such a great potential of abuse to substances which, when administered differently, would be much less addictive. A good knowledge of the kinetics and dynamics of drugs inducing dependence should allow a more rational approach in the treatment of toxicomania.

REFERENCES

1. **Jarvik, M. E. and Henningfield, J. E.,** Pharmacological treatment of tobacco dependence, *Pharmacol. Biochem. Behav.,* 30, 279, 1988.
2. **Oster, G., Huse, D. M., Delea, T. E., and Colditz, G. A.,** Cost-effectiveness of nicotine gum as an adjunct to physician's advice against cigarette smoking, *J. Am. Med. Assoc.,* 256, 1315, 1986.
3. U.S. Department of Health and Human Services. The Health Consequences of Smoking: Chronic Obstructive Lung Disease, A Report of the Surgeon General, Publication No. DHHS (PHS) 84-50205, Washington, DC, U.S. Government Printing Office, 1984.
4. **West, R. R. and Evans, D. A.,** Lifestyle changes in long-term survivors of acute myocardial infarction, *J. Epidemiol. Community Health,* 40, 103, 1986.
5. **Haefely, W.,** Biological basis of drug-induced tolerance, rebound and dependence. Contribution of recent research on benzodiazepines, *Pharmacopsychiatry,* 19, 353, 1987.
6. **Busto, U. and Sellers, E. M.,** Pharmacokinetics determinants for drug abuse and dependence. A conceptual perspective, *Clin. Pharmacokinetics,* 11, 144, 1986.
7. **Benowitz, N. L., Porchet, H. C., Sheiner, L. B., and Jacob, P., III,** Nicotine absorption and cardiovascular effects with smokeless tobacco use: comparison with cigarettes and nicotine gum, *Clin. Pharmacol. Ther.,* 44, 23, 1988.
8. **Benowitz, N. L., Jacob, P., III, Jones, R. T., and Rosenberg, J.,** Interindividual variability in the metabolism and cardiovascular effects of nicotine in man, *J. Pharmacol. Exp. Ther.,* 221, 368, 1982.

9. **Benowitz, N. L. and Jacob, P., III,** Nicotine renal excretion rate influences nicotine intake during cigarette smoking, *J. Pharmacol. Exp. Ther.*, 234, 153, 1985.
10. **Oldendorf, W. H.,** Lipid solubility and drug penetration in the blood-brain barrier, *Proc. Soc. Exp. Biol. Med.*, 147, 813, 1974.
11. **Hanson, E. and Schmiterlow, C. G.,** Physiological disposition and fate of ^{14}C-labeled nicotine in mice and rats, *J. Pharmacol. Exp. Ther.*, 137, 91, 1962.
12. **Stalhandski, T. and Slanina, P.,** Effect of nicotine treatment on the metabolism of nicotine in the mouse liver *in vitro, Acta Pharmacol. Toxicol.*, 28, 75, 1970.
13. **Mazière, M., Comar, D., Marazano, C., and Berger, G.,** Nicotine-^{11}C: synthesis and distribution kinetics in animals, *Eur. J. Nucl. Med.*, 1, 255, 1976.
14. **Ashton, H., Marsh, V. R., Millman, J. E., Rawlins, M. D., Telford, R., and Thompson, J. W.,** Biphasic dose-related responses of the CNV (contingent negative variation) to i.v. nicotine in man, *Br. J. Clin. Pharmacol.*, 10, 579, 1980.
15. **Rosenberg, J., Benowitz, N. L., Jacob, P., III, and Wilson, K. M.,** Dispositions kinetics and effects of intravenous nicotine, *Clin. Pharmacol. Ther.*, 28, 516, 1980.
16. **West, R. J. and Russell, M. A. H.,** Cardiovascular and subjective effects of smoking before and after 24 h of abstinence from cigarettes, *Psychopharmacology*, 92, 118, 1987.
17. **Comroe, J. H.,** The pharmacological actions of nicotine, *Ann. N.Y. Acad. Sci.*, 90, 48, 1960.
18. **Su, C.,** Actions of nicotine and smoking on circulation, *Pharmacol. Ther.*, 17, 129, 1982.
19. **Benowitz, N. L.,** Clinical pharmacology of nicotine, *Annu. Rev. Med.*, 37, 21, 1986.
20. **Ingenito, A. J., Barrett, J. P., and Procita, L.,** Direct central and reflexly mediated effects of nicotine on the peripheral circulation, *Eur. J. Pharmacol.*, 17, 375, 1972.
21. **Porsius, A. J. and Van Zwieten, P. A.,** The central actions of nicotine on blood pressure and heart rate after administration via the left vertebral artery of anesthetized nerves, *Arzneim. Forsch.*, 28, 1628, 1978.
22. **Porchet, H. C., Benowitz, N. L., Sheiner, L. B., and Copeland, J. R.,** Apparent tolerance to the acute effect of nicotine results in part from distribution kinetics, *J. Clin. Invest.*, 80, 1466, 1987.
23. **Marks, M. J., Burch, J. B., and Collins, A. C.,** Effect of chronic nicotine infusion on tolerance development and nicotinic receptors, *J. Pharmacol. Exp. Ther.*, 226, 817, 1983.
24. **Marks, M. J., Stitzel, J. A., and Collins, A. C.,** Time course study of the effects of chronic nicotine infusion on drug response and brain receptors, *J. Pharmacol. Exp. Ther.*, 235, 619, 1985.
25. **Porchet, H. C., Benowitz, N. L., and Sheiner, L. B.,** Pharmacodynamic model of tolerance: application to nicotine, *J. Pharmacol. Exp. Ther.*, 244, 231, 1988.
26. **Schwartz, R. D. and Kellar, K. J.,** Nicotinic cholinergic receptors binding sites in the brain: regulation in vivo, *Science*, 220, 214, 1983.
27. **Benowitz, N. L., Kuyt, F., and Jacob, P., III,** Influence of nicotine on cardiovascular and hormonal effects of cigarette smoking, *Clin. Pharmacol. Ther.*, 36, 74, 1984.
28. **Benowitz, N. L., Jacob, P., III, and Savanapridi, C.,** Determinants of nicotine intake while chewing nicotine polacrilex gum, *Clin. Pharmacol. Ther.*, 41, 467, 1987.
29. **Lam, W., Sze, P. C., Sacks, H. S., and Chalmers, T. C.,** Meta-analysis of randomized controlled trials of nicotine chewing-gum, *Lancet*, 2, 27, 1987.
30. **Tonnesen, P., Fryd, V., Hansen, M., Helsted, J., Gunnersen, A. B., Forchammer, H., and Stockner, M.,** Effect of nicotine chewing gum in combination with group counseling on the cessation of smoking, *N. Engl. J. Med.*, 318, 15, 1988.
31. **Abelin, T., Buehler, A., Müller, P., Vesanen, K., and Imhof, P. R.,** Controlled trial of transdermal nicotine patch in tobacco withdrawal, *Lancet*, 1, 7, 1989.
32. **Müller, P., Abelin, T., Ehrsam, R., Imhof, P., Howald, H., and Mauli, D.,** The use of transdermal nicotine in smoking cessation, *Lung*, (Suppl), 445, 1990.

Chapter 7

BEHAVIORAL NEUROPHARMACOLOGY OF NICOTINE

Charles Ksir

TABLE OF CONTENTS

I. INTRODUCTION

Human cigarette smoking appears to be capable of producing two seemingly contradictory psychopharmacological effects. When cigarette smokers are asked to list their reasons for smoking, they commonly answer that smoking relaxes them or helps them to deal with stress. Pomerleau and Pomerleau have recently reviewed evidence for the anxiety-reducing effect of smoking in humans.[35] Paradoxically, nicotine is most often classified as a mild CNS stimulant, and cigarette smoking has been shown to enhance performance on tests of alertness and attention when humans are asked to perform repetitive tasks.[51] Both types of effect have also been reported in animal activity experients with nicotine, and a selective review of the animal research will provide some insights into resolution of the apparent paradox.

II. EFFECTS ON LOCOMOTOR ACTIVITY

Morrison and Stephenson[32] provided an important part of the answer regarding the stimulant and depressant actions of nicotine. Rats given 0.4 or 0.8 mg/kg nicotine (base weight) s.c. immediately before placement in an activity test chamber showed less activity than control rats given saline injections. This difference in activity lasted for approximately 15 min and was seen clearly in rats that had never before experienced nicotine. When the same test was repeated at daily intervals (no tests on weekends), the relative decrease in activity produced by nicotine was seen only for the first 3 d. By the fifth test day and for the remainder of the experiment the nicotine rats showed greater activity than the saline rats. Once again this difference lasted for about 15 min of the 30-min test period. Thus, Morrison and Stephenson concluded that nicotine's initial effect is a depressant one, that tolerance develops to the depressant effect after a few days, and that the initial depressant effect is replaced by a stimulant effect.

Stolerman et al.[47] examined the activity-reducing effect of nicotine given 5 min before a 3-min behavioral test in a novel Y-maze. They demonstrated both acute tolerance (from a single injection of 0.32 mg/kg given 30 min to 4 h before the test) and tolerance to chronic exposure given by repeated injection, through subcutaneous reservoirs, or in the drinking water. After chronic exposure for 8 d, tolerance was still reported 80 d later. Under none of these conditions was more activity observed in the test rats than in the saline-treated controls (no evidence of nicotine-induced stimulation of activity), probably owing to the brevity of the test and the shortness of the post-injection time.

Battig et al.[2] gave 0.2 mg/kg nicotine 35 min prior to a 9-min test in a maze and reported increased activity in response to nicotine, especially in the Roman High-Avoidance strain of rats. The increased activity became greater as the tests were repeated daily for 12 d.

Clark and Kumar[8] used 80-min test sessions beginning immediately after the injection to study both the initial depressant effect and the later increase in activity. They pointed out that the decrease in activity seen within the first 20 min after the administration of 0.4 or 0.8 mg/kg was associated with a flaccid appearance, a loss of righting reflex in some cases, and a tendency for the rats to lie outstretched on the cage floor, panting. This observation raises the question whether this initial activity decrease should be seen as reflecting CNS sedation or, perhaps more appropriately, as motor ataxia associated with near-toxic doses. Tolerance developed to the activity-decreasing effect of 0.4 mg/kg: increased activity was seen during the first 20 min in a group of rats tested at weekly intervals but given daily injections between tests. This locomotor stimulant action was then present throughout the 80 min session after 1 to 4 weeks of nicotine exposure and also 3 weeks after nicotine was withdrawn. Mecamylamine, a ganglionic nicotinic receptor blocking agent that also acts in the brain, prevented both the depressant action and the stimulant action of nicotine, whereas hexamethonium, a ganglionic agent that does not readily enter the CNS, had no effect on either type of response to nicotine. These results imply that both the initial depressant effect and the later stimulant effect of nicotine occur as a result of nicotine's action in the CNS.

Ksir et al.[24] allowed rats to habituate to the test chambers for 1 h prior to the administration of nicotine. Under those conditions, nicotine-naive rats demonstrated an increase in activity during the first 20 min after 0.4 mg/kg, and rats given 0.1 or 0.2 mg/kg were not significantly affected. When these same doses were repeated daily for 5 d, by the 5th day all three doses increased activity in a dose-related manner, and the response to 0.4 mg/kg was almost twice as large as on the first day. This experiment also demonstrated that the behavioral stimulation produced by nicotine in nicotine-exposed rats was blocked by mecamylamine but not by hexamethonium. In a subsequent experiment from this laboratory, it was reported that 5 d of exposure to 0.2 mg/kg nicotine increased the locomotor response to each dose of nicotine over the range 0.1 to 0.8 mg/kg, without changing the shape of the dose-effect curve, which had a peak at 0.4 mg/kg in both nicotine-naive and nicotine-exposed rats.

Another experiment from our laboratory[55] pointed out the importance of habituating the rats to the test environment prior to testing with nicotine. Rats that were given saline injections before being placed in the test chamber were initially quite active and gradually reduced their activity over the first 40 min. In comparison to this high background of activity, 0.4 mg/kg nicotine given before rats were introduced to the test cages produced a clear decrease in activity. However, as reported before, when rats were allowed to habituate to the environment for 1 h, saline-injected rats were only moderately active for the first 10 min after injection. Relative to this lower baseline, 0.4 mg/kg nicotine produced a clear increase in activity. In rats given daily injections,

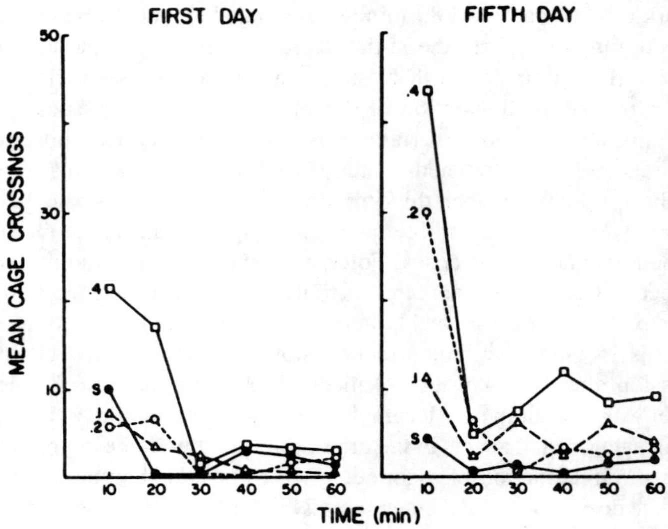

FIGURE 1. Locomotor activity measured in photocell test cages in 10-min blocks after the injection of saline (S) or nicotine (0.1, 0.2, or 0.4 mg/kg, s.c.). Each group of rats was given the same dose each day for 5 d, and data are shown for the first and fifth days (n = 6 for 0.1 and 0.4 mg/kg, n = 5 for 0.2 mg/kg). (From Ksir, C., et al. *Neuropharmacology,* 24, 527, 1985. With permission.)

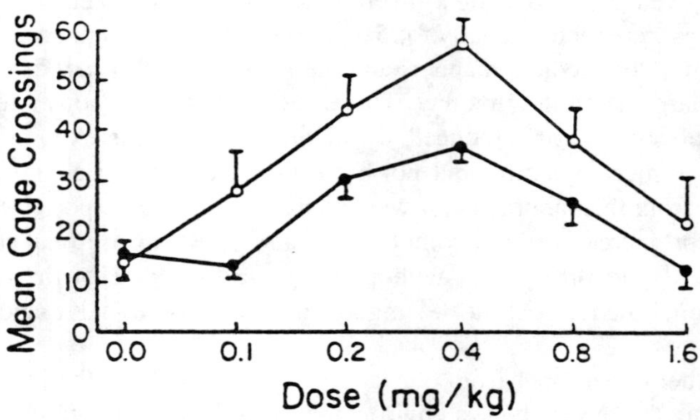

FIGURE 2. Cage crossings (±SEM) for groups of rats given different doses of nicotine before and after all animals received 5 d of exposure to 0.2 mg/kg/day nicotine. $N = 12$ for all groups, except $N = 6$ for the two 1.6 mg/kg points. Filled circles: before daily nicotine exposure. Open circles: after daily nicotine exposure. (From Ksir, C., et al., *Psychopharmacology,* 92, 25, 1987. With permission.)

by the 5th day nicotine produced dose-related increases in activity whether or not the rats were allowed to habituate to the cage before the injections were given.

Thus, there are at least two important determinants of the effect of nicotine. In addition to the effect of nicotine exposure, there is the separate influence of test cage adaptation. We interpreted this to be a *rate-dependent* effect: that is, in rats that were demonstrating a high level of activity, 0.4 mg/kg nicotine decreased activity, whereas in rats that were much less active (having already habituated to the test chamber), the same dose of nicotine increased activity. Such rate-dependent effects have been reported previously for amphetamine and many other behaviorally active drugs.[22,41] Balfour has reported that rats tested on an elevated, open maze were not so active as rats tested in a deep, enclosed maze and that nicotine increased the activity of the rats on the open maze, while having no effect on the activity of rats in the enclosed maze.[1] Although this effect was interpreted by Balfour in terms of the effect of nicotine on freezing reactions to the stress of an open maze, the data are consistent with the more general interpretation of a rate-dependent effect. In further support of the rate-dependent interpretation, there have been two reports that nicotine is more likely to produce stimulation when given during the daytime (when rats are typically less active) and more likely to decrease activity when given at night (when rats are typically more active).[3,21] However, Battig et al.[2] compared two strains of rats, the Roman High-Avoidance (RHA) and Roman Low-Avoidance (RLA), and they found that the RHA rats were the more active under control conditions and also showed a greater increase in activity when given nicotine, contrary to the rate-dependent prediction. Perhaps rate-dependency is useful when comparing activity levels under different conditions within a given strain, but other features of the strain differences between RHA and RLA rats may override rate-dependency.

In interpreting the effect of repeated daily testing with nicotine, it is important to consider the possibility of Pavlovian conditioning. Several laboratories have reported that amphetamine and cocaine have increasing stimulatory effects with repeated testing, and that a major factor is the pairing of the drug injection with distinctive stimuli associated with the test session. Rats given cocaine or amphetamine exposures in their home cages often do not show an enhanced response when tested for the first time in a distinctive environment. Although Clark and Kumar[8] and Ksir et al.[24,25] found enhanced reponding in rats given their nicotine exposures in the home cages and exposed to the test cage for the first time on the post-exposure behavioral test, it is possible that conditioning occurred to some other stimulus, such as handling or the injection procedure. Thus, the best control for conditioning effects is a ''pseudoconditioning'' procedure in which one group of rats receives saline injections in association with the test cage and nicotine in association with the home cage. Two experiments of this type have been reported. Walter and Kuchinsky[50] found that pseudoconditioned rats did not respond more to ni-

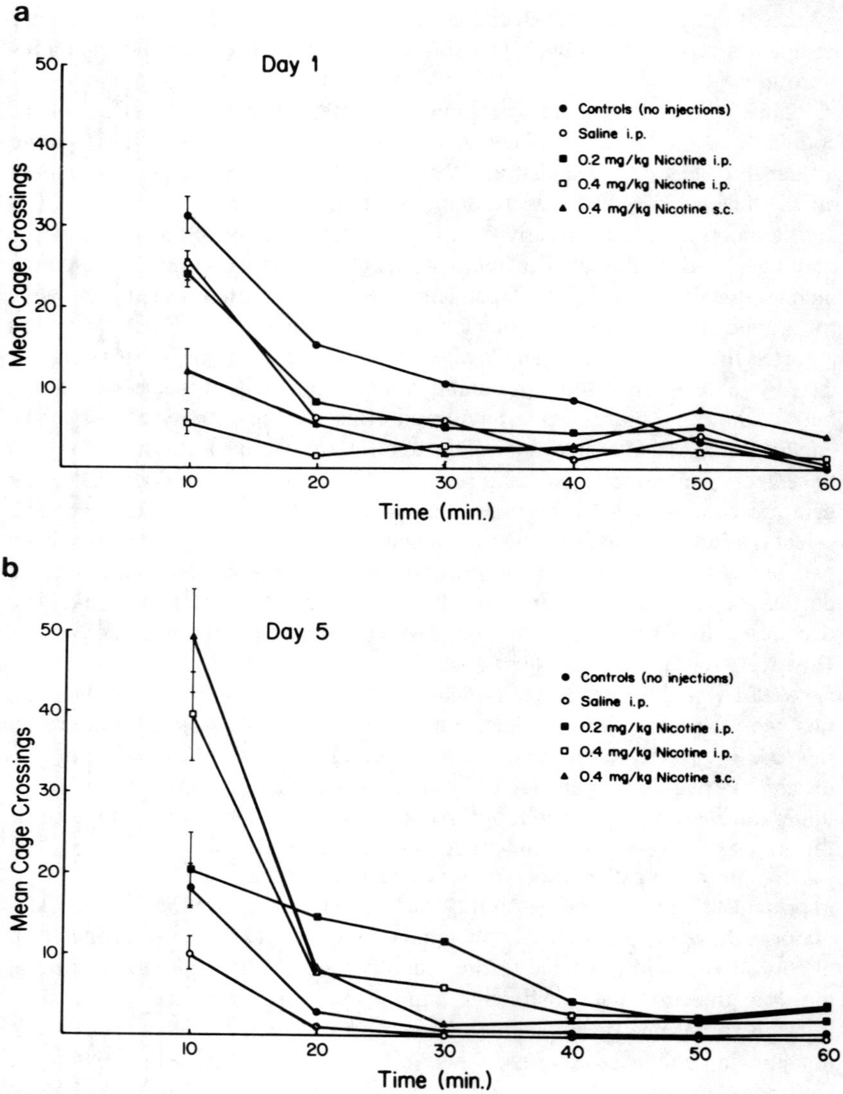

PART 1

FIGURE 3. (Part 1) Effect of test cage habituation on locomotor response to nicotine. Rats not allowed to habituate, but injected immediately before placement in the test cage. (a) Response to 0.2 or 0.4 mg/kg nicotine on day 1; (b) response to 0.2 or 0.4 mg/kg nicotine on day 5. (Part 2) Rats allowed to habituate for 60 min prior to injection. (c) Responses to 0.1, 0.2, or 0.4 mg/kg on day 1; (d) responses to 0.1, 0.2, or 0.4 mg/kg on day 5.

c

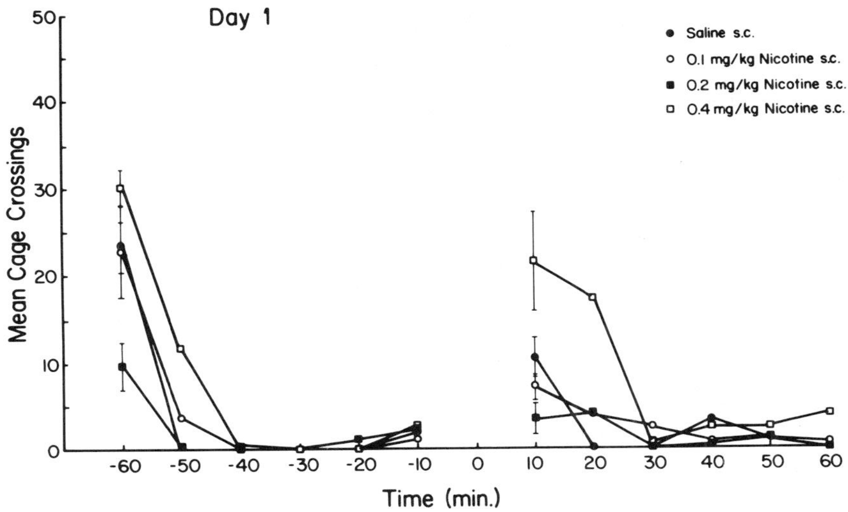

d

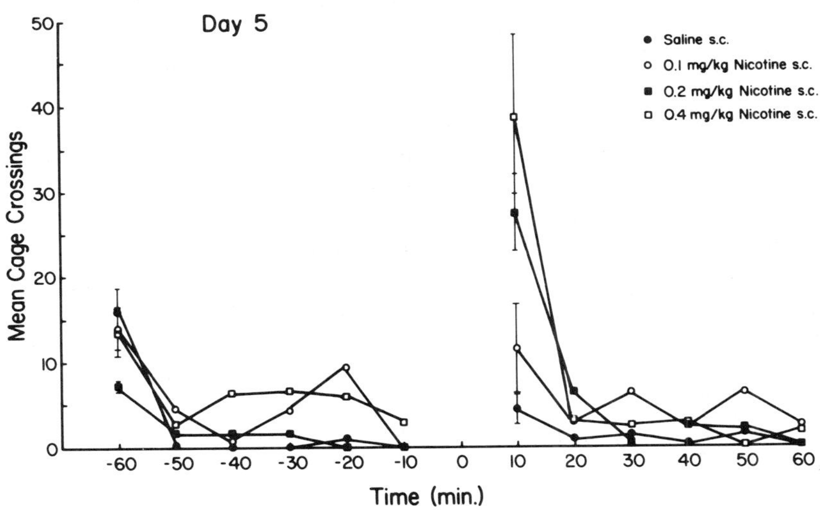

FIGURE 3 (PART 2)

cotine than did nicotine-naive rats, whereas Hakan and Ksir[17] found that pseudoconditioned rats showed as great an increase in activity as did the "conditioning" group. Further, Walker and Kuchinsky found that rats exposed to nicotine in the presence of a distinctive stimulus showed increased

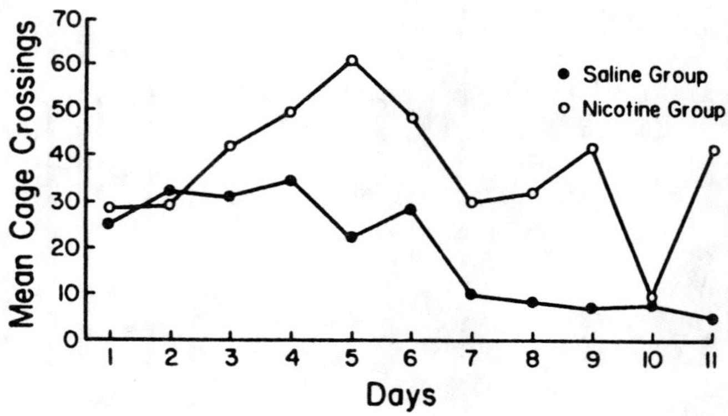

FIGURE 4. Mean cage crossings during the first 10 min after injection on days 1 to 11 for groups of rats given saline or 0.2 mg/kg nicotine. On days 1 to 9, injections were given in the presence of a complex auditory and visual stimulus. On day 10, both groups received saline in the presence of the previously conditioned complex stimulus. On day 11 the nicotine group was again given nicotine, but the complex stimulus was not presented. (From Hakan, R. L. and Ksir, C., *Pharmacol. Biochem. Behav.*, 29, 661, 1988. With permission.)

activity in response to that stimulus in the absence of nicotine, whereas Hakan and Ksir reported that presentation of a complex stimulus paired with nicotine resulted in behavior no different from that of saline-exposed rats, and nicotine in the absence of the conditioning stimulus produced as great a response as without the stimulus.

Thus, two experiments using similar strategies have reported opposite results and reached opposite conclusions about the role of Pavlovian conditioning in the enhanced response to nicotine after repeated nicotine administration. There were three differences in the procedures that may have been critical. First, in their experiment, Walter and Kuchinsky used a dose of 0.6 mg/kg, three times the dose used in the Hakan and Ksir experiment and large enough to produce initial hind-limb ataxia in most of their rats. Second, in the test-cage "pairing" experiments, Hakan and Ksir always habituated the rats to the test cages for an hour prior to injection, a procedure that would reduce the association between the cages and the injection. Walter and Kuchinsky always gave the injections just before introducing the rats to the test cages. Third, in the "complex stimulus" experiment, Hakan and Ksir employed a combined auditory and visual stimulus coincident with the injection, whereas Walter and Kuchinsky also added a distinctive odor cue. It appears that, given the right conditions, it is possible to demonstrate conditioning of nicotine's stimulating effects, but clearly sensitization to the stimulant effect of nicotine does not depend upon Pavlovian conditioning under a variety of circumstances. Under most of the conditions used in various laboratories, the enhanced stimulant response produced by nicotine following subchronic exposure appears to be a direct result of the repeated presence of nicotine itself. Whether the altered response to nicotine is to be attributed to pharmacokinetic or pharmacodynamic alterations in nicotine pharmacology remains a question for future research.

III. NICOTINE AND CHOLINERGIC MECHANISMS

Prior to the decade of the 1980s it was not clear how nicotine exerted its effects on the central nervous system and behavior. Classically, three major categories of receptor for acetylcholine had been identified in the periphery: the muscarinic receptor found on mammalian smooth muscle innervated by the parasympathetic system, the nicotinic receptor found at the neuromuscular junction, and the ganglionic type of nicotinic receptor. The nicotinic receptor had been most widely studied at the neuromuscular junction and in electric fish, such as the Torpedo, which is richly supplied with this type of receptor. The potent antagonist α-bungarotoxin (BGT) allowed pharmacological analyses to be accomplished in various tissues. Although BGT was known to bind to brain tissue, behavioral effects of systemic nicotine were not well blocked by intracerebral BGT. This led to the speculation that many of the behavioral effects of nicotine might be indirect, through the effect of nicotine on the adrenal medulla or through other mechanisms.

In the early 1980s the availability of radioactively labeled nicotine and acetylcholine with high specific activity made it possible for several laboratories to study nicotinic cholinergic binding in brain tissue using *in vitro* binding techniques. In particular, reports by Kellar's group[44] and by Collins' group[28] first began to point out that the most abundant form of CNS receptor for nicotine was not of the BGT (neuromuscular) type, but rather of a type similar to the ganglionic receptor. Their observations were made graphically evident by autoradiographic experiments demonstrating that the distributions of BGT and nicotine binding sites are quite distinct.[11,29] Study of this receptor type was made more difficult because no potent competitive antagonist exists that can be labeled and bound to the receptor with high affinity.

It is possible to study the pharmacology of the CNS "nicotine" site, however, using mecamylamine, a specific antagonist that is not competitive with nicotine for the same site but is capable of blocking nicotine's effects, both at peripheral ganglia and in the brain. For example, several laboratories have reported that both the decreased locomotor activity seen in rats and mice after acute doses of nicotine and the increased locomotor activity seen in rats after repeated drug exposure can be blocked by pretreatment with mecamylamine (see, for example, References 8, 9, 12, and 24). In addition, in rats trained to press a lever to obtain food after they have received saline injections, and a different lever after they have received nicotine injections (drug discrimination procedure), mecamylamine decreases the number of drug-appropriate responses when rats are subsequently given nicotine (blocks the discriminative stimulus properties of nicotine, as shown by Rosecrans and Meltzer[38]).

Also, mecamylamine appears to block the reinforcing properties of nicotine in rats trained to press a lever for small intravenous nicotine infusions, as shown, for example, by Corrigall and Coen.[13] Thus, it now appears that the interesting behavioral effects of nicotine are mediated through the type of receptor that has a much higher affinity for nicotine than for BGT, and that is noncompetitively antagonized by mecamylamine.[30]

In a further effort to relate this receptor to a behavioral effect of nicotine, it was noted that exposure of rats to large doses of nicotine for several weeks could result in increased numbers of nicotinic cholinergic binding sites.[42] Since several laboratories had previously reported that repeated exposure to lower doses of nicotine for a few days resulted in increased locomotor stimulation by nicotine, an experiment was conducted to see whether similar low doses and durations of exposure could also upregulate nicotinic sites;[24] they did. Further, both the enhanced locomotor response to nicotine and increased receptor numbers were found in rats that had been exposed to once-daily nicotine injections for 5 d and then received saline for 7 d. However, in rats given 21 d of saline treatment following nicotine exposure, both the behavioral response and the number of binding sites were not different from those observed for drug-naive rats. In a subsequent study it was found that a threshold daily exposure dose of approximately 0.03 mg/kg was necessary to produce either receptor upregulation or enhanced response to a test dose of nicotine.[25] It is worth noting that 0.03 mg/kg is incapable of producing measurable direct behavioral effects by itself. These results are consistent with the hypothesis that at least some of the enhanced stimulant effect produced by nicotine after repeated exposure is attributable to enhanced binding at some location in the CNS.

Tremendous strides have been made recently in the molecular biology of the nicotinic receptor, beginning with the complete sequencing of the subunits making up the neuromuscular (BGT) type of receptor. Antibodies raised to this type of receptor also reacted with neuronal receptors from chick brain, leading quickly to further immunological characterization of these neuronal nicotinic receptors.[27] Through the use of complementary DNA techniques, messenger RNAs for a variety of types of subunit have now been located in brain tissue, leading to the speculation that these subunits may combine in various ways to form a variety of subtypes of nicotinic cholinergic receptor. These advances also make possible the use of *in situ* hybridization techniques to describe the regional distributions of the various subunit mRNAs.[49] These techniques are promising, but their application to the study of behavioral phenomena has not yet appeared in the literature.

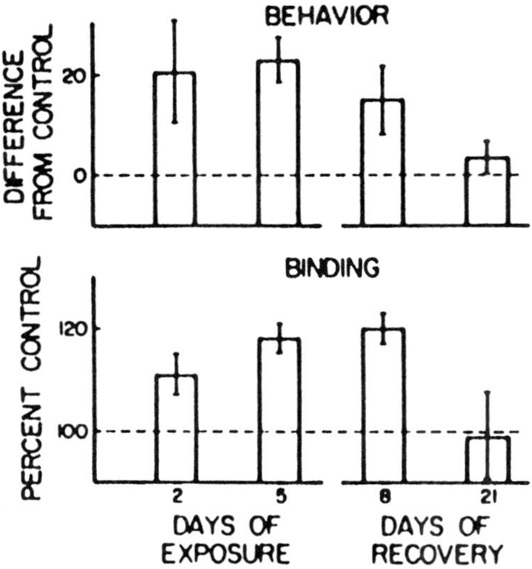

FIGURE 5. Comparison of the behavioral response to 0.2 mg/kg nicotine (locomotor activity) and specific binding to nicotinic cholinergic receptors in cortex. Two groups were given either 2 or 5 d of nicotine exposure (0.2 mg/kg/d), and behavioral testing and chemical assays were done on the 4th or 6th day. The other two groups were given 5 d of exposure to nicotine, followed by saline injections for either 7 or 20 d, and behavioral testing and chemical assays were done on the 8th or 21st day following the last nicotine injection. (From Ksir, C., et al., *Neuropharmacology,* 24, 527, 1985. With permission.)

IV. NICOTINE INTERACTIONS WITH DOPAMINE SYSTEMS

Neuroscientists have focused a great deal of attention over the past two decades on the two major brain dopamine pathways: the nigrostriatal pathway, which is characterized as originating in the substantia nigra pars compacta (SNC) A9 cell group in the midbrain and terminating in the caudate, and the mesolimbic pathway, which begins in the midbrain ventral tegmental area (VTA) A10 cell group and terminates in several forebrain structures, including, most prominently, the nucleus accumbens. The nigrostriatal pathway is known to be involved in movement disorders such as Parkinson's disease and the pseudo-Parkinsonism associated with chronic antipsychotic drug therapy. The mesolimbic system is known to be involved in the increased ambulation produced in rats by amphetamine and cocaine and is widely believed to be related to the reinforcing effects of drugs of abuse and possibly to psychosis.

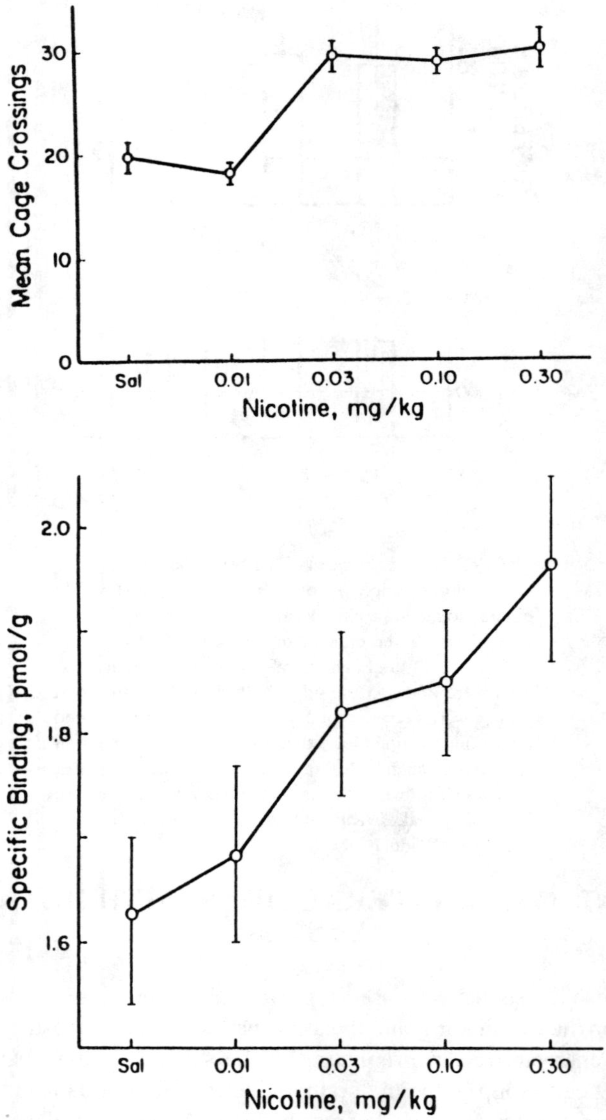

FIGURE 6. Comparison of the behavioral response to 0.2 mg/kg nicotine and specific binding to nicotinic cholinergic receptors in cortex. Groups of rats were exposed for 5 d to either saline or nicotine (0.01 to 0.30 mg/kg). Behavioral tests and chemical assays were done on the 6th day. (From Ksir, C., et al., *Psychopharmacology,* 92, 25, 1987. With permission.)

Noting some similarity between the behavioral effects of nicotine and those of low doses of amphetamine or cocaine, which are known to act through dopaminergic mechanisms, various methods have been used to examine the actions of nicotine on these dopamine pathways. For example, Westfall[52] studied the release of newly synthesized norepinephrine and dopamine from the *in vitro* striatal slice preparation, finding that micromolar concentrations of nicotine resulted in calcium-dependent release of dopamine. This release was also blocked by mecamylamine, implying that the effect is mediated through nicotinic receptors.[53] Lichtensteiger et al.[26] reported that subcutaneous injections of 0.3 to 1.0 mg/kg nicotine in urethane-anesthetized rats produced increases in the firing rates of SNC neurons. Clarke et al.[7] also reported increases in the firing rate of SNC neurons when nicotine was given either subcutaneously or intravenously to chloral hydrate anesthetized rats; the increased firing could be prevented by pretreatment with mecamylamine. Mereau et al.[31] reported that anesthesia was a critical variable in electrophysiological studies of dopamine neurons. SNC neurons were found to be activated by nicotine in rats given either general anesthesia or paralyzed and given local anesthesia. VTA neurons were slowed by nicotine in rats given general anesthesia but potently stimulated by nicotine in paralyzed, locally anesthetized rats. In fact, in locally anesthetized rats there was a greater effect of nicotine on VTA neurons than on SNC neurons. Intracellular recordings done in midbrain slices revealed that nicotine caused a direct depolarization of VTA neurons that were presumed on the basis of several properties to contain dopamine.[4]

Imperato et al.[19] used microdialysis to measure the release of dopamine and its metabolites from the caudate or the nucleus accumbens in freely moving rats. Although dopamine was released from both sites following 0.6 mg/kg nicotine s.c., the percentage increase in dopamine was greater in the nucleus accumbens than in the caudate. The authors related this finding to the fact that preferential release of dopamine from the accumbens relative to the caudate has also been reported for other substances of abuse, including amphetamine, cocaine, opiates, and ethanol. Rowell et al.[40] studied minced nucleus accumbens tissue *in vitro* and reported that nicotine in concentrations below 1 μM caused the release of [^{3}H]-dopamine that had been recently taken up by the tissue.

That dopaminergic interactions are important for at least some of the behavioral effects of nicotine is suggested by studies in which the selective neurotoxin 6-OHdopamine is used to damage the dopamine terminal projections in the nucleus accumbens. In one experiment[23] from our laboratory, such dopamine-depleting lesions resulted in a significant reduction in the locomotor response to nicotine in nicotine-sensitized rats (see Figure 7). Similarly, Singer et al.[46] found that dopamine-depleting lesions of the nucleus accumbens reduced the rate of self-injection of nicotine by rats, indicating a reduction in the reinforcing property of nicotine. Also, Rosecrans' group has

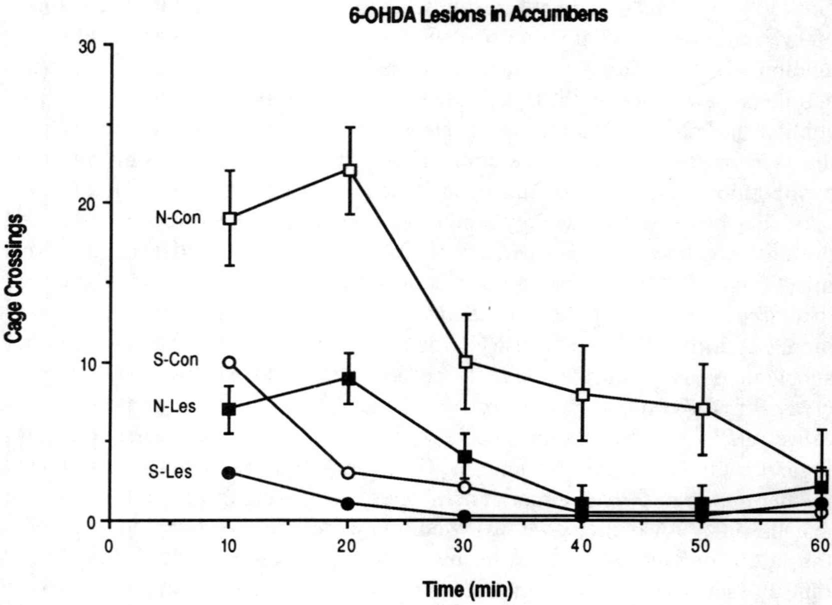

FIGURE 7. Locomotor activity after saline (S) or 0.2 mg/kg nicotine (N) in rats given bilateral injections of 6-OHDA into the nucleus accumbens (Les) and in control rats given sham injections (Con).

reported that neonatal dopamine depletion alters the discriminative stimulus properties of nicotine in adult rats.[37]

The combined results from behavioral, electrophysiological, excised tissue, and *in vivo* microdialysis experiments are confusing in one important sense: it is not clear to what extent systemically administered nicotine is acting directly on the cell bodies in the VTA, as implied by the intracellular VTA recordings of Calabresi et al.,[4] or at the terminals in the nucleus accumbens, as implied by the release of dopamine from isolated accumbens tissue reported by Rowell et al.[40] Nicotinic cholinergic receptors have been found to be abundant in both sites.[7,29] Selective neurochemical lesions that result in loss of dopamine terminals in either caudate or accumbens also result in losses of nicotinic receptors, implying that nicotinic receptors are located presynaptically on dopamine terminals.[10,43]

Immunohistochemical studies using antibodies to the nicotinic cholinergic receptor have reported localization of receptor-like reactivity in VTA and SNC cells as well as in the associated terminal regions.[15,48] Reports from three separate laboratories imply that at least the locomotor stimulation produced in rats by nicotine injections is probably related to effects at the VTA cell bodies rather than at the terminals. In these studies either nicotine or the nicotinic agonist cytisine was injected directly into the VTA in freely moving rats, resulting in increases in locomotor activity.[33,34,36] In two of the studies, injections were also given into the nucleus accumbens, and locomotor stimulation was not seen.

Several laboratories have recently examined the effects of either continuous or repeated (subchronic) exposure to nicotine on dopamine systems. Kirch et al.[20] infused nicotine continuously for 21 d into rats, using osmotic minipumps, and then examined postmortem concentrations of dopamine, norepinephrine, serotonin, and their metabolites in striatum, hippocampus, and hypothalamus. Although they did not find changes in the concentrations of dopamine or HVA, the levels of the dopamine metabolite DOPAC were decreased in striatum and hypothalamus, implying a decrease in turnover after 21 d of continuous exposure to either 3 or 12 mg/kg/d nicotine. Fung[16] used osmotic minipumps to expose rats to 1.5 mg/kg/d nicotine for 14 d. Studies were then conducted *in vitro* on nucleus accumbens slices from these and control (saline-exposed) rats. The slices were perfused with [^{3}H]-tyrosine, and the formation and release of [^{3}H]-dopamine were measured. When the slices were challenged with 10 μM nicotine, slices from saline-exposed rats did not release more dopamine, whereas slices from nicotine-exposed rats did release more dopamine. Damsma et al.[14] gave injections of 0.35 mg/kg nicotine s.c. once per day for 12 to 15 d consecutively. The rats were then anesthetized and studied using *in vivo* microdialysis, with the probe implanted in the nucleus accumbens. Dopamine was released in the dialysate in response to nicotine in both drug-naive and chronically exposed rats, and the values did not differ. Thus, these three studies of the effect of chronic nicotine on dopamine systems, using three different techniques, have resulted in three outcomes: dopamine turnover was reduced in one study; dopamine release was enhanced in another; and there was no effect on dopamine release in the third. Presumably future research will resolve differences in procedural variables so that some consistency can be discovered.

Overall, the research indicating direct effects of nicotine on dopamine systems is convincing, and current major theorizing is based on the concept that the mesolimbic dopamine system mediates both the locomotor stimulating effect of nicotine[68] and the reinforcing property of nicotine that presumably is responsible for tobacco dependence.[5]

V. EFFECTS ON OTHER NEURAL SYSTEMS

Although the major focus has been on the interactions of nicotine with dopamine systems, there has been limited but promising exploration of other actions of nicotine. Nicotine is believed to exist presynaptically on serotonin neurons,[43] and Rowell has reviewed evidence that nicotine is capable of releasing noradrenaline, dopamine, acetylcholine, and serotonin from their terminals.[39] Nicotine releases prolactin into the blood stream through a central mechanism.[18,45] Also, nicotine-induced release of GABA has been reported.[54]

There is great potential for future research based on newly developed techniques for localizing the nicotinic receptor subunits and determining their locations with reference to markers for other neurotransmitter systems, and based on pharmacological analyses of CNS effects of nicotine using the cornucopia of selective antagonists now being made available for various neurotransmitters.

REFERENCES

1. **Balfour, D. J. K.,** Influence of nicotine on the release of monoamines in the brain, in *Nicotinic Receptors in the CNS: Their Role in Synaptic Transmission,* Nordberg, A., Fuxe, K., Holmstedt, B., and Sundwall, A., Eds., Elsevier, Amsterdam, 1989, 165.
2. **Battig, K., Driscoll, P. J. S., and Uster, H. J.,** Effects of nicotine on the exploratory locomotion patterns of female Roman high- and low-avoidance rats, *Pharmacol. Biochem. Behav.,* 4, 435, 1976.
3. **Bovet, D., Bovet-Nitti, F., and Oliverio, A.,** Action of nicotine on spontaneous and acquired behavior in rats and mice, *Ann. N.Y. Acad. Sci.,* 142, 261, 1967.
4. **Calabresi, P., Lacey, M. G., and North, R. A.,** Nicotine excitation of rat ventral tegmental neurones *in vitro* studied by intracellular recording, *Br. J. Pharmacol.,* 98, 135, 1989.
5. **Clarke, P. B. S.,** Mesolimbic dopamine activation—the key to nicotine reinforcement?, in *The Biology of Nicotine Dependence,* Bock, G. and Marsh, J., Eds., John Wiley & Sons, New York, 1990, 153.
6. **Clarke, P. B. S., Fu, D. S., Jakubovic, A., and Fibiger, H. C.,** Evidence that mesolimbic dopaminergic activation underlies the locomotor stimulant action of nicotine in rats, *J. Pharmacol. Exp. Ther.,* 246(2), 701, 1988.
7. **Clarke, P. B. S., Hommer, D. W., Pert, A., and Skirboll, L. R.,** Electrophysiological actions of nicotine on substantia nigra single units, *Br. J. Pharmacol.,* 85, 827, 1985.
8. **Clarke, P. B. S. and Kumar, R.,** The effects of nicotine on locomotor activity in non-tolerant and tolerant rats, *Br. J. Pharmacol.,* 78, 329, 1983.
9. **Clarke, P. B. S. and Kumar, R.,** Characterization of the locomotor stimulant action of nicotine in tolerant rats, *Br. J. Pharmacol.,* 80, 587, 1983.
10. **Clarke, P. B. S. and Pert, A.,** Autoradiographic evidence for nicotinic receptors on nigrostriatal and mesolimbic dopaminergic neurons, *Brain Res.,* 348, 355, 1985.
11. **Clarke, P. B. S., Schwartz, R. D., Paul, S. M., Pert, C. B., and Pert, A.,** Nicotinic binding in rat brain: autoradiographic comparison of ^{3}H-acetylcholine, ^{3}H-nicotine, and ^{125}I-α-bungarotoxin, *J. Neurosci.,* 5, 1307, 1985.
12. **Collins, A. C., Evans, C. B., Miner, L. L., and Marks, M. J.,** Mecamylamine blockade of nicotine respones: evidence for two brain nicotinic receptors, *Pharmacol. Biochem. Behav.,* 24, 1767, 1986.
13. **Corrigall, W. A. and Coen, K. M.,** Nicotine maintains robust self-administration in rats on a limited-access schedule, *Psychopharmacology,* 99, 473, 1989.
14. **Damsma, G., Day, J., and Fibiger, H. C.,** Lack of tolerance to nicotine-induced dopamine release in the nucleus accumbens, *Eur. J. Pharmacol.,* 168, 363, 1989.
15. **Deutsch, A. Y., Holliday, J., Roth, R. H., Chun, L. L. Y., and Hawrot, E.,** Immunohistochemical localization of a neuronal nicotinic acetylcholine receptor in mammalian brain, *Proc. Natl. Acad. Sci. U.S.A.,* 84, 8697, 1987.

16. **Fung, Y. K.,** Effects of chronic nicotine pretreatment on (+)-amphetamine and nicotine-induced synthesis and release of [^{3}H]tyrosine in rat nucleus accumbens, *J. Pharm. Pharmacol.,* 41, 66, 1989.
17. **Hakan R. L. and Ksir, C.,** Nicotine induced locomotor activity in rats: the role of Pavlovian conditioning, *Pharmacol. Biochem. Behav.,* 29, 661, 1988.
18. **Hulihan-Giblin, B. A., Lumpkin, M. D., and Kellar K. J.,** Acute effects of nicotine on prolactin release in the rat: agonist and antagonist effects of a single injection of nicotine, *J. Pharmacol. Exp. Ther.,* 252, 15, 1990.
19. **Imperato, A., Mulas, A., and Di Chiara, G.,** Nicotine preferentially stimulates dopamine release in the limbic system of freely moving rats, *Eur. J. Pharmacol.,* 132, 337, 1986.
20. **Kirch, D. G., Gerhardt, G. A., Shelton, R. C., Freedman, R., and Wyatt, R. J.,** Effect of chronic nicotine administration on monoamine and monoamine metabolite concentrations in rat brain, *Clin. Neuropharmacol.,* 10(4), 376, 1987.
21. **Kita, T., Nakashima, T., and Kurogochi, Y.,** Circadian variation of nicotine-induced ambulatory activity in rats, *Jpn. J. Pharmacol.,* 41, 55, 1986.
22. **Ksir, C.,** Rate-convergent effects of drugs, in *Advances in Behavioral Pharmacology,* Thompson, T., Dews, P. B., and McKim, W. A., Eds., Academic Press, San Diego, 1981, 39.
23. **Ksir, C. and Cline, E. J.,** Lesions of nucleus accumbens block the stimulant effects of nicotine in rats, *Soc. Neurosci. Abstr.,* 13, 447, 1987.
23a. **Ksir, C. and Haken, R. L.,** unpublished.
24. **Ksir, C., Hakan, R., Hall, D. P., Jr., and Kellar, K. J.,** Exposure to nicotine enhances the behavioral stimulant effect of nicotine and increases binding of [^{3}H]acetylcholine to nicotinic receptors, *Neuropharmacology,* 24(6), 527, 1985.
25. **Ksir, C., Hakan, R. L., and Kellar, K. J.,** Chronic nicotine and locomotor activity: influences of exposure dose and test dose, *Psychopharmacology,* 92, 25, 1987.
26. **Lichtensteiger, W., Felix, D., Lienhart, R., and Hefti, F.,** A quantitative correlation between single unit activity and fluorescence intensity of dopamine neurones in zona compacta of substantia nigra, as demonstrated under the influence of nicotine and physostigmine, *Brain Res.,* 117, 85, 1976.
27. **Lindstrom, J., Schoepfer, R., Conroy, W. G., and Whiting, P.,** Structural and functional heterogeneity of nicotinic receptors, in *The Biology of Nicotine Dependence,* Bock, G. and Marsh, J., Eds., John Wiley & Sons, New York, 1990, 23.
28. **Marks, M. J. and Collins, A. C.,** Characterizations of nicotine binding in mouse brain and comparison with the binding of a-bungarotoxin and quinuclidinyl benzilate, *Mol. Pharmacol.,* 22, 554, 1982.
29. **Marks, M. J., Stitzel, J. A., Romm, E., Wehner, J. M., and Collins, A. C.,** Nicotinic binding sites in rat and mouse brain: comparison of acetylcholine, nicotine, and α-bungarotoxin, *Mol. Pharmacol.,* 30, 427, 1986.
30. **Martin, B. R., Onaivi, E. S., and Martin, T. J.,** What is the nature of mecamylamine's antagonism of the central effects of nicotine? *Biochem. Pharmacol.,* 38(20), 3391, 1989.
31. **Mereau, G., Yoon, K.-W.P., Boi, V., Gessa, G. L., Naes, L., and Westfall, T. C.,** Preferential stimulation of ventral tegmental area dopaminergic neurons by nicotine, *Eur. J. Pharmacol.,* 141, 395, 1987.
32. **Morrison, C. F. and Stephenson, J. A.,** The occurrence of tolerance to a central depressant effect of nicotine, *Br. J. Pharmacol.,* 46, 151, 1972.
33. **Museo, E. and Wise, R. A.,** Locomotion induced by ventral tegmental microinjections of a nicotinic agonist, *Pharmacol. Biochem. Behav.,* 35, 737, 1990.
34. **Pert, A. and Chiueh, C. C.,** Effects of intracerebral nicotinic agonists on locomotor activity: involvement of mesolimbic dopamine, *Soc. Neurosci. Abstr.,* 12, 917, 1986.
35. **Pomerleau, O. F. and Pomerleau, C. S.,** Behavioural studies in humans: anxiety, stress and smoking, in *The Biology of Nicotine Dependence,* Bock, G. and Marsh, J., Eds., John Wiley & Sons, New York, 1990, 225.

36. **Reavill, C. and Stolerman, I. P.,** Locomotor activity in rats after administration of nicotinic agonists intracerebrally, *Br. J. Pharmacol.*, 99, 273, 1990.
37. **Rosecrans, J. A. and Chance, W. T.,** Cholinergic and non-cholinergic aspects of the discriminative stimulus properties of nicotine, in *Discriminative Stimulus Properties of Drugs,* Lal, H., Ed., Plenum Press, New York, 1977, 155.
38. **Rosecrans, J. A. and Meltzer, L. T.,** Central sites of mechanisms of action of nicotine, *Neurosci. Biobehav. Rev.*, 5, 497, 1981.
39. **Rowell, P. P.,** Current concepts on the effects of nicotine on the central nervous system, in *Tobacco Smoking and Nicotine,* Martin, W. R., Van Loon, G. R., Iwamoto, E. T., and Davies, L., Eds., Plenum Press, New York, 1987, 191.
40. **Rowell, P. P., Carr, L. A., and Garner, A. C.,** Stimulation of [^{3}H]dopamine release by nicotine in rat nucleus accumbens, *J. Neurochem.*, 49(5), 1449, 1987.
41. **Sanger, D. J. and Blackman, D. E.,** Rate-dependent effects of drugs: a review of the literature, *Pharmacol. Biochem. Behav.*, 4, 73, 1976.
42. **Schwartz, R. D. and Kellar, K. J.,** Nicotinic cholinergic receptor binding sites in the brain: regulation *in vivo, Science,* 220, 214, 1983.
43. **Schwartz, R. D., Lehmann, J., and Kellar, K. J.,** Presynaptic nicotinic cholinergic receptors labelled by [^{3}H]acetylcholine on catecholamine and serotonin axons in brain, *J. Neurochem.*, 42, 1495, 1984.
44. **Schwartz, R. D., McGee, R., and Kellar, K. J.,** Nicotinic cholinergic receptors labeled by [^{3}H]acetylcholine in rat brain, *Mol. Pharmacol.*, 22, 56, 1982.
45. **Sharp, B. M. and Beyer, H. S.,** Rapid desensitization of the acute stimulatory effects of nicotine on rat plasma adrenocorticotropin and prolactin, *J. Pharmacol. Exp. Ther.*, 238, 486, 1986.
46. **Singer, G., Wallace, M., and Hall, R.,** Effects of dopaminergic nucleus accumbens lesions on the acquisition of schedule induced self injection of nicotine in the rat, *Pharmacol. Biochem. Behav.*, 17, 579, 1982.
47. **Stolerman, I. P., Fink, R., and Jarvik, M. E.,** Acute and chronic tolerance to nicotine measured by activity in rats, *Psychopharmacologia (Berlin),* 30, 329, 1973.
48. **Swanson, L. W., Simmons, D. M., Whiting, P. J., and Lindstrom, J.,** Immunohistochemical localization of neuronal nicotinic receptors in the rodent central nervous system, *J. Neurosci.*, 7, 3334, 1987.
49. **Wada, E., Wada, K., Boulter, J., Deneris, E., Heinemann, S., Patrick, J., and Swanson, L. W.,** Distribution of alpha2, alpha3, alpha4, and beta2 neuronal nicotinic receptor subunit mRNAs in the central nervous system: a hybridization histochemical study in the rat, *J. Comp. Neurol.*, 284, 314, 1989.
50. **Walter, S. and Kuchinsky, K.,** Conditioning of nicotine effects on motility and behaviour in rats, *Naunyn-Schmiedebergs Arch. Pharmakol.*, 339, 208, 1989.
51. **Wesnes, K. and Warburton, D. M.,** Effects of scopolamine and nicotine on human rapid information processing and performance, *Psychopharmacology,* 82, 147, 1984.
52. **Westfall, T. C.,** Effect of nicotine and other drugs on the release of ^{3}H-norepinephrine and ^{3}H-dopamine from rat brain slices, *Neuropharmacology,* 13, 693, 1974.
53. **Westfall, T. C., Mereau, G., Vickery, L., Perry, H., Naes, L., and Yoon, K.-W. P.,** Regulation by nicotine of midbrain dopamine neurons, in *Nicotinic Receptors in the CNS: Their Role in Synaptic Transmission,* Nordberg, A., Fuxe, K., Holmstedt, B., and Sundwall, A., Eds., Elsevier, Amsterdam, 1989, 173.
54. **Wonnacott, S., Drasdo, A., Sanderson, E., and Rowell, P. P.,** Presynaptic nicotine receptors and the modulation of transmitter release, in *The Biology of Nicotine Dependence,* Bock, G. and Marsh, J., Eds., John Wiley & Sons, New York, 1990, 102.
55. **Ksir, C. and Hakan, R. L.,** unpublished observations.

Chapter 8

NEUROPHARMACOLOGY OF AMPHETAMINE: CHANGES IN THE ACTIVITY OF STRIATAL NEURONS DURING BEHAVIOR

George V. Rebec

TABLE OF CONTENTS

I. INTRODUCTION

The behavioral effects of amphetamine have intrigued psychopharmacologists ever since this drug became widely available in the 1930s. Not only does amphetamine increase motor activity, it also elicits feelings of energy, heightened mental awareness, and euphoria.[1-3] These latter effects undoubtedly contribute to its alarming potential for abuse. In fact, the recent availability of the methylated version of amphetamine in a pure, smokable form known as ''ice'' threatens to surpass cocaine as the major drug-abuse problem of the 1990s.[4] This threat highlights another behavioral feature of amphetamine: its ability to elicit a paranoid psychosis with chronic or sustained use. The symptoms of this psychosis, which include delusions and hallucinations as well as specific postures and motor response patterns, are remarkably similar to those of idiopathic paranoid schizophrenia.[5-9] The resemblance is sufficiently close to suggest that both conditions share many of the same underlying neural mechanisms.[10,11] Thus, research on how amphetamine exerts its behavioral effects has implications for understanding the neural basis of drug abuse and certain forms of mental illness as well.

Because amphetamine elicits characteristic motor responses in virtually all mammals, these behaviors have been studied extensively in animals, typically rodents, as models of the human amphetamine response.[12-16] Even amphetamine psychosis, which obviously cannot be documented in animals, appears to have a counterpart in the repetitive or stereotyped nature of the motor responses produced in animals by this drug.[10,14-20] Indeed, stereotyped movements are the hallmark of amphetamine administration, and like the drug-induced psychosis, many of these movements are enhanced with repeated use. They also are blocked or attenuated by neuroleptic drugs, which are used widely in the treatment of psychotic behavior. Disruption of amphetamine-induced stereotypy by neuroleptics is highly correlated with their clinical potency, and the stereotypy test has become one of many standard behavioral screens in the development of new neuroleptic drugs.[21,22]

The stereotyped responses produced by amphetamine typically include motor components of investigative behavior, which in the rat are manifest as forward locomotion and rearing as well as sniffing and head movements. The invariant and repetitive nature of these behaviors distinguishes them from normal exploratory responses of nondrugged animals.[23,24] At relatively high doses, a constricted form of stereotypy emerges in which locomotion and rearing give way to focused head and oral movements.[12,14,16]

Numerous brain areas have been implicated in the behavioral response to amphetamine, but the structure that has received overwhelming attention in this regard is the striatum, a large forebrain area with dorsal and ventral regions that have been linked, respectively, to the extrapyramidal motor system and the limbic system.[25] The dorsal striatum, sometimes known in the rat as the neostriatum or caudate-putamen complex, serves as a sensorimotor integration area of the basal ganglia.[26-28] Damage to the dorsal striatum or its afferents has been reported to interfere with several aspects of the amphetamine behavioral response, especially the highly focused pattern of stereotypy.[14,19,29-31] The ventral striatum, which has extensive anatomical connections to various limbic structures, includes the nucleus accumbens and olfactory tubercle.[32-34] This striatal region has been linked to the nonfocused motor responses produced by amphetamine, such as locomotion and rearing.[30,31]

In order to understand the neuronal mechanisms by which the striatum mediates the amphetamine behavioral response, it is necessary to assess the effects of this drug on striatal physiology. Efforts to make such an assessment were carried out initially in anesthetized or immobilized animals secured in a stereotaxic frame, a procedure that made it possible to obtain stable recordings of cellular activity for prolonged periods. The stereotaxic preparation provided basic, though complex, information about the actions of amphetamine on striatal neurons. In dorsal striatum, the neuronal response to amphetamine was found to vary with both region and dose.[35] Anteromedial neurons, for example, typically responded with an inhibition of firing rate that progressively shifted to a pronounced and prolonged excitatory response with increasing dose.[36,37] In fact, the same neuron inhibited by systemic injection of 1.0 mg/kg d-amphetamine was excited by a dose of 5.0 mg/kg. This dose-dependent pattern was reversed in anterolateral areas such that low doses usually excited and high doses inhibited neuronal activity.[38] Ventral striatal neurons, in contrast, were inhibited by amphetamine regardless of dose, though the duration and intensity of the inhibition was clearly dose-dependent.[37]

These results were intriguing because in dorsal striatum they paralleled the dose-dependent shift in behavior from general locomotor activation to intense, focused stereotypy. In ventral striatum, on the other hand, the increase in inhibition with increasing dose appeared to reflect the dose-dependent change in behavioral intensity. It was possible, using this and other relevant data from the stereotaxic preparation, to make inferences regarding the role of dorsal and ventral striatal neurons in the amphetamine behavioral response. Such inferences, however, had to be treated with caution, because even though they were derived from a detailed investigation of amphetamine effects on striatal neurons, the results were limited by the state of the animal. At the very least, stereotaxic recordings are confounded by the prior use of anesthetic agents or immobilizing drugs, to say nothing of the changes in physiology that must accompany such treatment.

The recent application of electrophysiological techniques to tissue-culture and brain-slice preparations has provided important new information on the cellular mechanisms of drug action, but the advances have not diminished the need to understand the behavioral significance of drug-induced changes in neuronal activity. In fact, such advances now make it more necessary than ever to study basic physiological changes in the context of ongoing behavior.

Drug effects that can be demonstrated in an isolated neuronal system or in an animal restrained in a stereotaxic device may not occur or may be altered dramatically in a behaving animal. Ultimately, if amphetamine effects in the striatum are to be understood in terms of their behavioral consequences, they must be assessed during behavior.

Such assessments are now possible because of recent breakthroughs in the miniaturization of electronic circuits and in the development of sophisticated computer software for data acquisition and analysis. The result has been a steadily growing literature on the actions of amphetamine on striatal neurons in awake, behaving animals. As more information becomes available on this topic, new insights are likely to emerge into the neuronal substrates of the amphetamine behavioral response. This chapter examines the data generated by this increasingly popular line of research and the implications of such data for understanding the mechanisms by which this drug exerts its behavioral effects.

II. METHODOLOGICAL CONSIDERATIONS

The number of factors that can influence the action of a systemically injected drug on striatal activity in behaving animals is enormous. Some of these factors reflect the complex neuronal circuitry associated with the striatum; others relate to the confounding influence of behavior itself. All of them influence the conclusions that can be drawn about amphetamine effects on striatal neurons. Before discussing such effects, therefore, it would be useful to review some of the most important issues that confront this line of research and the various means available for dealing with them.

A. NEURONAL CIRCUITRY

Following systemic injection, many drugs are distributed throughout the brain where they are likely to have simultaneous actions in numerous structures, many of which either directly or indirectly influence striatal neurons. Amphetamine is believed to exert its behavioral effects primarily by facilitating the synaptic action of dopamine, the most abundant monoamine neurotransmitter in the striatum.[19,29,30,40] More than 70% of total brain dopamine is concentrated in this structure, strongly suggesting that dopamine is a major contributor to amphetamine-induced changes in the activity of striatal neurons. Dopamine, however, does not act in isolation, and any postsynaptic effect of this neurotransmitter must be considered in terms of how it interacts with other neuronal systems and processes that also may be altered either by amphetamine or by the behavioral state of the animal.

The striatum receives its supply of dopamine from several brainstem nuclei, including those of the substantia nigra pars compacta (SNC) and the ventral tegmental area (VTA). Dopaminergic neurons in the SNC innervate dorsal striatum, whereas VTA neurons send dopaminergic projections to ventral striatum, as well as to many limbic and cortical areas.[41-45] The firing pattern of neurons in both SNC and VTA is controlled, at least in part, by local — primarily dendritic — release of dopamine onto inhibitory autoreceptors.[35,46-50] In addition, SNC neurons also receive feedback from both dorsal and ventral striatum, and following amphetamine administration, this feedback circuit could play a crucial role in further controlling impulse flow in SNC and thus dopamine release in dorsal striatum.[35,47,51] Apart from dopamine, however, amphetamine also facilitates the synaptic action of other monoamines, including serotonin and norepinephrine.[39,40,52,53] Serotonergic neurons in the raphé complex send direct projections to both dorsolateral and ventral striatum, and although relatively few noradrenergic terminals exist in striatum, they are plentiful throughout the reticular formation, thalamus, and cerebral cortex, all of which are influenced by amphetamine and which, in turn, influence striatal activity.[35] In fact, cortical and thalamic neurons account for the majority of striatal afferents.[54-57] Thus, amphetamine-induced changes in striatal activity are likely to reflect the composite influence of this drug on several different afferent systems. Add to this complexity the equally complex neuronal circuitry within the striatum itself, in which neurons receive information from their neighbors via nonmonoaminergic axon collaterals and interneurons,[58-63] and it becomes virtually impossible to assume that dopamine plays the sole or even the primary role in the neuronal changes produced by amphetamine.

In order to obtain information about the involvement of specific neurotransmitter systems in the action of amphetamine, additional manipulations are necessary. Direct application of the drug to the recording site is an effective means of eliminating the confounding influence of amphetamine action in areas outside the neostriatum. However, this maneuver, though easily accomplished in the stereotaxic preparation, presents its share of difficulties in the behaving animal. The solution may lie in a miniaturized device that can accommodate a multibarreled glass pipette that includes a recording barrel and several barrels for direct application of drugs via iontophoresis. Preliminary evidence indicates that this system works well in behaving rats.[64,65] Until such a device becomes widely available, however, other approaches can be used to begin the process of identifying those neurochemical systems that contribute to the amphetamine neuronal response — for example, inactivation of dopaminergic afferents either with neurotoxic lesions or the administration of selective dopamine antagonists.

A further complication in the interpretation of drug effects in the striatum arises from the fact that striatal neurons are not anatomically and biochemically homogeneous and are not likely to receive the same inputs or be equally sensitive to these inputs. The striatum, for example, contains many known neurotransmitters, some of which, including dopamine, show regional differences in distribution, release, and receptor density.[66-72,129] Moreover, the striatum receives, in rough topographical order, direct inputs from virtually all cortical areas and either direct or indirect inputs from many brainstem regions as well.[57,73,74] These complexities underscore the need for obtaining as much information as possible about the neuronal signals detected by the recording electrode. A recording system that samples impulses generated by a large number of cells without differentiating their signals is likely to provide severely limited information on drug effects. If, for example, a drug increases the activity of one cell type but inhibits the activity of another, a recording system that fails to distinguish between them might suggest that the drug produced no effect, or, depending on the number of each cell type in the vicinity of the recording electrode and the magnitude of each response, one could erroneously assume that the drug produced a uniform excitation or inhibition of striatal activity. For accurate characterization of drug effects in the striatum, therefore, it is critical to record well-isolated action potentials from individual neurons. Indeed, some would argue that before drug-induced changes in multiple-unit records become credible in any brain region, the changes must be verified with single-unit recordings.[75,76]

The primary morphological cell type in the striatum is the medium spiny neuron, which has a relatively small cell body (12 to 18 μm in diameter) and an axon that comprises the major projection system to the substantia nigra pars reticulata (SNR).[74,77-79] More than 90% of striatal neurons are believed to be in the medium spiny class. Relatively large, aspiny interneurons and numerous other morphological cell types account for the remainder of the striatal neuronal population. Unfortunately, these varied cell types are not associated with unique electrophysiological signals. In the stereotaxic preparation, single-unit recordings have identified two groups of cells in the striatum:[80-82] Type I neurons, which display an initial negative-going wave and peak-to-peak action potentials of 300 to 600 μV, and Type II neurons, which fire with an initially positive wave and peak-to-peak action potentials that often exceed 1000 μV. Recordings at some laboratories show a much higher frequency for Type II neurons,[80,81] suggesting that they correspond to medium spiny neurons. In fact, extracellular recordings from neostriatal neurons subsequently identified with intracellular injections of horseradish peroxidase confirm a link between Type II cells and the medium spiny morphology.[83] In other investigations, however, Type I cells have been encountered more frequently than Type II,[82] and the possibility remains that the electrophysiological distinction of cell types reflects differing electrode and signal-filtering parameters.[80,84] Wave form alone, therefore, may not be a reliable indicator of morphological cell type in striatum.

Other procedures, however, can be used in conjunction with single-unit recording techniques to gain additional information about striatal neurons. It is possible, for example, to insert a stimulating electrode into an appropriate afferent or efferent pathway in order to establish anatomical neuronal connections. Many striatal cells can be activated by orthodromic stimulation of cortical afferents, and this finding has been used in the stereotaxic preparation to further differentiate Type I and Type II cells.[80,85] With stimulating electrodes in the SNR, it is possible to use antidromic activation to identify a major group of striatal efferents. A sizable number of cells in the neostriatum can be activated antidromically with SNR stimulation,[86] and this procedure has been used successfully by Ryan et al.[87] to isolate a group of striatonigral projection neurons in freely moving rats. Such combined stimulation and recording studies are important in that they provide information that can be crucial for interpreting drug effects. It is useful to know, for example, whether amphetamine excites or inhibits striatal output cells, because the answer suggests a mechanism by which the striatum controls the behavioral response to this drug. In addition, electrical stimulation is likely to identify very slowly firing or phasically active neurons that might otherwise be missed in recordings that rely on sampling only tonically firing, spontaneously active neurons.

Of course, one cannot assume that all projection neurons are capable of antidromic activation. Numerous factors, many of which are only poorly understood, actually may prevent antidromic invasion of the soma.[86,88,89] Extreme caution, therefore, is required before results obtained from a limited number of antidromically activated neurons are generalized to an entire projection pathway. Moreover, neither antidromic or orthodromic activation guarantees a uniform population of cells or a uniform drug response. Striatal cells receive numerous inputs, including complex intrastriatal connections, that may differ markedly even among cells that receive similar cortical afferents or that project to SNR. Thus, electrical stimulation procedures should be viewed not as a solution to the problem of identifying cell types within the striatum, but as another mechanism for focusing drug effects on specific groups of cells in a structure in which such identification is extremely difficult.

B. BEHAVIORAL FEEDBACK

Another major methodological consideration is the extent to which a drug-induced change in neuronal activity reflects feedback from behavior rather than a direct synaptic effect of the drug. It is conceivable — highly likely, in fact — that a brain region such as the striatum, which receives considerable sensorimotor information, is acutely responsive to behavioral changes. If amphetamine alters striatal activity mainly because of its ability to alter behavior, then investigators who focus exclusively on the direct synaptic actions of this drug are likely to make erroneous conclusions about underlying neuronal mechanisms. It is important, therefore, to include procedures in the experimental paradigm that address the issue of behavioral feedback.

One such procedure involves an examination of neuronal activity when comparable behaviors are displayed before and after drug administration.[40,75,90] In effect, this approach attempts to "clamp," or hold constant, the influence of behavior in order to reveal the direct effect of the drug on neuronal activity. In one version of behavioral clamping, a conditioning or training paradigm can be used to insure that pre- and post-drug comparisons of neuronal activity are made during the same behavioral state. An effective training paradigm used by West et al.[91] to characterize amphetamine effects is treadmill locomotion. Animals quickly learn to walk the treadmill, which is set to move at a predetermined rate. Neuronal activity then can be monitored during an easily quantifiable behavior that remains essentially the same both before and after amphetamine administration. This paradigm is especially appropriate for examining the effects of relatively low doses of amphetamine, which do not produce the components of focused stereotypy that would interfere with treadmill locomotion. Even low doses of this drug, however, can elicit behaviors, such as sniffing and head bobbing, that may confound pre- and post-drug behavioral comparisons. Thus, behavioral observations still are required with the treadmill paradigm, and the investigator must be able to demonstrate whether or not these additional behaviors contribute to any change in firing rate that may occur after amphetamine administration.

It is also important to stress that recording unit activity exclusively during a given behavioral response, such as locomotion, severely limits data interpretation because only neurons active during the behavioral paradigm are likely to be examined for a drug response. Cells that may respond during other behaviors or that may be inhibited by behavioral activation are likely to be missed. In addition, prior conditioning or training may influence the baseline firing rate of certain neurons, which itself could influence how neurons respond to amphetamine.

An alternative clamping method, which has been used successfully by Haracz et al.,[92] relies on a videotaping system to identify comparable open-field behaviors before and after amphetamine administration. When such behaviors have been identified, neuronal activity, stored on the same videotape, then can be compared. This procedure takes advantage of the fact that many of the same behavioral responses produced by amphetamine also occur in the absence of the drug. Locomotion, rearing, sniffing, and head movements, for example, occur during normal exploration in the rat, and to the extent that these behaviors also occur following amphetamine administration, they provide a basis for behavioral clamping.

Although many behaviors before and after amphetamine appear similar, they are not necessarily appropriate for clamping in every case. An obvious difference is that post-amphetamine behaviors often appear in an exaggerated form or in a stereotyped pattern that rarely, if ever, appears pre-amphetamine. Moreover, certain post-amphetamine behaviors sometimes occur simultaneously (e.g., locomotion and sniffing), whereas these same behaviors may appear individually during the pre-amphetamine period. Thus, the selection of pre-amphetamine behavioral episodes that match post-amphetamine behavioral episodes requires careful scrutiny of videotape records. The observer must be attentive not only to specific movements but also to their duration and intensity. Even with such restrictions, it is possible to identify numerous pre- and post-amphetamine behavioral episodes that match and that then can be analyzed for differences in neuronal activity. To the extent that such differences appear, they suggest a direct drug effect rather than an effect secondary to a drug-induced behavioral change.

This method of behavioral clamping also offers the opportunity to monitor neuronal activity during the execution of different behaviors and in that way to determine the types of behavioral responses that are associated with changes in firing rate. Such information can be useful in the interpretation of drug effects. If, for example, amphetamine increases the firing rate of neurons whose activity is closely associated with movement, then a further examination of these cells may identify the mechanisms by which amphetamine exerts its effects on striatal activity and behavior.

No form of clamping, however, provides a complete behavioral control. It is impossible, for example, to rule out completely a relationship between neuronal activity and covert behavioral processes such as changes in attention or sensory responsiveness. Nor does clamping address all the possible peripheral variables, including changes in autonomic function, that could influence striatal activity independently of a direct drug effect. Despite these limitations, however, clamping provides a valuable mechanism for interpreting drug effects independently of any obvious behavioral changes.

III. INTERPRETING AMPHETAMINE EFFECTS

The first reports of amphetamine-induced changes in striatal neuronal activity in awake, behaving animals failed to show any evidence of the region-specific, dose-dependent shift in firing rate that had been described for the stereotaxic preparation. The drug was reported, instead, to elicit a simple, monotonic increase in striatal activity with dose.[93] Moreover, under certain conditions, the drug-induced excitation that was observed in ambulant animals could be dissociated from the behavioral response, suggesting that striatal mechanisms alone cannot account for the behavioral effects of amphetamine.[94,95] The inferences that can be drawn from these data, however, are severely limited. In these initial studies, the results were based on gross recordings of multiple-unit activity; no attempt was made to isolate the firing pattern of individual neurons, which likely could respond to amphetamine in a way very different from that represented by the composite group response of a multiple-unit record. In fact, single-unit evidence, published in a preliminary form at approximately the same time as the multiple-unit data, indicates that at certain doses amphetamine is capable of producing both excitations and inhibitions in the striatum of freely moving animals.[96] These early studies also lacked an adequate control for behavioral feedback, though efforts were made to rule out artifacts in the neuronal signal from gross changes in movement.

Despite these limitations, however, the data raised serious questions about the adequacy of the stereotaxic preparation for inferring neuronal correlates of the amphetamine behavioral response. As a case in point, Warenycia and McKenzie[97] showed that multiple-unit recordings from the same animals under both freely moving and anesthetized conditions revealed very different effects of amphetamine. Thus, whatever the shortcomings of the multiple-unit method, it seems clear that the freely moving preparation is detecting the influence of systems that are either not functioning or are somehow altered in anesthetized animals.

The freely moving preparation soon became the method of choice for assessing striatal mechanisms in the behavioral effects of amphetamine. Single-unit recordings now dominate this research, and various approaches are in use to control for behavioral feedback. Although these developments promise to make data interpretation easier, they do not guarantee a simple explanation of how the striatum participates in the amphetamine behavioral response.

A. AMPHETAMINE ELICITS BOTH EXCITATIONS AND INHIBITIONS

Different striatal responses to amphetamine have been reported to occur in different behavioral situations. In a training paradigm, such as running on a treadmill, the neuronal response to the drug is said to be uniformly excitatory,[91,98] whereas in an open-field arena, in which the animal is free to engage in a wide variety of behaviors, amphetamine may elicit increases as well as decreases in neuronal activity.[87,99-101] The reason for such diversity may have less to do with a differential action of amphetamine in these behavioral situations than with the ways in which the behavioral situation interacts with the type of cell that is being recorded.

When striatal neurons are recorded from rats trained to run on a treadmill, the electrode is likely to detect neurons excited by movement but to miss those that respond with an inhibition or those that fire relatively slowly. That the neuronal sample is biased toward movement-excited cells becomes clear when the treadmill operates on a regular cycle (e.g., 30 sec on, 30 sec off); firing rate typically is slow during the off-phase of the cycle, as the animals rest quietly, and then increases with the movement associated with treadmill onset.[91] These movement-excited cells are, in turn, excited by amphetamine, an effect that does not appear to be caused simply by behavioral feedback. Following systemic injection of 1.0 mg/kg d-amphetamine, West et al.[91] reported that the firing rate of all neurons sampled during treadmill running increased markedly (up to 500%) compared to the same behavior following an injection of saline. Other behaviors that might occur after amphetamine, such as head movements, were examined for correlations with unit activity, but none was found, further suggesting that the amphetamine-induced increase in striatal activity in the treadmill paradigm reflects a direct effect of the drug on striatal neurons.

This paradigm says nothing, however, about the effects of amphetamine on neurons that fail to increase firing rate with movement. In an open-field situation, behavior can vary considerably and so too can the types of neuronal responses that one is likely to encounter. When animals are actively moving about in the open-field arena, many, though not all, striatal neurons exhibit distinct changes in firing rate compared to periods of quiet rest.[99,100] Some cells increase or decrease firing rate in close temporal association with specific behaviors, such as limb or neck movements, while other cells respond to more general changes in behavioral activity. It also is possible to find still other cells that do not respond reliably to any obvious behavioral episode, but increase or decrease firing rate in conjunction with an apparent change in behavioral arousal. In their initial investigation of the effects of systemic amphetamine on these different categories of cells, Gardiner et al.[99] reported that 1.0 mg/kg d-amphetamine produced both excitations and inhibitions. The drug-induced changes in neuronal activity roughly paralleled the time course and the intensity of the behavioral activation; they began shortly after injection, peaked within 30 to 40 min, and returned to baseline after approximately 60 min. At higher doses of amphetamine, however, the pattern of neuronal activity was not simply a more intense version of the low-dose response. In several cases, systemic administration of 5.0 mg/kg d-amphetamine 60 min after administration of 1.0 mg/kg actually reversed the direction of the low-dose neuronal response. Surprisingly, this dose-dependent shift in firing rate did not occur in all neurons tested, even though there was a uniform shift in behavior from nonfocused activity (e.g., locomotion, sniffing, and rearing) to focused stereotypy (e.g., a decline in locomotion accompanied by the appearance of intense head movements and occasional licking). It appears, therefore, that amphetamine does not elicit uniform increases in firing rate throughout the striatum, but rather that the neuronal response to this drug reflects a complex interaction of several factors, including ongoing behavior and dose.

To the extent that amphetamine produces divergent effects on striatal neurons, it is important to identify the underlying neuronal mechanisms, and a series of experiments has been initiated to explore this issue from several different perspectives. In one line of research, attempts are made to relate the response of striatal neurons during spontaneous movement to the direction and magnitude of the neuronal response to amphetamine. Data emerging from this work indicate that amphetamine-induced excitations occur almost exclusively in cells excited by motor behaviors, whereas nonmotor-related neurons typically are inhibited by the drug. Parallel experiments have focused on the role of dopaminergic mechanisms in the striatal response to amphetamine. Both types of studies are reviewed in this section — a subsequent section being devoted to an examination of the effects of amphetamine on an electrophysiologically identified subpopulation of striatal cells: projection neurons to SNR.

1. Motor- vs Nonmotor-Related Neurons

Ample evidence now indicates that the majority of spontaneously active striatal neurons change their firing rate in association with movement.[91,92,99-101] Typically, the response is manifest as a frank increase above the rate recorded during periods of quiet rest. Relatively few cells are inhibited during motor activity. The excitation usually appears as a burst of activity coinciding with specific behaviors (e.g., turning of the head), though tonic increases in firing rate beginning shortly before movement onset and ending shortly after movement offset also have been observed. Other striatal neurons fail to show any consistent change in firing rate with movement; some of these cells appear to change activity in response to arousing stimuli, but this effect has been observed among a subpopulation of motor-related neurons as well.[92,99,100] Both motor- and nonmotor-related cells are found throughout the anterior striatum, including dorsal and ventral regions,[92,100] although there may be a tendency for locomotor-related neurons to cluster in lateral areas.[91,99] In those studies in which the waveform and firing pattern of individual neurons have been analyzed, both motor- and nonmotor-related cells typically were found to display biphasic, positive-negative action potentials characteristic of Type II neurons showing a relatively slow and irregular spontaneous rate during quiet rest (0.5 to 8.0 spikes/sec).[92,99]

A series of studies have tested the hypothesis that striatal cells activated during movement are more likely to increase their firing rate to amphetamine than neurons showing activity unrelated to movement.[92,99-101] The results indicate that indeed neurons excited by movement continue to display excitation after systemic injection of 1.0 mg/kg d-amphetamine, whereas nonmotor-related cells usually are inhibited by the drug. Statistical analyses indicate that this difference in response direction is highly significant.[92,100] The magnitude of each response also is substantial: drug-induced increases routinely exceed the resting baseline rate by more than 200%, and inhibitions reach approximately 10% of the baseline rate.

Because amphetamine increases movement, it is conceivable that the drug-induced acceleration of firing rate is secondary to the change in behavior. An extensive analysis of firing rates during pre- and post-amphetamine behaviors matched for intensity and duration has revealed, however, a highly significant increase in neuronal activity following amphetamine administration.[92] Such a response to amphetamine when behavior is clamped or matched in this fashion suggests a primary drug effect rather than a secondary neuronal response to behavior. This conclusion is supported by comparisons of pre- and post-amphetamine firing rates during treadmill running, which also revealed an amphetamine-induced excitation of movement-related neuronal activity (see above). In fact, the uniform increases in firing rate to amphetamine that occur in animals performing a motor task can be explained by the selective inclusion of motor-related cells in the neuronal sample. Because these are the cells most likely excited by amphetamine, it is not surprising that this drug should elicit increases in firing rate in rats performing a motor task. Data from nonmotor-related cells, however, make it clear that amphetamine does not produce uniformly excitatory effects in the striatum of ambulant animals. Rather, amphetamine elicits both excitations and inhibitions of striatal activity, and the drug response that a given neuron is likely to display is strongly related to its motor-response characteristics.

The neurochemical mechanisms underlying these amphetamine-induced neuronal responses undoubtedly reflect the complex neuronal circuitry of the striatum, which includes both cortical and brainstem afferents as well as intrastriatal lateral inhibitory networks, all regulated by many different neurotransmitters and neuromodulators (see above). Dopamine, however, is known to be released by amphetamine in the striatum of freely moving animals,[40,102-104] and it seems likely, given the importance of this neurotransmitter in the amphetamine-induced behavioral response,[14,19,20,29] that dopamine contributes in some form to the neuronal changes produced by this drug. Unfortunately, a consensus has not yet emerged on the postsynaptic action of dopamine in the striatum. Both excitatory and inhibitory influences of dopamine have been postulated, and both postulates are supported by substantial experimental evidence.[41,48,49,105-110] Although many theories have been proposed to reconcile these opposing actions of dopamine[28,74,105,109,111-113] and many new lines of research have been attempted to address the problem,[114-123] some conflict still exists. The issue is complicated further by the fact that the large majority of data on the postsynaptic action of dopamine has been obtained from the stereotaxic preparation, which may not adequately represent the neuronal functions that obtain in a behavioral situation. An examination of the amphetamine response by itself, therefore, provides little useful information on dopamine effects in the striatum. Recently, however, the striatal response to amphetamine has been assessed in freely moving animals in which dopaminergic function has been impaired. This approach begins to address the issue more directly, and the results, summarized below, favor the view that dopamine participates in amphetamine-induced excitations, but cannot rule out dopaminergic mechanisms in the inhibitory response to this drug.

2. Manipulations of Dopaminergic Activity

Neuroleptic drugs are believed to attenuate the behavioral response to amphetamine by blocking dopamine receptors.[14,16,19,20,124] Thus, to the extent that dopamine contributes to the changes in striatal activity produced by amphetamine, neuroleptic drugs also should reverse these neuronal changes. This hypothesis has been tested with haloperidol, a classical neuroleptic that blocks virtually all components of the amphetamine behavioral response, and clozapine, an atypical neuroleptic that reverses relatively few amphetamine-induced behaviors.[125-127]

Neuronal activity was recorded from ambulant rats in an open-field situation.[92,101] The animals first received systemic d-amphetamine (1.0 mg/kg) followed 30 min later by a similar injection of either haloperidol (0.1 to 1.0 mg/kg) or clozapine (5.0 to 30.0 mg/kg). In some cases, animals received a second neuroleptic 30 min after the first. As expected, both neuroleptics altered amphetamine-induced changes in behavior and striatal neuronal activity, but Haracz et al.[92] found that these neuroleptics were far from equivalent. An analysis of behavior revealed that whereas haloperidol suppressed amphetamine-induced locomotion, rearing, sniffing, and head bobbing, clozapine attenuated only the rearing and sniffing produced by amphetamine and actually enhanced repetitive head movements. Moreoever, although clozapine did not exert a detectable change in standard ratings of amphetamine-induced locomotion, the expression of this behavior changed markedly following clozapine administration. A distinct locomotor ataxia, confined mainly to the hindlimbs and characterized by an unsteady gait and abdominal contact with the cage floor, appeared in clozapine-treated rats. Haloperidol, in contrast, suppressed movement in all limbs. Clozapine-induced disruptions of hindlimb function have been reported in other contexts as well,[128] suggesting a unique behavioral response that may serve as a predictor of the clinical effects of atypical neuroleptics.[92,128]

Despite these behavioral differences, both haloperidol and clozapine have been reported to reverse amphetamine-induced increases in the striatum of ambulant rats.[92,101] Each neuroleptic blocked this response and often caused unit activity to decline below the pre-amphetamine baseline rate. These findings support the view that dopamine plays a role in the excitatory effects of amphetamine on striatal neurons, a view consistent with preliminary data showing that locally applied dopamine elicits striatal excitations in freely moving animals.[65]

Histological analysis of data obtained from haloperidol-treated rats suggests a stronger effect in lateral striatal areas than in medial striatum, including nucleus accumbens.[92] It is of considerable interest that this medial-lateral response gradient parallels the density of D2 dopamine receptors: high in lateral, but low in medial striatal areas.[130-132] Thus, the relatively high affinity of haloperidol for this receptor implicates D2 blockade in the neuronal inhibition produced by this drug. It also is noteworthy that lateral areas are the primary striatal target of sensorimotor information from cortex.[28,57,91,133] Thus, a haloperidol-induced disruption of sensorimotor processing in lateral striatum may contribute to the motor side effects of this drug. Although clozapine-induced neuronal effects have not been analyzed for regional differences, this drug appears to reverse amphetamine-induced excitations throughout the striatum.[92]

A sharply different story emerges when haloperidol and clozapine are compared in terms of their ability to block the inhibitions produced by amphetamine. In this case, haloperidol further suppresses the activity of neurons already suppressed by amphetamine, whereas clozapine either partially or completely reverses this inhibitory response.[92,101] The failure of haloperidol to reverse amphetamine-induced inhibitions in striatum (and actually enhance this response) is surprising in view of the strong blockade that this neuroleptic exerts on the behavioral response to amphetamine. In fact, these neuronal data are clearly paradoxical in that haloperidol and amphetamine are known to exert functionally opposite effects on dopaminergic transmission. It seems likely, therefore, that apart from any involvement of dopamine, at least one of these drug-induced inhibitions requires the participation of nondopaminergic mechanisms, and not unlikely that both do.

Many such mechanisms are possible, and a review of striatal function suggests some intriguing ones. In the case of amphetamine, an inhibition may arise indirectly from stimulation of lateral inhibitory circuits linked to activated motor-related neurons. Indeed, activity in one area of striatum is known to cause an inhibition of firing rate in surrounding areas,[134] an effect most likely mediated by a dense network of intrastriatal, GABAergic collaterals.[58,135] Thus, to the extent that amphetamine exerts a more powerful excitatory effect on motor- than on nonmotor-related neurons, the latter cells would be vulnerable to GABA-mediated lateral inhibition, thereby masking any direct excitatory action of the drug. The supposition that GABAergic mechanisms can have this masking effect is supported by evidence that muscimol, a GABA agonist, blocks amphetamine-induced excitations in the striatum of ambulant rats.[94] Haloperidol, in contrast, may suppress striatal activity by an entirely different mechanism, including an interaction with sigma receptors. This neuroleptic has a high affinity for these receptors,[136] but their role in regulating striatal activity is not yet known. Although preliminary evidence indicates that BMY-14802, a selective sigma antagonist, can attenuate at least some amphetamine-induced excitations in the striatum of freely moving rats,[137] it remains to be established whether sigma receptors play a role in the inhibitory effects of haloperidol, especially the potentiation of the inhibitory response to amphetamine.

A more direct assessment of the dopaminergic contribution to the amphetamine neuronal response has been made in animals with neurotoxic lesions of striatal dopaminergic terminals. Such lesions, which depleted striatal dopamine in some cases by more than 75%, completely blocked the excitatory effects of amphetamine on striatal multiple-unit activity in freely moving rats.[138] Although multiple-unit records provide no information on any inhibitory responses to amphetamine that may occur (see above), it is significant that striatal dopamine appears to be necessary for amphetamine-induced excitations. These results do not rule out nondopaminergic mechanisms in this response, however, since bilateral removal of frontoparietal cortex, which caused a significant reduction in striatal glutamate, also has been reported to abolish the excitatory effects of amphetamine.[95] Thus, although dopamine may play a direct role in the neuronal excitations produced by this drug, it seems clear that the action of amphetamine on striatal neurons cannot be understood independently of other neuronal systems, including a major projection from cerebral cortex.

One way in which cortical afferents can influence the striatal response to amphetamine is to modulate the postsynaptic action of dopamine. In the stereotaxic preparation, striatal neurons have been reported to respond to direct applications of dopamine with either excitations or inhibitions (see above), but in the presence of glutamate, which is presumably released from corticostriatal terminals, dopamine potentiates glutamate-induced excitations.[111] Thus, dopamine may be capable of exerting either excitatory or inhibitory effects on striatal neurons, depending on the level of cortical activation they receive. This bidirectional action of dopamine may explain the opposing responses of striatal neurons to amphetamine in ambulant animals. If, for example, motor-related cells receive more cortical activation than nonmotor-related neurons, then amphetamine-induced dopamine release, either directly or indirectly, may be expected to excite the former and inhibit the latter. The hypothesis that dopamine may have such a divergent effect, proposed independently in a network model of catecholamine function,[139] is consistent with data presented in the following section on amphetamine-induced changes in the activity of striatal output neurons in behaving animals.

B. OUTPUT NEURONS

A major striatal efferent system projects to SNR, which exerts a significant level of control over various descending motor pathways. By antidromically activating the striatonigral projection in ambulant rats, Ryan et al.[87] were able to record from many of the striatal neurons that comprise this important output system and to assess their response to amphetamine. These cells fired infrequently during the pre-drug baseline period, even when the animals actively engaged in a variety of spontaneous behaviors. Many nonantidromically activated neurons also were recorded in this study, and they showed a similarly slow baseline firing rate. Following systemic amphetamine administration (0.25, 1.0, 2.5, or 5.0 mg/kg d-amphetamine), both types of neurons typically responded with an inhibition of activity, which was evident at all doses. Amphetamine-induced excitations were recorded in less than 25% of the neuronal sample.

These results appear to conflict with data summarized in the previous section, which indicate that although inhibitions occur with amphetamine, excitations are the predominant response. However, the unusually slow baseline firing rates reported by Ryan et al.[87] suggest that their amphetamine data were derived from a different population of neurons. The basal activity of these cells was considerably lower than the pre-drug firing rate of neurons in which amphetamine was reported to produce a majority of excitatory responses.[92,100,101] In fact, most of the excitatory responses reported by Ryan et al.[87] occurred among cells that fired most rapidly. It is conceivable, therefore, that amphetamine acts significantly in the striatum of freely moving animals to inhibit slowly firing cells and to excite neurons that have a relatively active basal rate. Such divergence may be inherent in the complex circuitry of the striatum, which allows rapid-firing neurons to inhibit their less active neighbors,[134] or in a bidirectional action of dopamine mediated by the level of cortical activation (see above).

The inability of Ryan et al.[87] to activate many rapid-firing cells with antidromic stimulation does not necessarily rule them out as SNR projection neurons. In fact, the anatomy (e.g., small diameter) and electrophysiology (e.g., slow conduction time) of these output cells may preclude antidromic activation when they are tonically active.[86,88] Thus, to the extent that both slow- and fast-firing cells are projection neurons, a further divergence in their firing rates produced by amphetamine may become a significant factor in the emergence of relatively invariant, stereotyped behaviors.

Further insight into the neuronal mechanisms underlying the behavioral effects of amphetamine can be gained by monitoring the response of SNR efferents. The activity of these neurons is controlled, in part, either directly or indirectly by the descending striatal projection,[74,105,140,141] and the information that SNR efferents convey to sites even farther downstream in the neuronal circuitry should reflect this striatal influence. Olds[142] reported that a subpopulation of SNR efferents respond to systemic amphetamine with increases in activity that closely parallel the drug-induced increase in motor behavior. This effect persisted even when the behavioral response was blocked by light anesthesia, arguing against a simple effect of behavioral feedback. A role for the striatum was confirmed in subsequent experiments in which intrastriatal applications of haloperidol significantly blocked both the behavioral response to amphetamine and the concomitant increase in SNR efferent activity.[143] Of course, interneurons and other neuronal connections intrinsic to the SNR may alter the amphetamine-induced striatal output signal before it reaches SNR efferents, but at present it seems safe to suggest that this signal exerts significant control over the output of SNR neurons during the amphetamine behavioral response.

IV. CONCLUSIONS

Although numerous issues remain to be addressed concerning the mechanisms of action of amphetamine in the striatum of ambulant animals, several conclusions can be drawn from the data at hand. First, amphetamine does not exert uniform effects throughout the striatum but rather excites or inhibits neurons depending, at least in part, on their motor-response characteristics. Neurons that increase activity in close association with movement further increase their rate following amphetamine administration, whereas cells showing nonmotor-related activity tend to be inhibited by the drug. Baseline firing rate also may determine the direction of the amphetamine-induced neuronal response as slowly firing cells appear more likely to respond with an inhibition than very active neurons. The net effect would be a divergence in firing rate that could exert a crucial influence on descending motor pathways and thus may play an important role in shaping the drug-induced behavioral response. Second, although dopaminergic mechanisms may contribute to both increases and decreases in neuronal activity produced by amphetamine, currently available evidence suggests that dopamine is likely to be involved either directly or indirectly in amphetamine-induced excitations. These increases in neuronal activity, for example, are blocked by dopamine antagonists as well as by destruction of striatal dopaminergic axon terminals. Dopaminergic mechanisms, however, do not exert exclusive control over this response since it can be modified by other manipulations, including anesthesia (see below). The mechanisms underlying amphetamine-induced inhibitions seem more difficult to identify since these responses are resistant to haloperidol, a potent D2 receptor antagonist, but not to clozapine, an atypical neuroleptic with a broad mechanism of action. Complex lateral inhibitory networks, which are known to operate in the striatum, perhaps via GABAergic axon collaterals or interneurons, may contribute to this inhibitory response.

Perhaps the most significant conclusion to be made from research on the striatal electrophysiology of amphetamine is that previous data obtained from the stereotaxic preparation may not accurately reflect the drug-induced neuronal changes that occur during behavior. As outlined in previous sections, amphetamine-induced effects on striatal neurons in ambulant animals differ markedly from the changes in neuronal activity produced by the drug in anesthetized or immobilized animals. This finding is supported by data showing that a motor-related neuron, which available evidence suggests should be excited by amphetamine, responds with an inhibition when the animal undergoes chloral hydrate anesthesia.[101] A state-dependent reversal of the neuronal response also applies to the neuroleptic drugs, which in the stereotaxic preparation routinely accelerate striatal activity,[85,144-146] but in freely moving animals are very effective in inhibiting amphetamine-induced excitations.[92,101,147] Taken together, these results suggest that striatal activity is controlled by factors that are highly sensitive to the state of the animal. One such factor is likely to be the level of afferent input. To the extent that dopamine either excites or inhibits striatal neurons as a function of the level of this input,[92] a bidirectional synaptic action of dopamine may be critically involved in the state-dependent effects of amphetamine and neuroleptics on striatal neurons activity. Further research on these and other synaptic mechanisms in the striatum of the freely moving preparation promises to shed new light on the neuronal systems and processes by which these drugs exert their behavioral effects.

ACKNOWLEDGMENT

Preparation of this chapter was supported, in part, by the National Institute on Drug Abuse (DA 02451). Faye Caylor provided expert secretarial assistance.

REFERENCES

1. **Angrist, B., Corwin, J., Bartlik, B., and Cooper, T.,** Early pharmacokinetics and clinical effects of oral D-amphetamine in normal subjects, *Biol. Psychiatry,* 22, 1357, 1987.
2. **Griffith, J. D., Cavanaugh, J., Held, J., and Oates, J. A.,** Dextroamphetamine: evaluation of psychotomimetic properties in man, *Arch. Gen. Psychiatry,* 26, 97, 1972.
3. **Lieberman, J. A., Kane, J. M., and Alvir, J.,** Provocative tests with psychostimulant drugs in schizophrenia, *Psychopharmacology,* 91, 415, 1987.
4. **Cho, A.,** Ice: a new dosage form of an old drug, *Science,* 249, 631, 1990.
5. **Angrist, B., Sathananthan, G., Wilk, S., and Gershon, S.,** Amphetamine psychosis: behavioral and biochemical aspects, *J. Psychiatr. Res.,* 11, 13, 1974.
6. **Bell, D. S.,** The experimental reproduction of amphetamine psychosis, *Arch. Gen. Psychiatry,* 29, 35, 1973.
7. **Ellinwood, E. H.,** Amphetamine psychosis: I. description of the individuals and process, *J. Nerv. Ment. Dis.,* 144, 273, 1967.
8. **Sato, M.,** Acute exacerbation of methamphetamine psychosis and lasting dopaminergic supersensitivity — a clinical survey, *Psychopharmacol. Bull.,* 22, 751, 1986.
9. **Schiorring, E.,** Psychopathology induced by "speed drugs," *Pharmacol. Biochem. Behav.,* 14 (Suppl. 1), 109, 1981.
10. **Ellinwood, E. H., Sudilovsky, A., and Nelson, L. M.,** Evolving behavior in the clinical and experimental amphetamine (model) psychosis, *Am. J. Psychiatry,* 130, 1088, 1973.
11. **Snyder, S. H.,** Amphetamine psychosis: a "model" schizophrenia mediated by catecholamines, *Am. J. Psychiatry,* 130, 61, 1973.
12. **Ellinwood, E. H. and Kilbey, M. M.,** Species differences in response to amphetamine, in *Psychopharmacogenetics,* Eleftheriou, B. E., Ed., Plenum Press, New York, 1975, 323.
13. **Randrup, A. and Munkvad, I.,** Biochemical, anatomical, and physiological investigations of stereotyped behavior induced by amphetamine, in *Amphetamines and Related Compounds,* Costa, E. and Garattini, S., Eds., Raven Press, New York, 1970, 695.
14. **Rebec, G. V. and Bashore, T. R.,** Critical issues in assessing the behavioral effects of amphetamine, *Neurosci. Biobehav. Rev.,* 8, 153, 1984.
15. **Ridley, R. M. and Baker, H. F.,** Stereotypy in monkeys and humans, *Psychol. Med.,* 12, 61, 1982.
16. **Segal, D. S. and Janowsky, D. S.,** Psychostimulant-induced behavioral effects: Possible models of schizophrenia, in *Psychopharmacology: A Generation of Progress,* Lipton, A., Di Mascio, A., and Killam, K. F., Eds., Raven Press, New York, 1978, 1113.
17. **Ellinwood, E. H. and Sudilovsky, A.,** Chronic amphetamine intoxication: behavioral model of psychosis, in *Psychopathology and Psychopharmacology,* Cole, J. O., Freedman, A. M., and Friedhoff, A. J., Eds., Johns Hopkins Press, Baltimore, MD, 1973, 51.
18. **Randrup, A. and Munkvad, I.,** Pharmacology and physiology of stereotyped behavior, *J. Psychiatr. Res.,* 11, 1, 1974.
19. **Kokkinidis, L. and Anisman, H.,** Amphetamine psychosis and schizophrenia: a dual model, *Neurosci. Biobehav. Rev.,* 5, 449, 1981.
20. **Robinson, T. E. and Becker, J. B.,** Enduring changes in brain and behavior produced by chronic amphetamine administration: a review and evaluation of animal models of amphetamine psychosis, *Brain Res. Rev.,* 11, 157, 1986.
21. **Angrist, B.,** Pharmacological models of schizophrenia, in *Handbook of Schizophrenia,* Vol. 2, *Neurochemistry and Neuropharmacology of Schizophrenia,* Henn, F. A. and DeLisi, L. E., Eds., Elsevier, Amsterdam, 1987, 391.
22. **Miller, R.,** The time course of neuroleptic therapy for psychosis: role of learning processes and implications for concepts of psychotic illness, *Psychopharmacology,* 92, 405, 1987.

23. **Schiorring, E.,** An open-field study of stereotyped locomotor activity in amphetamine-treated rats, *Psychopharmacology (Berlin),* 66, 281, 1979.
24. **Mueller, K., Hollingsworth, E. M., and Cross, D. R.,** Another look at amphetamine-induced stereotyped locomotor activity in rats using a new statistic to measure locomotor stereotypy, *Psychopharmacology,* 97, 74, 1989.
25. **Heimer, L., Alheid, G. F., and Zaborszky, L.,** Basal ganglia, in *The Rat Nervous System,* Vol. 1, Paxinos, G., Ed., Academic Press, San Diego, 1985, 37.
26. **Lidsky, T. I., Manetto, C., and Schneider, J. S.,** A consideration of sensory factors involved in motor functions of the basal ganglia, *Brain Res. Rev.,* 9, 133, 1985.
27. **Penney, J. B., Jr. and Young, A. B.,** Speculations on the functional anatomy of basal ganglia disorders, *Annu. Rev. Neurosci.,* 6, 73, 1983.
28. **Schneider, J. S.,** Basal ganglia-motor influences: role of sensory gating, in *Basal Ganglia and Behavior: Sensory Aspects of Motor Functioning,* Schneider, J. S. and Lidsky, T. I., Eds., Huber, Stuttgart, 1987, 103.
29. **Groves, P. M. and Rebec, G. V.,** Biochemistry and behavior: some central actions of amphetamine and antipsychotic drugs, *Annu. Rev. Psychol.,* 27, 91, 1976.
30. **Iverson, S. D.,** Neural substrates mediating amphetamine responses, in *Cocaine and Other Stimulants,* Ellinwood, E. H. and Kilbey, M. M., Eds., Plenum Press, New York, 1977, 31.
31. **Evenden, J. L. and Ryan, C. N.,** Behavioral responses to psychomotor stimulant drugs: localization in the central nervous system, *Pharmacol. Ther.,* 36, 151, 1988.
32. **Mogenson, G. J., Jones, D. L., and Yim, C. Y.,** From motivation to action: functional interface between the limbic system and the motor system, *Prog. Neurobiol.,* 14, 69, 1980.
33. **Haber, S. N., Groenewegen, H. J., Grove, E. A., and Nauta, W. J. H.,** Efferent connections of the ventral pallidum: evidence of a dual striato pallidofugal pathway, *J. Comp. Neurol.,* 235, 322, 1985.
34. **Oades, R. D. and Halliday, G. M.,** Ventral tegmental (A10) system: neurobiology. 1. Anatomy and connectivity, *Brain Res. Rev.,* 12, 117, 1987.
35. **Rebec, G. V.,** Electrophysiological pharmacology of amphetamine, in *Neurobiology of Drug Abuse,* Marwah, J., Ed., S. Karger, Basal, 1987, 1.
36. **Rebec, G. V. and Segal, D. S.,** Dose-dependent biphasic alterations in the spontaneous activity of neurons in the rat neostriatum produced by d-amphetamine and methylphenidate, *Brain Res.,* 150, 353, 1978.
37. **Rebec, G. V. and Zimmerman, K. S.,** Opposite effects of d-amphetamine on spontaneous neuronal activity in the neostriatum and nucleus accumbens, *Brain Res.,* 201, 485, 1980.
38. **Rebec, G. V. and Curtis, S. D.,** Reciprocal changes in the firing rate of neostriatal and dorsal raphe neurons following local infusions or systemic injections of d-amphetamine: evidence for neostriatal heterogeneity, *J. Neurosci.,* 3, 2240, 1983.
39. **Kuczenski, R.,** Biochemical actions of amphetamine and other stimulants, in *Stimulants: Neurochemical, Behavioral, and Clinical Perspectives,* Creese, I., Ed., Raven Press, New York, 1983, 31.
40. **Rebec, G. V.,** Changes in brain and behavior produced by amphetamine: a perspective based on microdialysis, voltammetry, and single-unit electrophysiology in freely moving animals, in *Biochemistry and Physiology of Substance Abuse,* Vol. III, Watson, R. R., Ed., CRC Press, Boca Raton, FL, 1991, 93.
41. **Moore, R. Y. and Bloom, F. E.,** Central catecholamine neuron systems: anatomy and physiology of the dopamine systems, *Annu. Rev. Neurosci.,* 1, 129, 1978.
42. **Faull, R. L. M. and Mehler, W. R.,** The cells of origin of nigrotectal, nigrothalamic and nigrostriatal projections in the rat, *Neuroscience,* 3, 989, 1978.
43. **Fallon, J. H. and Moore, R. V.,** Catecholamine innervation of the basal forebrain. IV. Topography of the dopamine projection to the basal forebrain and neostriatum, *J. Comp. Neurol.,* 180, 545, 1978.

44. **Ungerstedt, U.,** Stereotaxic mapping of the monoamine pathways in the rat brain, *Acta Physiol. Scand.,* (Suppl. 367), 1, 1971.
45. **Agnati, L. F. and Fuxe, K.,** Computer assisted morphometry and microdensitometry of transmitter identified neurons with special reference to the mesostriatal dopamine pathway, *Acta Physiol. Scand.,* (Suppl. 532), 1, 1985.
46. **Groves, P. M., Wilson, C. J., Young, S. J., and Rebec, G. V.,** Self-inhibition by dopaminergic neurons: an alternative to the ''neuronal feedback loop'' hypothesis for the mode of action of certain psychotropic drugs, *Science,* 190, 522, 1975.
47. **Groves, P. M., Staunton, D. A., Wilson, C. J., and Young, S. J.,** Sites of action of amphetamine intrinsic to catecholaminergic nuclei: catecholaminergic presynaptic dendrites and axon terminals, *Prog. Neuropsychopharmacol.,* 3, 325, 1979.
48. **Chiodo, L. A.,** Dopamine-containing neurons in the mammalian central nervous system: electrophysiology and pharmacology, *Neurosci. Biobehav. Rev.,* 12, 49, 1988.
49. **Grace, A. A.,** The regulation of dopamine neuron activity as determined by in vivo and in vitro intracellular recordings, in *Neurophysiology of Dopaminergic Systems — Current Status and Clinical Perspectives,* Chioco, L. A. and Freeman, A. S., Eds., Lakeshore, Detroit, MI, 1987, 1.
50. **Cheramy, A., Leviel, V., and Glowinski, J.,** Dendritic release of dopamine in the substantia nigra, *Nature (London),* 289, 537, 1981.
51. **Wang, R. Y.,** Dopaminergic neurons in the rat ventral tegmental area. III. Effects of D- and L-amphetamine, *Brain Res. Rev.,* 3, 153, 1981.
52. **Tessel, R. E.,** Noradrenergic processes in the behavioral actions of psychomotor stimulants, *Drug Develop. Res.,* 20, 359, 1990.
53. **Fuxe, K. and Ungerstedt, U.,** Histochemical, biochemical and functional studies on central monoamine neurons after acute and chronic amphetamine administration, in *Amphetamines and Related Compounds,* Costa, E. and Garatini, S., Eds., Raven Press, New York, 1970, 257.
54. **Kemp, J. M. and Powell, T. P. S.,** The site of termination of afferent fibres in the caudate nucleus, *Philos. Trans. R. Soc. London,* B., 262, 413, 1971.
55. **Kitai, S. T., Kocsis, J. D., and Wood, J.,** Origin and characteristics of the cortico-caudate afferents: an anatomical and electrophysiological study, *Brain Res.,* 118, 137, 1976.
56. **Wilson, C. J., Chang, H. T., and Kitai, S. T.,** Origins of post synaptic potentials evoked in spiny neostriatal projection neurons by thalamic stimulation in the rat, *Exp. Brain Res.,* 51, 217, 1983.
57. **Grofova, I.,** Extrinsic connections of the neostriatum, in *The Neostriatum,* Divac, I. and Obert, R. G. E., Eds., Pergamon Press, Oxford, 1979, 37.
58. **Park, M. R., Lighthall, J. W., and Kitai, S. T.,** Recurrent inhibition in the rat neostriatum, *Brain Res.,* 194, 359, 1980.
59. **Jimenez-Castellanos, J. and Graybiel, A. M.,** Compartmental origins of striatal efferent projections in the cat, *Neuroscience,* 32 (2), 297, 1989.
60. **Graybiel, A. M.,** Neurotransmitters and neuromodulators in the basal ganglia, *Trends Neurosci.,* 13 (7), 244, 1990.
61. **Gerfen, C. R.,** The neostriatal mosaic: striatal patch-matrix organization is related to cortical lamination, *Science,* 246, 385, 1989.
62. **Loopuijt, L. D. and Van Der Kooy, D.,** Organization of the striatum: collateralization of its efferent axons, *Brain Res.,* 348, 86, 1985.
63. **Wilson, C. J. and Groves, P. M.,** Fine structure and synaptic connections of the common spiny neuron of the rat neostriatum: a study employing intracellular injection of horseradish peroxidase, *J. Comp. Neurol.,* 194, 599, 1980.
64. **West, M. O., Chapin, J. K., and Woodward, D. J.,** Drug actions on neural circuits in awake, behaving rats: new techniques and strategies, in *Modern Methods in Pharmacology,* Vol. 3, *Electrophysiological Techniques in Pharmacology,* Geller, H. M., Ed., Alan R. Liss, New York, 1986, 35.

65. **Shimizu, N., West, M. O., Lee, R.-S., Chapin, J. K., and Woodward, D. J.,** Dopaminergic involvement in striatal neuronal activity during locomotor behavior and its enhancement by cocaine, *Soc. Neurosci. Abstr.,* 13, 979, 1987.
66. **Graybiel, A. M. and Ragsdale, C. W., Jr.,** Biochemical anatomy of the striatum, in *Chemical Neuroanatomy,* Emson, P. C., Ed., Raven Press, New York, 1983, 427.
67. **Haber, S. N.,** Neurotransmitters in the human and nonhuman primate basal ganglia, *Human Neurobiol.,* 5, 159, 1986.
68. **Widmann, R. and Sperk, G.,** Topographical distribution of amines and major amine metabolites in the rat striatum, *Brain Res.,* 367, 244, 1986.
69. **Gerfen, C. R.,** The neostriatal mosaic: the reiterated processing unit, in *Neurotransmitter Interactions in the Basal Ganglia,* Sandler, M., et al., Eds., Raven Press, New York, 1987, 19.
70. **Beal, M. F. and Martin, J. B.,** Topographical dopamine and serotonin distribution and turnover in rat striatum, *Brain Res.,* 358, 10, 1985.
71. **Kemel, M.-L., Desban, M., Glowinski, J., and Gauchy, C.,** Distinct presynaptic control of dopamine release in striosomal and matrix areas of the cat caudate nucleus, *Proc. Natl. Acad. Sci. U.S.A.,* 86, 9006, 1989.
72. **Basse-Tomusk, A. and Rebec, G. V.,** Regional distribution of ascorbate and 3,4-dihydroxyphenylacetic acid (DOPAC) in rat striatum, *Brain Res.,* 538, 29, 1991.
73. **Dray, A.,** The physiology and pharmacology of mammalian basal ganglia, *Prog. Neurobiol.,* 14, 221, 1980.
74. **Groves, P. M.,** A theory of the functional organization of the neostriatum and the neostriatal control of voluntary movement, *Brain Res. Rev.,* 5, 109, 1983.
75. **Deadwyler, S. A.,** Electrophysiological investigations of drug influences in the behaving animal, in *Modern Methods in Pharmacology,* Vol. 3, *Electrophysiological Techniques in Pharmacology,* Geller, H. M., Ed., Alan R. Liss, New York, 1986, 1.
76. **Lemon, R.,** *Methods for Neuronal Recording in Conscious Animals,* John Wiley & Sons, New York, 1984.
77. **Pasik, P., Pasik, T., and DiFiglia, M.,** The internal organization of the neostriatum in mammals, in *The Neostriatum,* Divac, I. and Obert, R. G. E., Eds., Pergamon Press, Oxford, 1979, 5.
78. **Chang, H. T., Wilson, C. J., and Kitai, S. T.,** A golgi study of rat neostriatal neurons: light microscopic analysis, *J. Comp. Neurol.,* 208, 107, 1982.
79. **Smith, A. D. and Bolam, J. P.,** The neural network of the basal ganglia as revealed by the study of synaptic connections of identified neurones, *Trends Neurosci.,* 13 (7), 259, 1990.
80. **Nisenbaum, E. S., Orr, W. B., and Berger, T. W.,** Evidence for two functionally distinct subpopulations of neurons within the rat striatum, *J. Neurosci.,* 8, 4138, 1988.
81. **Skirboll, L. R. and Bunney, B. S.,** The effects of acute and chronic haloperidol treatment on spontaneously firing neurons in the caudate nucleus of the rat, *Life Sci.,* 25, 1419, 1979.
82. **Hu, X. T. and Wang, R. Y.,** Comparison of effects of D1 and D2 dopamine receptor agonists on neurons in the rat caudate putamen: an electrophysiological study, *J. Neurosci.,* 8, 4340, 1988.
83. **Wilson, C. J. and Groves, P. M.,** Spontaneous firing patterns of identified spiny neurons in the rat neostriatum, *Brain Res.,* 220, 67, 1981.
84. **Hu, X. T., Wachtel, S. R., Galloway, M. P., and White, F. J.,** Lesions of the nigrostriatal dopamine projection increase the inhibitory effects of D1 and D2 dopamine agonists on caudate-putamen neurons and relieve D2 receptors from the necessity of D1 receptor stimulation, *J. Neurosci.,* 10, 2318, 1990.
85. **Nisenbaum, E. S., Stricker, E. M., Zigmond, M. J., and Berger, T. W.,** Long-term effects of dopamine-depleting brain lesions on spontaneous activity of Type II striatal neurons: relation to behavioral recovery, *Brain Res.,* 398, 221, 1986.

86. **Ryan, L. J., Young, S. J., and Groves, P. M.,** Substantia nigra stimulation evoked antidromic responses in rat neostriatum, *Exp. Brain Res.,* 63, 449, 1986.
87. **Ryan, L. J., Young, S. J., Segal, D. S., and Groves, P. M.,** Antidromically identified striatonigral projection neurons in the chronically implanted behaving rat: relations of cell firing to amphetamine-induced behaviors, *Behav. Neurosci.,* 103, 3, 1989.
88. **Kocsis, J. D. and VanderMaelen, C. P.,** A supernormal period in central axons following single cell stimulation, *Exp. Brain Res.,* 36, 381, 1979.
89. **Steriade, M. and Deschenes, M.,** The thalamus as a neuronal oscillator, *Brain Res. Rev.,* 8, 1, 1984.
90. **Ranck, J. B., Jr., Kubie, J. L., Fox, S. E., Walfson, S., and Muller, R. L.,** Single-unit recording in behaving animals: bridging the gap between neural events and sensory behavioral variables, in *Behavioral Contributions to Brain Research,* Robinson, T., Ed., Oxford University Press, Oxford, 1982, 130.
91. **West, M. O., Michael, A. J., Knowles, S. E., Chapin, J. K., and Woodward, D. J.,** Striatal unit activity and the linkage between sensory and motor events, in *Basal Ganglia and Behavior: Sensory Aspects of Motor Functioning,* Schneider, J. S. and Lidsky, T. I., Eds., Hans Huber, Toronto, 1987, 27.
92. **Haracz, J. L., Tschanz, J. T., Wang, Z., White, I. M., and Rebec, G. V.,** Effects of amphetamine, haloperidol, and clozapine on behavior and striatal single-unit activity in freely moving rats, *Neuroscience,* in press.
93. **Hansen, E. L. and McKenzie, G. M.,** Dexamphetamine increases in striatal neuronal firing in freely moving rats, *Neuropharmacology,* 18, 547, 1979.
94. **McKenzie, G. M. and Hansen, E. L.,** GABA agonists dissociate striatal unit activity from drug-induced stereotyped behaviour, *Neuropharmacology,* 19, 957, 1980.
95. **Warenycia, M. W., McKenzie, G. M., Murphy, M., and Szerb, J. C.,** The effects of cortical ablation on multiple unit activity in the striatum following dexamphetamine, *Neuropharmacology,* 26 (8), 1107, 1987.
96. **Trulson, M. E. and Jacobs, B. L.,** Effects of D-amphetamine on striatal unit activity and behavior in freely moving cats, *Neuropharmacology,* 18, 735, 1979.
97. **Warenycia, M. W. and McKenzie, G. M.,** Immobilization of rats modifies the response of striatal neurons to dexamphetamine, *Pharmacol. Biochem. Behav.,* 21, 53, 1984.
98. **West, M., Carelli, R., Shimizu, N., Chapin, J., and Woodward, D.,** Single unit correlations with specific movements in lateral striatum of freely moving rats, *Soc. Neurosci. Abstr.,* 13, 979, 1987.
99. **Gardiner, T. W., Iverson, D. A., and Rebec, G. V.,** Heteregeneous responses of neostriatal neurons to amphetamine in freely moving rats, *Brain Res.,* 463, 268, 1988.
100. **Haracz, J. L., Tschanz, J. T., Greenberg, J., and Rebec, G. V.,** Amphetamine-induced excitations predominate in single neostriatal neurons showing motor-related activity, *Brain Res.,* 489, 365, 1989.
101. **Rebec, G. V., Haracz, J. L., Tschanz, J. T., Wang, Z., and White, I.,** Responses of motor- and nonmotor-related neostriatal neurons to amphetamine and neuroleptic drugs, in *Basal Ganglia III,* Bernardi, G., Ed., Plenum Press, New York, 1990, 447.
102. **Kuczenski, R. and Segal, D.,** Concomitant characterization of behavioral and striatal neurotransmitter response to amphetamine using in vivo microdialysis, *J. Neurosci.,* 9, 2051, 1989.
103. **Sharp, T., Zetterstrom, T., Ljungberg, T., and Ungerstedt, U.,** A direct comparison of amphetamine-induced behaviours and regional brain dopamine release in the rat using intracerebral dialysis, *Brain Res.,* 402, 322, 1987.
104. **Robinson, T. E., Jurson, P. A., Bennett, J. A., and Bentgen, K. M.,** Persistent sensitization of dopamine neurotransmission in ventral striatum (nucleus accumbens) produced by prior experience with (+)-amphetamine: a microdialysis study in freely moving rats, *Brain Res.,* 462, 211, 1988.

105. **Bunney, B. S., Sesack, S. R., and Silva, N. L.,** Midbrain dopaminergic systems: neurophysiology and electrophysiological pharmacology, in *Psychopharmacology: The Third Generation of Progress,* Meltzer, H. Y., Ed., Raven Press, New York, 1987, 113.
106. **Feger, J., Deniau, J. M., deChamplain, J., and Feltz, P.,** A survey of electrophysiology and pharmacology of neostriatal input-output relations, in *The Neostriatum,* Divac, I. and Oberg, R. G. E., Eds., Pergamon Press, Oxford, 1979, 71.
107. **Herrling, P. L. and Hull, C. D.,** Iontophoretically applied dopamine depolarizes and hyperpolorizes the membrane of cat caudate neurons, *Brain Res.,* 192, 441, 1980.
108. **Kitai, S. T., Preston, R. J., Bishop, G. A., and Kocsis, J. D.,** Striatal projection neurons: morphological and electrophysiological studies, in *Advances in Neurology,* Vol. 24, Poirier, L. J., Sourkes, T. L., and Bedard, P. J., Eds., Raven Press, New York, 1979, 45.
109. **Siggins, G. R.,** Electrophysiological role of dopamine in striatum: excitatory or inhibitory? in *Psychopharmacology: A Generation of Progress,* Lipton, M. A., Di Mascio, A., and Killam, K. F., Eds., Raven Press, New York, 1978, 143.
110. **Connor, J. D.,** Electrophysiology of the nigro-caudate dopamine pathway, *Pharmacol. Ther. B.,* 1 (3), 357, 1975.
111. **Chiodo, L. A. and Berger, T. W.,** Interactions between dopamine and amino acid-induced excitation and inhibition in the striatum, *Brain Res.,* 375, 198, 1986.
112. **Johnson, S. W., Palmer, M. R., and Freedman, R.,** Effects of dopamine on spontaneous and evoked activity of caudate neurons, *Neuropharmacology,* 22 (7), 843, 1983.
113. **Schoener, E. P. and Elkins, D. P.,** Neuronal response to dopamine in rat neostriatum, *Neuropharmacology,* 23 (6), 611, 1984.
114. **Abercrombie, E. D. and Jacobs, B. L.,** Dopaminergic modulation of sensory responses of striatal neurons: single unit studies, *Brain Res.,* 358, 27, 1985.
115. **Akaike, A., Ohno, Y., Sasa, M., and Takaori, S.,** Excitatory and inhibitory effects of dopamine on neuronal activity of the caudate nucleus neurons in vitro, *Brain Res.,* 418, 262, 1987.
116. **Boyajian, C. L., Bickford-Wimer, P., Kim, M. B., Freedman, R., and Cooper, D. M. F.,** Pertussis toxin lesioning of the nucleus caudate-putamen attentuates adenylate cyclase inhibition and alters neuronal electrophysiological activity, *Brain Res.,* 495, 66, 1989.
117. **Calabresi, P., Benedetti, M., Mercuri, N. B., and Bernardi, G.,** Endogenous dopamine and dopaminergic agonists modulate synaptic excitation in neostriatum: intracellular studies from naive and catecholamine-depleted rats, *Neuroscience,* 27 (1), 145, 1988.
118. **Calabresi, P., Mercuri, N. B., Stefani, A., and Bernardi, G.,** Synaptic intrinsic control of membrane excitability of neostriatal neurons. I. An *in vivo* analysis, *J. Neurophysiol.,* 63 (4), 651, 1990.
119. **Calabresi, P., Mercuri, N. B., and Bernardi, G.,** Synaptic and intrinsic control of membrane excitability of neostriatal neurons. II. An *in vitro* analysis, *J. Neurophysiol.,* 63 (4), 663, 1990.
120. **Rutherford, A., Garcia-Munoz, M., and Arbuthnott, G. W.,** An afterhyperpolarization recorded in striatal cells *"in vitro": effect of dopamine administration, *Exp. Brain Res.,* 71, 399, 1988.
121. **Mercuri, N., Bernardi, G., Calabresi, P., Cotugno, A., Levi, G., and Stanzione, P.,** Dopamine decreases cell excitability in rat striatal neurons by pre- and postsynaptic mechanisms, *Brain Res.,* 358, 110, 1985.
122. **Orr, W. B., Stricker, E. M., Zigmond, M. J., and Berger, T. W.,** Effects of dopamine depletion on the spontaneous activity of type I striatal neurons: relation to local dopamine concentration and motor behavior, *Synapse,* 1, 461, 1987.
123. **Rolls, E. T., Thorpe, S. J., Boytim, M., Szabo, I., and Perrett, D. I.,** Responses of striatal neurons in the behaving monkey. 3. Effects of iontophoretically applied dopamine on normal responsiveness, *Neuroscience,* 12, 1201, 1984.

124. **Rebec, G. V. and Anderson, G. D.,** Regional neuropharmacology of the antipsychotgic drugs: implications for the dopamine hypothesis of schizophrenia, *Behav. Assess.*, 8, 11, 1986.
125. **Ljungberg, T. and Ungerstedt, U.,** A rapid and simple behavioral screening method for simultaneous assessment of limbic and striatal blocking effects of neuroleptic drugs, *Pharmacol. Biochem. Behav.*, 23, 479, 1985.
126. **Robertson, A. and MacDonald, C.,** Atypical neuroleptics clozapine and thioridazine enhance amphetamine-induced stereotypy, *Pharmacol. Biochem. Behav.*, 21, 97, 1984.
127. **Tschanz, J. T. and Rebec, G. V.,** Atypical antipsychotic drugs block selective components of amphetamine-induced stereotypy, *Pharmacol. Biochem. Behav.*, 31, 519, 1989.
128. **Ellenbroek, B. and Cools, A. R.,** The paw test: an animal model for neuroleptic drugs which fulfills the criteria for pharmacological isomorphism, *Life Sci.*, 42, 1205, 1988.
129. **Yamamoto, B. K. and Pehek, E. A.,** A neurochemical heterogeneity of the rat striatum as measured by in vivo electrochemistry and microdialysis, *Brain Res.*, 506, 236, 1990.
130. **Joyce, J. N. and Marshall, J. F.,** Striatal topography of D-2 receptors correlates with indexes of cholineric neuron localization, *Neurosci. Lett.*, 53, 127, 1985.
131. **Farde, L., Wiesel, F.-A., Stone-Elander, S., Halldin, C., Nordstrom, A.-L., Hal, I. H., and Sedvall, G.,** D_2 Dopamine receptors in neuroleptic-naive schizophrenic patients: a positron emission tomography study with [^{11}C]raclopride, *Arch. Gen. Psychiat.*, 47, 213, 1990.
132. **Loopuijt, L. D.,** Distribution of dopamine D-2 receptors in the rat striatal complex and its comparison with acetylcholinesterase, *Brain Res. Bull.*, 22, 805, 1989.
133. **Dube, L., Smith, A. D., and Bolam, J. P.,** Identification of synaptic terminals of thalamic or cortical origin in contact with distinct medium-size spiny neurons in the rat neostriatum, *J. Comp. Neurol.*, 267, 455, 1988.
134. **Rebec, G. V. and Curtis, S. D.,** Reciprocal zones of excitation and inhibition in the neostriatum, *Synapse*, 2, 633, 1988.
135. **Katayama, Y., Miyazaki, S., and Tsubokawa, T.,** Electrophysiological evidence favoring intracaudate axon collaterals of GABAergic caudate output neurons in the cat, *Brain Res.*, 216, 180, 1981.
136. **Taylor, D. P. and Dekleva, J.,** Potential antipsychotic BMY-14802 selectively binds to sigma sites, *Drug Develop. Res.*, 11, 65, 1987.
137. **Rebec, G. V., Wang, Z., and Haracz, J. L.,** BMY-14802 attenuates amphetamine-induced excitations in the neostriatum of freely moving rats, *Soc. Neurosci. Abstr.*, in press.
138. **Warenycia, M. W. and McKenzie, G. M.,** Activation of striatal neurons by dexamphetamine is antagonized by degeneration of striatal dopaminergic terminals, *J. Neural Transm.*, 70, 217, 1987.
139. **Servan-Schreiber, D., Printz, H., and Cohen, J. D.,** A network model of catecholamine effects: gain, signal-to-noise ratio, and behavior, *Science*, 249, 892, 1990.
140. **Dray, A.,** The striatum and the substantia nigra: a commentary on their relationships, *Neuroscience*, 4, 1407, 1979.
141. **Scarnati, E., Campana, E., and Pacitti, C.,** The functional role of the nucleus accumbens in the control of the substantia nigra: electrophysiological investigations in intact and striatum-globus pallidus lesioned rats, *Brain Res.*, 265, 249, 1983.
142. **Olds, M. E.,** Amphetamine-induced increase in motor activity is correlated with higher firing rates of non-dopamine neurons in substantia nigra and ventral tegmental area, *Neuroscience*, 24 (2), 477, 1988.
143. **Olds, M. E.,** The response of non-dopamine neurons in substantia nigra and ventral tegmental area to amphetamine and apomorphine during hypermotility: the striatal influence, *Brain Res.*, 452, 237, 1988.

144. **Rebec, G. V., Bashore, T. R., Zimmerman, K. S., and Alloway, K. S.,** "Classical" and "atypical" antipsychotic drugs: differential antagonism of amphetamine- and apomorphine-induced alterations of spontaneous neuronal activity in the neostriatum and nucleus accumbens, *Pharmacol. Biochem. Behav.,* 11, 529, 1979.
145. **Rebec, G. V., Bashore, T. R., Zimmerman, K. S., and Alloway, K. D.,** Neostriatal and mesolimbic neurons: dose-dependent effects of clozapine, *Neuropharmacology,* 19, 281, 1980.
146. **Napier, T. C., Coyle, S., and Breese, G. R.,** Ontogeny of striatal unit activity and effects of single or repeated haloperidol administration in rats, *Brain Res.,* 333, 35, 1985.
147. **Warenycia, M. W. and McKenzie, G. M.,** Pharmacological activity profiles of dopamine D-1 and D-2 receptor agonists and antagonists on striatal neuronal activity and the response to dexamphetamine in freely moving rats, *Gen. Pharmacol.,* 20 (3), 295, 1989.

Chapter 9

NEUROCHEMISTRY OF THE BENZODIAZEPINES

Lawrence G. Miller and David J. Greenblatt

TABLE OF CONTENTS

I. INTRODUCTION

Benzodiazepines have been in clinical use since the early 1960s. This class of drugs has been demonstrated to be effective in the treatment of anxiety and insomnia, and benzodiazepines have largely replaced barbiturates and other anxiolytic/hypnotic agents because they are less toxic and less likely to be abused. Nonetheless, benzodiazepines are abused, and the widespread use of this class of drugs suggests that abuse may be relatively common. Several recent reviews have addressed clinical and behavioral components of benzodiazepine abuse.[1-5] This chapter will address the neurochemistry of the benzodiazepines, and in particular, how it relates to their abuse.

II. THE BENZODIAZEPINE RECEPTOR

In 1977, two groups of investigators identified a specific benzodiazepine receptor in brain.[6,7] Subsequent electrophysiological and neurochemical studies established that the benzodiazepine receptor was coupled to a GABA-gated chloride channel, usually termined the $GABA_A$ receptor.[8] Benzodiazepines do not act directly, but rather are allosterically coupled to the GABA site; benzodiazepine binding promotes GABA binding which, in turn, promotes chloride flux. Chloride influx then leads to neuronal hyperpolarization, with an inhibitory effect on individual neurons. It should be pointed out that, depending upon the location of a neuron in its network, the net result might be functional inhibition or excitation.

As an allosterically coupled receptor, the benzodiazepine site allows an unusual range of modulation of GABA. Benzodiazepine agonists (GABA-positive) promote GABA binding and chloride influx; antagonists (GABA-neutral) bind to the benzodiazepine site but have little or no effect on GABA binding and chloride flux; inverse agonists (GABA-negative) inhibit GABA binding and thus promote chloride efflux. Binding studies indicate that all three types of benzodiazepine ligands bind to the benzodiazepine receptor, but that the binding domains may be somewhat different.

During the past several years, the benzodiazepine receptor has been studied as a component of the $GABA_A$ receptor. Biochemical studies indicate that the purified $GABA_A$ receptor in mammalian brain has two major subunits with an apparent size of 53 kDa (α) and 56 kDa (β).[9] The overall molecular weight is estimated to be 220 to 355 kDa based on these studies,[10] and 220 to 400 kDa based on irradiation studies.[11] Photoaffinity labeling indicates that the benzodiazepine site is located on the α-subunit, with the GABA site on the β-subunit, although it is possible that some cross-binding exists.[12,13] These data suggest that the functional $GABA_A$ receptor binding both GABA and benzodiazepines is a hetero-oligomer. Although the apparent molecular weight of the receptor suggests that the receptor includes at least four subunits, the oligomeric structure remains uncertain.

Substantial additional information has been obtained from the study of deduced sequences from cloned $GABA_A$ receptor genes. This topic has been recently reviewed in detail.[14] Initial studies isolated two cDNAs termed alpha and beta.[15] Sequences of these subunits established that the $GABA_A$ receptor was a member of the ligand-gated ion channel family of receptors. This family is exemplified by the nicotinic acetylcholinergic receptor and includes the strychnine-sensitive glycine receptor. The common structure is that of a hetero-oligomeric complex, with multiple membrane-spanning regions forming the walls of the ion channel. The neurotransmitter binding site is present on the same protein complex.

Subsequent studies soon demonstrated that α- and β-subunits reconstituted in *Xenopus oocytes* exhibited some, but not all, characteristics of the native receptor protein (e.g., Levitan et al.[16]). In particular, benzodiazepine potentiation of GABA effects was not reliably present. Concurrently, additional α- and β-subunits were demonstrated in bovine and other species, indicating potentially diverse receptor subunit combinations. Anatomical studies indicate marked variations in distributions of mRNAs based on *in situ* hybridization.[16a] For example, the α_1-subunit is widely distributed in cortex and cerebellum, with substantial levels in hippocampus. In contrast, the α_2-subunit is less abundant with moderate levels only in hippocampus. Between classes of subunits, the α and β mRNAs also have different distributions.

Identification of a third class of subunits, γ, together with expression studies suggest that this class is necessary to benzodiazepine modulation of GABA binding.[17] The mechanism for this effect remains uncertain. Further, the deduced γ polypeptide has not yet been identified in purified receptor preparations. However, support for the role of γ-subunits in benzodiazepine binding is provided by expression studies indicating altered benzodiazepine binding characteristics in cells expressing α_1- as compared to α_3-subunits.[18] In addition, a fourth class, δ, has been identified, but its role in receptor function remains uncertain.

In summary, data from biochemical and molecular biological studies establish the existence of multiple $GABA_A$ receptor mRNAs, and make it highly likely that multiple $GABA_A$ receptors exist. Benzodiazepine binding is likely to vary among receptor subtypes, apparently due to differences in α- and, perhaps, γ-subunits. Differential anatomical distribution of receptor subunits provides support for the role of specific receptor subtypes, and perhaps specific benzodiazepine ligands, in different functional effects of benzodiazepines.

It should also be noted that substantial evidence exists in support of endogenous benzodiazepine ligands.[19] These include a polypeptide (diazepam-binding inhibitor), a benzodiazepine agonist (desmethyldiazepam), and an inverse agonist. The contributions of these compounds to benzodiazepine actions remain uncertain.

III. BENZODIAZEPINE ABUSE

Neurochemical studies of benzodiazepine abuse have focused largely on benzodiazepine dependence, and, in turn, on benzodiazepine withdrawal as an indicator of dependence. As such, these studies involve effects of chronic benzodiazepine administration. Although the majority of neurochemical studies of benzodiazepines involved acute exposure, a growing literature has addressed chronic drug effects. Most of the recent studies indicate alterations in the $GABA_A$ receptor associated with chronic benzodiazepine administration. Several studies reported decreased benzodiazepine receptor binding,[20,21] while others found decrements in $GABA_A$ receptor function.[21,22] A variety of benzodiazepine agonists have been associated with receptor alterations, including diazepam, lorazepam, flurazepam, alprazolam, and clonazepam.[20-24]

A more modest literature concerns the neurochemical correlates of benzodiazepine withdrawal. Studies from our laboratory indicate that 2 to 4 d after discontinuation of lorazepam administration (1 week), mice exhibit increased motor activity which returns to control levels at 7 d after discontinuation.[25] This behavioral alteration is accompanied by an increase in benzodiazepine receptor binding in several brain regions and an increase in GABA-dependent chloride uptake in cortex (Figure 1). Similar results have been obtained with alprazolam and clonazepam.[24,26] These data indicate that a behavioral withdrawal syndrome occurs in mice exposed to a dose and duration of benzodiazepines sufficient to produce tolerance. This syndrome is associated with an increase in benzodiazepine binding and $GABA_A$ receptor function, with subsequent return to control values for both behavioral and neurochemical effects.

To examine the mechanism for increased benzodiazepine binding during withdrawal, we evaluated benzodiazepine receptor half-life using the protein-modifying reagent *N*-ethyoxycarbonyl-2-ethoxy-1,2-dihydroquinoline (EEDQ). This compound appears to irreversibly bind a majority of benzodiazepine receptors, so that receptor recovery can be determined and receptor half-life estimated.[27] Benzodiazepine receptor half-life was reduced in mice treated with EEDQ 1 d after lorazepam discontinuation compared to vehicle-treated mice (Figure 2; Miller et al.[28]). This accelerated recovery of benzodiazepine receptors is consistent with the increase in binding observed during withdrawal. These data suggest an increase in benzodiazepine synthesis associated with withdrawal, although a reduction in the rate of receptor degradation is also consistent with these results.

Possible alterations in benzodiazepine receptor synthesis after drug discontinuation might reflect alterations at the genome. Using probes based on receptor subunit cDNAs, as described above, we found a decrease in mRNAs for the α_1- and γ_2-subunits after 14 d of lorazepam administration.[29] However, no alteration was observed for either subunit after 7 d, an interval sufficient for the development of tolerance and receptor downregulation. Thus, this

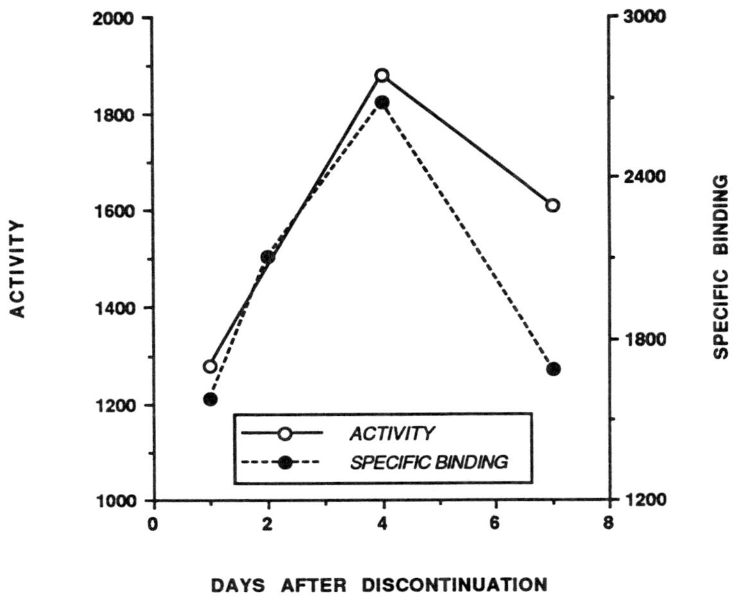

FIGURE 1. Behavior and benzodiazepine binding in cortex after lorazepam discontinuation. Results illustrate open-field activity and benzodiazepine binding *in vivo* in cortex at days 1 to 7 after lorazepam discontinuation (2 mg/kg/day for 7 d). Both activity and binding are significantly increased ($p<0.05$) compared to vehicle at day 4. Results are adapted from Miller et al.[21,25]

finding may reflect an additional, long-term effect of benzodiazepine administration; decrements in mRNAs persists at 28 d of administration. Similar data indicating reductions in the α_1-subunit after prolonged drug administration was recently reported with diazepam.[30] Preliminary studies from our laboratory indicate that mRNA levels return to control values 1 week after drug discontinuation (14 d of lorazepam). Thus, at present, insufficient data are available to indicate the role of gene expression in alterations of benzodiazepine/GABA function during withdrawal.

It is also possible that interactions of benzodiazepines with other neurotransmitter systems are involved in benzodiazepine dependence and withdrawal. Substantial behavioral data indicate benzodiazepine interactions with the opioid and adenosine receptor systems. Limited neurochemical data support these associations. In our own studies, chronic morphine, but not naloxone administration, augmented benzodiazepine binding and $GABA_A$ receptor function in cortex.[31] Conversely, acute administration of endogenous peptides with opioid effects, MIF-1 and Tyr-MIF-1, augmented $GABA_A$ receptor binding and function.[32,33] With regard to adenosine receptor effects, acute caffeine administration increased benzodiazepine receptor binding but not $GABA_A$ receptor function.[34] Data from these and other studies support

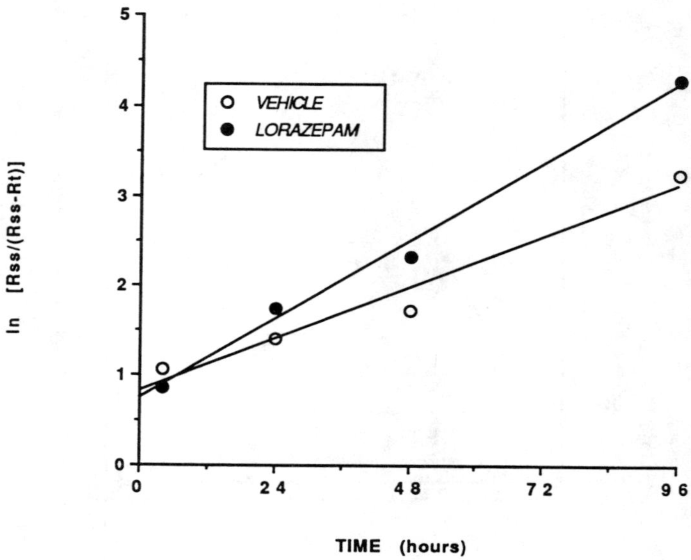

FIGURE 2. Kinetics of benzodiazepine receptor recovery in cortex after EEDQ administration. The slopes of the regression lines are significantly different ($p<0.05$), indicating different half-lives in vehicle- and lorazepam-treated mice. (Adapted from Miller, L. G., Lumpkin, M., Greenblatt, D. J., and Shader, R. I., *Fed. Proc. Fed. Am. Soc. Exp. Biol. J.*, in press, 1991.)

interactions between the benzodiazepine/GABA systems and neurotransmitter systems implicated in central reward effects. However, available data do not address specific mechanisms for benzodiazepine dependence.

IV. CONCLUSIONS

In summary, benzodiazepines act in the central nervous system at a specific receptor incorporated in the $GABA_A$ receptor complex. It is likely that multiple benzodiazepine receptor subtypes exist based on combinations of receptor subunit polypeptides. Chronic benzodiazepine administration is associated with behavioral tolerance and receptor downregulation. Benzodiazepine withdrawal is associated with increases in benzodiazepine binding and $GABA_A$ receptor function, and with accelerated benzodiazepine receptor recovery after inactivation. Prolonged benzodiazepine administration is associated with decreases in mRNAs for several $GABA_A$ receptor subunits. In conjunction, these data indicate neurochemical alterations associated with benzodiazepine withdrawal as a marker for benzodiazepine dependence. However, the data are insufficient to indicate a neurochemical mechanism for benzodiazepine withdrawal or dependence. It is also possible that interactions between the benzodiazepine/GABA system and other neurotransmitter systems contribute to benzodiazepine dependence.

ACKNOWLEDGMENTS

The authors wish to acknowledge the important contributions of Dr. Richard I. Shader to these studies, and the assistance of Dr. Jamie Barnhill, Wendy Galpern, Inwha Kang, Fred Lopez, and Beth Roy. Supported in part by grants DA-05258, MH-34223, and AG-01006 from the U.S. Public Health service. Dr. Miller is the recipient of a Faculty Development Award in Clinical Pharmacology from the Pharmaceutical Manufacturers Association Foundation.

REFERENCES

1. **Woods, J. H., Katz, J. L., and Winger, G.,** Abuse liability of benzodiazepines, *Pharmacol. Rev.,* 39, 254, 1987.
2. **Ator, N. A. and Griffiths, R. R.,** Self-administration of barbiturates and benzodiazepines: a review, *Pharmacol. Biochem. Behav.,* 27, 391, 1987.
3. **Noyes, R., Garvey, M. J., Cook, B. L., and Perry, P. J.,** Benzodiazepine withdrawal: a reivew of the evidence, *J. Clin. Psychiatry,* 49, 382, 1988.
4. **Roy-Byrne, P. P. and Hommer, D.,** Benzodiazepine withdrawal: overview and implications for the treatment of anxiety, *Am. J. Med.,* 84, 1041, 1988.
5. **Tyrer, P.,** Dependence as a limiting factor in the clinical use of minor tranquilizers, *Pharmacol. Ther.,* 36, 173, 1988.
6. **Möhler, H. and Okada, T.,** Benzodiazepine receptor: demonstration in the central nervous system, *Science,* 198, 849, 1977.
7. **Braestrup, C. and Squires, R. E.,** Specific benzodiazepine receptors in rat brain characterized by high-affinity (^{3}H)-diazepam binding, *Proc. Natl. Acad. Sci. U.S.A.,* 74, 3805, 1977.
8. **Haefely, W., Kyburz, E., Gerecke, M., and Möhler, H.,** Recent advances in the molecular pharmacology of benzodiazepine receptors and in the structure-activity relationships of their agonists and antagonists, *Adv. Drug Res.,* 14, 165, 1985.
9. **Sigel, E., Stephenson, F. A., Mamalaki, C., and Barnard, E. A.,** A γ-aminobutyric acid/benzodiazepine receptor complex of bovine cerebral cortex: purification and partial characterization, *J. Biol. Chem.,* 258, 6965, 1983.
10. **Stephenson, F. A., Watkins, A. E., and Olsen, R. W.,** Physicochemical characterization of detergent-solubilized γ-aminobutyric acid and benzodiazepine receptor protein from bovine brain, *Eur. J. Biochem.,* 123, 291, 1982.
11. **Chang, L. R. and Barnard, E. A.,** The benzodiazepine/GABA receptor complex: molecular size in brain synaptic membranes and in solution, *J. Neurochem.,* 39, 1507, 1982.
12. **Sweetnam, P., Nestler, E., Gallombardo, P., Brown, S., Duman, R., Bracha, H. S., and Tallman, J.,** Comparison of the molecular structure of GABA/benzodiazepine receptor purified from rat and human cerebellum, *Mol. Brain. Res.,* 2, 223, 1987.
13. **Bureau, M. and Olsen, R. W.,** GABA/benzodiazepine receptor protein carries binding sites for both ligands on two major peptide subunits, *Biochem. Biophys. Res. Commun.,* 153, 1006, 1988.
14. **Olsen, R. W. and Tobin, A. J.,** Molecular biology of $GABA_A$ receptors, *Fed. Proc. Fed. Am. Soc. Exp. Biol. J.,* 4, 1469, 1990.

15. **Schofield, P. R., Darlison, M. G., Fujita, N., Burt, D. R., Stephenson, F. A., Rodriguez, H., Rhee, L. M., Ramachandian, J., Reale, V., Glencourse, T. A., Seeburg, P. H., and Barnard, E. A.,** Sequence and functional expression of the $GABA_A$ receptor shows a ligand-gated receptor superfamily, *Nature (London),* 328, 221, 1987.

16a. **Shivers, B. D., Killison, I., Sprengel, R., Sontheimer, H., Köhler, M., Schofield, P. R., and Seeburg, P. H.,** Two novel $GABA_A$ receptor subunits exist in distinct neuronal subpopulations, *Neuron,* 3, 327, 1989.

16. **Levitan, E. S., Schofield, P. R., Burt, D. R., Rhee, L. M., Wisden, W., Köhler, M., Fujita, N., Rodriguez, H. F., Stephenson, F. A., Darlison, M. G., Barnard, E. A., and Seeburg, P. H.,** Structural and functional basis for $GABA_A$ receptor heterogeneity, *Nature (London),* 335, 76, 1988.

17. **Pritchett, D., Sontheimer, H., Shivers, B. D., Ymer, S., Kettenmann, H., Schofield, P. R., and Seeburg, P.,** Importance of a novel $GABA_A$ receptor subunit for benzodiazepine pharmacology, *Nature (London),* 338, 582, 1989.

18. **Pritchett, D. B., Luddens, H., and Seeburg, P. H.,** Type I and type II $GABA_A$-benzodiazepine receptors produced in transfected cells, *Science,* 245, 1389, 1989.

19. **Haefely, W.,** Endogenous ligand of the benzodiazepine receptor, *Pharmacopsychiatry,* 21, 43, 1988.

20. **Tietz, E. I., Rosenberg, H. C., and Chiu, T. H.,** Autoradiographic localization of benzodiazepine receptor downregulation, *J. Pharmacol. Exp. Ther.,* 236, 284, 1986.

21. **Miller, L. G., Greenblatt, D. J., Barnhill, J. G., and Shader, R. I.,** Chronic benzodiazepine administration. I. Tolerance is associated with downregulation of benzodiazepine receptor binding and γ-aminobutyric $acid_A$ receptor function, *J. Pharmacol. Exp. Ther.,* 246, 170, 1988.

22. **Marley, R. J. and Gallager, D. W.,** Chronic diazepam treatment produces regionally specific changes in GABA-stimulated chloride influx, *Eur. J. Pharmacol.,* 159, 217, 1989.

23. **Miller, L. G., Woolverton, S., Greenblatt, D. J., Lopez, F., Roy, R. B., and Shader, R. I.,** Chronic benzodiazepine administration. IV. Rapid development of tolerance and receptor downregulation associated with alprazolam administration, *Biochem. Pharmacol.,* 38, 3773, 1989.

24. **Galpern, W. R., Lumpkin, M., Greenblatt, D. J., Shader, R. I., and Miller, L. G.,** Chronic benzodiazepine administration. VII. Behavioral tolerance and withdrawal and receptor alterations associated with clonazepam administration, *Psychopharmacology,* 104, 225, 1991.

25. **Miller, L. G., Greenblatt, D. J., Summer, W. R., and Shader, R. I.,** Chronic benzodiazepine administration. II. Discontinuation syndrome is associated with upregulation of γ-aminobutyric $acid_A$ receptor binding and function, *J. Pharmacol. Exp. Ther.,* 246, 177, 1988.

26. **Lopez, F., Miller, L. G., Greenblatt, D. J., Chesley, S., Schatzki, A., and Shader, R. I.,** Chronic administration of benzodiazepines. V. Rapid onset of behavioral and neurochemical alterations after alprazolam discontinuation, *Neuropharmacology,* 29, 237, 1990.

27. **Miller, L. G., Lumpkin, M., Galpern, W. R., Greenblatt, D. J., and Shader, R. I.,** Modification of $GABA_A$ receptor binding and function by *N*-ethoxycarbonyl-2,2-ethoxy-1,2-dihydroquinoline *in vitro* and *in vivo:* effects of aging, *J. Neurochem.,* 56, 1241, 1990.

28. **Miller, L. G., Lumpkin, M., Greenblatt, D. J., and Shader, R. I.,** Accelerated benzodiazepine receptor recovery following lorazepam discontinuation, *Fed. Proc. Fed. Am. Soc. Exp. Biol. J.,* 5, 93, 1991.

29. **Kang, I. and Miller, L. G.,** Decreased $GABA_A$ receptor subunit mRNA concentrations following chronic lorazepam administration, *Br. J. Pharmacol.,* 103, 1285, 1991.

30. **Heninger, C., Saito, N., Tallman, J. F., Duman, R. S., Garrett, K. M., Vitek, M. P., and Gallager, D. W.,** Effects of continuous diazepam administration on GABA-A subunit mRNA in rat brain, *Soc. Neurosci. Abstr.*, 831, 1989.
31. **Lopez, F., Miller, L. G., Thompson, M. L., Schatzki, A., Chesley, S., Greenblatt, D. J., and Shader, R. I.,** Chronic morphine administration augments benzodiazepine binding and $GABA_A$ receptor function, *Psychopharmacology*, 101, 545, 1990.
32. **Miller, L. G. and Kastin, A. J.,** MIF-1 and Tyr-MIF-1 augment GABA-stimulated benzodiazepine receptor binding, *Peptides*, 8, 751, 1987.
33. **Miller, L. G., Kastin, A. J., and Roy, R. B.,** MIF-1 and Tyr-MIF-1 augment muscimol-stimulated chloride uptake in cerebral cortical synaptoneurosomes, *Brain Res. Bull.*, 23413, 1989.
34. **Lopez, F., Miller, L. G., Greenblatt, D. J., Kaplan, G. B., and Shader, R. I.,** Interaction of caffeine with the $GABA_A$ receptor complex: alterations in receptor function but not ligand binding, *Eur. J. Pharmacol., (Mol. Pharmacol. Sect.),* 172, 453, 1989.

Chapter 10

NEUROBIOLOGY OF ANDROGEN ABUSE

Richard E. Harlan and Meredith M. Garcia

TABLE OF CONTENTS

I. INTRODUCTION

Use and abuse of steroid hormones by athletes has become of increasing concern at several levels of society. It is estimated that as many as one million athletes in this country alone may take anabolic steroids.[1] Recent reviews of the topic[2-6] have served to illustrate the profound lack of basic information on effects of pharmacological levels of these hormones on the nervous system. Despite evidence that anabolic steroids can be associated with psychological disturbances,[7] little is known about the mechanisms of action of these hormones or the possible deleterious consequences for the nervous system of long-term use. Endogenous steroid hormones are well documented to regulate several neuroendocrine systems, but very little is known about neuroendocrine effects of steroids of abuse. This chapter will survey the literature concerning effects of these suprapharmacological levels of steroids on the brain, and will propose possible mechanisms of action.

A. WHAT IS STEROID ABUSE?

Drug abuse can be defined as "use, usually by self-administration, of any drug in a manner that deviates from the approved medical or social patterns within a given culture."[8] As will be evident below, the doses and combinations of steroids used by athletes and body-builders are far in excess of what is ever prescribed for any medical treatment. Moreover, this level of steroid abuse clearly deviates from normal social patterns within our culture, since the steroids are obtained illegally[9-11] and are used despite increasing negative publicity in the news media.

B. WHAT IS THE EPIDEMIOLOGY OF STEROID ABUSE?

Information about prevalence of androgen abuse is, at best, incomplete. Most studies have concentrated on androgen abuse patterns in athletes. A small number of studies have attempted to estimate the prevalence of androgen abuse in specific segments of the general population. Pope et al.[12] estimated that 2% of men in college use anabolic steroids, based on more than 1000 respondents to a questionnaire. Use of anabolic steroids was more prevalent among athletic than among nonathletic responders. However, since only 17 of the respondents reported using anabolic steroids, meaningful analysis of these results is difficult. Johnson et al.[13] reported that 11% ($N = 95$) of 853 high school males responding to a questionnaire acknowledged anabolic steroid use. The following reasons for taking steroids were listed: to increase strength (64%), to increase size (50%), to improve physical appearance (27%), and because their friends were using steroids (10%). In a larger statistical study, Buckley et al.[14] found that 6.7% ($N = 226$) of 3403 male high school seniors reported current or previous anabolic steroid use. The user group tended to be older than the nonuser group ($p<0.001$), had a greater minority representation ($p<0.001$), and were less likely to have a parent who had

TABLE 1
Typical Androgen Abuse Patterns in Humans[a]

Usage classification (typical)	Drug[b]	Dose	Duration (weeks)
Light	Methandrostenolone[c]	15 mg/d po	6
Moderate	Methandrostenolone	20 mg/d po	10
	Testosterone cypionate[d]	200 mg/d im	10
	Nandrolone decanoate[e]	200 mg/d im	10
Heavy	Methandrostenolone	40 mg/d po	16
	Oxandrolone[f]	40 mg/d po	16
	Nandrolone decanoate	600 mg/week im	16
	Methenolone enanthate[g]	4 ml/week im	16
	Boldenone undecylenate[h]	8 ml/week im	16

[a] Data taken from Refs. 7 and 16.
[b] All drugs taken simultaneously.
[c] Trade name Dianabol; most commonly abused orally active steroid.
[d] Trade name Depo-testosterone.
[e] Trade name Deca-Durabolin.
[f] Trade name Anavar.
[g] Not legally available for human use in the U.S.; hence potency is unknown. The related compound methenolone acetate has the trade name Primobolan.
[h] Boldenone (Equipoise), one of several veterinary preparations used by the study subjects, is a steroid used chiefly for horses; a dose of 8 ml would represent either 200 or 400 mg of boldenone, since the drug is marketed in two strengths.

finished high school ($p<0.001$). Age of onset of use was quite early, with more than one third of the users reporting first use at age 15 or younger. Reasons for taking steroids were listed as follows: to improve athletic ability (47%), to improve appearance (27%), for social reasons (9%), and for injury prevention or treatment (11%).

C. WHAT ARE THE STEROIDS THAT ARE MOST COMMONLY ABUSED?

The most commonly abused steroids are all derivatives of androgens, and thus presumably are anabolic. Three general types of modifications to the steroid structure have been induced, to overcome the rapid clearance and metabolism of testosterone (see Reference 4 for a review): (a) alkylation at the 17α-position, for orally active steroids; (b) esterification at the 17β-hydroxyl group, for parenterally active steroids; (c) modifications of the ring structure, for either oral or parenteral administration. The pattern of steroid abuse is tailored to individuals and ranges from light to very heavy use (Table 1). Heavy usage typically involves a pattern called "stacking," where several different steroids are taken, some daily and some at intervals of 1 to 2 weeks (see Reference 15 for an example). Typically, a person will self-administer increasing doses of steroids for several weeks, followed by a period of several weeks without steroids; the cycle is often repeated for years. Since none of the steroids used by athletes are purely anabolic, i.e., without androgenic actions, the term androgenic-anabolic steroid will be used in this chapter.

II. PSYCHIATRIC EFFECTS OF ANDROGENIC-ANABOLIC STEROIDS

A small literature is extant on the psychological effects of high doses of androgens in humans. Two types of studies have been reported: clinical case studies and retrospective studies based on interviews of abusers of androgens. One of the earliest studies of psychiatric effects of androgens[17] reported on two castrated men, two hypogonadal men, and two cases of "psychic impotence." All six patients were "disturbed, anxious, and broken in spirit." The four clear-cut cases of organic problems displayed mental depression, vasomotor disturbances (hot flashes), emotional instability, and mental and physical fatiguability. The patients were given testosterone propionate (20 mg three times weekly), which resulted in not only cessation of the previous mental condition, but also appearance of elation and "rational aggressiveness." A larger study on the effects of testosterone in the treatment of depression was conducted several years later.[18] Thirty-one patients with depression (including the depressed phase of manic depression, involutional psychoses with depression, reactive depression, and one case of schizophrenia with depression) were treated with testosterone, 50 mg/d for 2 to 3 weeks, with or without electroconvulsive shock therapy. While the results were somewhat difficult to interpret, in the large majority of cases, testosterone treatment seemed to be as effective as shock treatment. In a more recent study,[19] the synthetic androgen, mesterolone, was reported to be as effective as the tricyclic antidepressant amitriptyline for treatment of depression in men. The tricyclic antidepressants are believed to exert their therapeutic effect by inhibiting re-uptake of catecholamines, thus increasing neurotransmitter concentration at the synapse. This study is an interesting contrast to a previous report[20] on treatment of depression in five men with a combination of methyltestosterone (15 mg daily) and imipramine (25 to 50 mg three times daily). In this latter study, four of the five men promptly converted from depression to a paranoid reaction, which cleared rapidly with withdrawal of the steroid. This report suggests a potential interaction between androgens and catecholamine turnover, a subject to which we will return.

Despite the suggestive involvement of androgens in adverse mental status, the doses of androgens administered in these studies were considerably lower than the doses self-administered by many athletes. Reports of severe psychological reactions to the massive doses of androgenic-anabolic steroids taken by athletes began to appear in the early 1980s. Annitto and Layman[21] reported a case study of a 17-year-old male who was diagnosed as schizophrenic, paranoid type. The patient was placed on haloperidol, which resulted in marked improvement. About a year later, the patient stopped the therapy, and symptoms returned. Upon representation, he revealed that during the 6 months prior to his original visit, he had been lifting weights and taking anabolic steroids (types or doses not listed). Although the evidence for a causal relationship between the diagnosed mental status and the self-administration of steroids was tenuous at best, this study served to alert physicians to the possible adverse mental effects of androgenic-anabolic steroids.

Since this report was published, several other reports have appeared. In many cases, a constellation of symptoms appears to be induced by androgenic-anabolic steroids, including euphoria (often alternating with depression and irritability), nervousness, hyperactivity, increased libido, sleep disorders, delusions of reference, grandiose delusions, psychomotor retardation or acceleration, paranoid ideation, and hallucinations of the auditory or visual systems.[7,20-24] It should be noted that not all symptoms in the cluster are always present. Pope and Katz[23] cite several examples of manic or hypomanic behavior, based on interviews of body-builders and football players taking large amounts of androgenic-anabolic steroids. One subject "developed the grandiose delusion that he could pick up a car and tip it over." Another man "bought a $17,000 sports car while taking methandrostenolone. When he stopped the drug, he realized that he could not afford the payments and sold the car; a year later, during another cycle (of steroid use), he impulsively bought a $20,000 sports car."

In the most extensive study to date, Pope and Katz[7] reported the results of interviews of 39 men and 2 women self-administering androgenic-anabolic steroids and actively involved in weight lifting or body-building. Androgen abuse patterns were classified as light, moderate, or heavy (see Table 1). Note that the moderate and heavy doses are 10 to 100 times the doses used in typical medical studies. While none of the subjects recalled any adverse psychological episodes during absence of steroid exposure, 5 subjects (12.2%) "met DSM-III-R criteria for psychotic symptoms during periods of steroid exposure." All 5 subjects experienced hallucinations or delusions. An additional 4 subjects "reported mild or equivocal psychotic symptoms (defined as 'subthreshold' psychotic symptoms) during periods of steroid exposure." All psychotic symptoms appeared during steroid use. Interestingly, all subjects experiencing psychotic episodes were taking multiple steroids. Furthermore, 5 subjects met criteria, and 8 additional subjects narrowly missed diagnosis for a manic episode during steroid exposure. While withdrawing from steroids, 5 subjects also developed major depression. Despite the limitations of this type of retrospective study, these results, combined with several case studies, clearly suggest adverse changes in mental status during exposure to large doses of androgenic-anabolic steroids.

Interestingly, many of the same psychiatric problems associated with excess androgen treatment have also been reported with chronic *underexposure* to endogenous androgens, due to castration or hypogonadism.[17] This similarity of psychiatric problems associated with under- or vast overexposure to androgens can be compared with the hippocampal-damaging effects of either chronic adrenalectomy[25] or chronic exposure to high doses of corticosteroids in adrenally intact rats.[26,27] Thus, deviations toward chronic under- or overexposure to androgenic-anabolic steroids may cause psychiatric problems.

The cluster of psychiatric symptoms seen in people experiencing high levels of androgens has been classified by several authors as psychosis, especially affective psychosis.[7,21,23] Within the realm of affective disorders, the constellation of symptoms of androgen-induced psychosis seems to fit most completely the symptoms described for manic disorder or the manic phase of manic-depressive illness, viz., heightened mood, faster speech, quicker thought and mental activity, more energy, decreased need for sleep, irritability, perceptual acuity, paranoia, heightened sexuality, and impulsivity. More severe cases of manias (acute mania) may include delusions, especially grandiose delusions and hallucinations.[28]

This classification suggests that the neurobiology of androgen-induced psychiatric problems may have some similarity to the neurobiology of psychosis, especially manic or manic-depressive illness. While the biochemistry and neurobiology of manic or manic-depressive illness is quite complex,[28] a central role of increased turnover of norepinephrine and especially dopamine in several neural systems appears to be a key feature.[29,30] This increased dopamine turnover probably occurs in several dopaminergic systems, although the ventral tegmental A10 dopaminergic system and its projections to the limbic system, nucleus accumbens, and cerebral cortex appears to be a prime suspect in the neurobiology of affective disorders. That androgen-induced psychosis may have as its foundation an alteration, especially a stimulation, of dopaminergic transmission in the brain is supported by the successful treatment of psychiatric problems associated with androgen abuse by administration of haloperidol,[7] a dopamine receptor blocker commonly used in the treatment of psychosis and schizophrenia.

It is the major hypothesis of this chapter that administration of large amounts of androgenic-anabolic steroids can lead to increased dopaminergic transmission in several brain systems; this increased dopaminergic activity then leads to a variety of neurobiological problems associated with androgen abuse.

A. DOPAMINERGIC SYSTEMS IN THE BRAIN

Several different dopaminergic systems have been identified in the brain (reviewed in References 31 to 33), including nigrostriatal, mesocortical, mesolimbic, tuberohypophyseal, retinal, incertohypothalamic, medullary periventricular, and periglomerular neurons in the olfactory bulb. Of these systems, we postulate that androgenic-anabolic steroids exert behavioral and psychiatric effects through the mesocortical, mesolimbic, and/or nigrostriatal systems, and exert neuroendocrine effects through the tuberohypophyseal system.

The tuberohypophyseal system consists of dopaminergic cell bodies in the medial basal hypothalamus, particularly the arcuate nucleus, with projections to the median eminence. Release of dopamine into the hypothalamic-hypophyseal portal system inhibits pituitary release of prolactin and modulates release of other pituitary trophic hormones. The nigrostriatal system consists of dopaminergic cell bodies in the substantia nigra pars compacta, pars reticulata, and pars lateralis, with extensive projections to the caudate and putamen, by way of a pathway which includes the lateral hypothalamus and medial portion of the internal capsule. This system is critically involved in the control of movement, since increased dopaminergic activity is associated with increased movement. The mesolimbic (mesocortical, according to Moore and Bloom[31]) system consists of dopaminergic cell bodies in the ventral tegmental area, medial to the substantia nigra. These neurons project in five pathways to: (a) the amygdala and the entorhinal, perirhinal, and piriform cortex; (b) the septum and the bed nucleus of the stria terminalis; (c) the frontal and anterior cingulate cortex; (d) the nucleus accumbens; (e) the olfactory tubercle, anterior olfactory nucleus, and the olfactory bulb. In primates, a mesocortical system has also been identified, with extensive projections to various portions of the neocortex which are involved in planning and execution of goal-directed motor behavior.[30] While this complex telencephalic projection system is involved in a variety of different functions, of most importance for this chapter are the projections to the frontal cortex and nucleus accumbens, which appear to be involved in emotional outlook, reinforcement, and goal-directed (motivated) behavior.[30,32,34] We hypothesize that high doses of androgenic-anabolic steroids increase dopaminergic activity in the projection pathways of the mesolimbic and mesocortical systems, thus leading to psychiatric problems and potential addiction to the steroids.

B. ARE ANDROGENIC-ANABOLIC STEROIDS ADDICTIVE?

Three studies have suggested that androgenic-anabolic steroids can be addictive for some people. First, Tennant et al.[1] reported a case in which a body-builder taking large doses of several steroids claimed that he was addicted to steroids and that he could not stop taking these drugs without becoming depressed and fatigued and experiencing an intense craving for steroids. The patient was given a naloxone challenge (0.2 mg) which precipitated an opiatelike withdrawal syndrome including nausea, chills, headache, dizziness, diaphoresis, piloerection, increased pulse rate, and blood pressure. The withdrawal symptoms were treated with clonidine, a common treatment for opiate withdrawal syndrome. Seven days after naloxone treatment, the patient telephoned to say that he could not stand the depression, fatigue, and craving, and was resuming the steroid regimen. Brower et al.[24] reported a similar apparent addiction case in a noncompetitive weight lifter taking very high doses of a variety of steroids as well as human chorionic gonadotropin (to stimulate endogenous production of androgens). The patient complained primarily of depression and outbursts of anger that he associated with steroid use; he sought professional help to discontinue use of the steroids. While taking steroids, he "felt energetic, he required only 4 to 5 h sleep each night, and he thought he looked and performed better." Without steroids "he felt depressed, low in energy, fatigued and weak, and he suffered from headaches." The patient was hospitalized, and his mental outlook improved. He participated in a chemical dependency program for 5 d, "when he signed out against medical advice for no clear reason." The authors argued that the patient met six of the nine criteria for substance dependence, as defined by the Diagnostic and Statistical Manual of the American Psychiatric Association (DSM-III-R).[35]

Finally, Kashkin and Kleber[36] reviewed the literature and concluded that androgen abuse can be classified as addictive, as defined by the following criteria, taken from the DSM-III-R[35]: "(1) the hormones are used over longer periods than desired; (2) attempts are made to stop use without success; (3) substantial time is spent obtaining, using, or recovering from the hormones; (4) use continues despite knowledge of significant psychological problems caused by the hormones; (5) characteristic withdrawal symptoms occur; and (6) the hormones are often taken to relieve these withdrawal symptoms." The case studies reviewed above, as well as the evidence for psychological effects of androgenic-anabolic steroids, support the contention that these steroids can be addictive.

A consensus is gradually emerging that many, if not all, substances with addictive potential have as a common action a stimulation of dopaminergic turnover or activity in the mesolimbic dopaminergic system.[37] For instance, cocaine has been well documented to increase dopaminergic activity in the nucleus accumbens by inhibiting the dopamine transporter,[38] thus increasing the concentration of dopamine in the synaptic cleft. Similar increases in dopamine activity in the nucleus accumbens have been reported for opiates, amphetamine, and alcohol.[37] If androgenic-anabolic steroids increase dopaminergic and norepinephrinergic transmission in several brain systems, as postulated above, then this may provide at least a partial explanation for the neurobiological cause of the addictive quality of these hormones.

III. NEUROENDOCRINE EFFECTS OF ANDROGENIC-ANABOLIC STEROIDS

Although testosterone has been recognized as a participant in neuroendocrine feedback systems for decades, very little information is available regarding neuroendocrine effects of the massive doses of androgenic-anabolic steroids taken by athletes. Alen et al.[39] reported neuroendocrine effects of androgenic-anabolic steroids in a group of 5 athletes, compared to a control group of 6 athletes not taking steroids. The experimental group self-administered a 26-week regimen of increasing doses of three androgenic-anabolic steroids (methandienone, stanozolol, and nandrolone) as well as testosterone. This was followed by a 16-week period without steroid self-administration. Blood samples were taken periodically (at 8, 14, 20, 26, 29, 32, 38, and 42 weeks); each blood sample was taken at 8:00 A.M., after 1 d of reduced athletic training and overnight fasting. Androgenic-anabolic steroids significantly reduced luteinizing hormone (LH) and follicle-stimulating hormone (FSH) levels, which recovered following cessation of steroid use. Serum prolactin levels showed a biphasic response, with an initial decrease, followed by an increase. However, the changes were not significantly different from the levels found in the control group. Serum testosterone levels tended to increase throughout the steroid exposure, due to self-administration of this steroid, but reached a significant increase only at the end of the 26-week period of self-administration. Following cessation of steroid administration, testosterone levels fell significantly to concentrations 50% below the levels found in the control group. Circulating estradiol levels increased significantly at all time-points measured during steroid administration, reaching a sevenfold increase by 26 weeks. Following cessation of steroid use, estradiol concentrations returned to levels comparable to or slightly below those found in the control group. Circulating adrenocorticotropin (ACTH) levels were significantly lower in the experimental group at all time-points measured during steroid use. ACTH levels gradually returned to control levels by 12 weeks following cessation of administration. Serum cortisol levels were not changed during or after steroid administration.

In a second study,[40] this same research group examined serum levels of thyroid hormones, gonadotropins and steroid hormones in a group of seven athletes self-administering androgenic-anabolic steroids over a 12-week period, followed by a 13-week period without steroids. During steroid use, serum levels of thyroid-stimulating hormone (TSH), thyroxine (T_4), and triiodothyronine (T_3) all decreased significantly, with recovery to baseline levels following cessation of steroid use. Circulating levels of both thyroxine-binding globulin (TBG) and sex hormone-binding globulin (SHBG) also declined significantly during steroid self-administration, with subsequent recovery to baseline following steroid withdrawal. The previous findings of decreased circulating LH and FSH, and increased levels of estradiol were also confirmed. Although mean serum levels of growth hormone (GH) were not significantly changed, in five of the seven subjects, circulating GH levels increased 5- to 60-fold at 8 or 12 weeks.

These results suggest several neuroendocrine effects of androgenic-anabolic steroids.

1. The steroids, or their metabolites (e.g., estradiol), exert a strong negative feedback effect on circulating LH and FSH levels. This effect has

been reported by other groups;[41,42] however, this effect has not always been seen.[2] These reduced levels of gonadotropins result in testicular atrophy and a prolonged reduction in testicular production of testosterone, as reported by several groups.[2]

2. The steroids may act as glucocorticoids, exerting a negative feedback effect on circulating ACTH levels. If this is the case, then some of the behavioral and psychiatric effects of androgenic-anabolic steroids may be mediated by glucocorticoid receptors, which have a very widespread distribution in the central nervous system.[43] Moreover, the decreased circulating ACTH levels may imply decreased production and/or release of corticotropin-releasing factor in the brain. Although one would predict that the decreased circulating levels of ACTH would result in decreased levels of cortisol, circulating cortisol levels were reported to be normal during and after androgenic-anabolic steroid use. The reasons for this are not currently known.
3. The large increase in circulating estradiol levels, reaching concentrations typical of females, may account for not only the increased occurrence in gynecomastia, but also some of the extrapyramidal and psychological effects of the steroids. The increased circulating estradiol levels presumably derive from aromatization of the self-administered androgenic-anabolic steroids.
4. The decreased levels of SHBG may result in increased delivery of free steroids to the brain and other target tissues. The cause of the decreased levels of SHBG and TBG are not known, but could involve damage to the liver produced by the large doses of androgenic-anabolic steroids.
5. The decreased levels of TSH, T_3, and T_4 may suggest decreased production and/or release of thyrotropin-releasing hormone (TRH) in the hypothalamus. It should be noted that normal T_3 levels following androgenic-anabolic steroid use were reported by another group.[42]

IV. POSSIBLE MECHANISMS OF ACTION OF ANDROGENIC-ANABOLIC STEROIDS ON THE NERVOUS SYSTEM

As indicated above, androgenic-anabolic steroids have a variety of effects on circulating hormone levels. Thus, it is difficult to ascribe the neurobiology of androgen abuse to a direct action of androgenic-anabolic steroids on the brain or to an effect that is secondary to alterations in circulating hormone levels. For example, the markedly increased circulating levels of estrogens in athletes taking androgenic-anabolic steroids may also exert direct effects on the nervous system. A very extensive literature exists on effects of steroid hormones on the nervous system (see References 44 and 45 for reviews). Review of this literature is beyond the scope of this chapter. We will focus on studies that relate to our hypothesis that androgenic-anabolic steroids affect dopaminergic transmission in several pathways in the brain. We will propose three possible mechanisms by which androgenic-anabolic steroids affect the brain: (1) by a direct action on neurons that express androgen receptors, (2) by aromatization to estrogens, which then act on the brain, (3) by action through neurons that express glucocorticoid receptors.

A. ANDROGENIC-ANABOLIC STEROIDS MAY ACT THROUGH NORMAL ANDROGEN RECEPTOR SYSTEMS IN THE BRAIN

Since the steroids that are abused by athletes are all androgens, the most straightforward mechanism of steroid abuse may be that the androgenic-anabolic steroids act through normal androgen receptor systems, which are linked to dopamine turnover in certain systems in the brain. This hypothesis is supported by three arguments: (1) normal androgen-regulated behaviors depend on dopamine transmission in the nigrostriatal, mesolimbic, and mesocortical systems; (2) normal androgen-regulated behaviors are a subset of motivated behaviors that depend on dopamine transmission in the mesolimbic and mesocortical systems; (3) normal circulating androgens increase dopamine turnover in the mesolimbic and mesocortical systems. Each of these arguments will be reviewed and possible mechanisms of androgenic regulation of dopamine transmission in the mesolimbic and mesocortical systems will be proposed.

An example of a behavior regulated by androgen, at least in some mammals, is masculine sexual behavior. Although this behavior is mediated by a variety of neural systems, a strong case can be made for a critical involvement of dopamine transmission in the mesolimbic, mesocortical, and nigrostriatal systems in this behavior. Bitran and Hull[46] have recently reviewed the evidence for dopamine involvement in masculine sexual behavior in the rat, and have noted that the behavior is facilitated by drugs increasing dopamine turnover (L-DOPA) and by drugs that act as dopamine receptor agonists (apomorphine, 3-PPP, LY163502); the behavior is inhibited by drugs that act as dopamine receptor blockers (haloperidol) and by lesioning of the nigrostriatal tract. Microinjection of apomorphine directly into the nucleus accumbens, but not into the caudate-putamen, increased masculine sexual performance.[47] There is also evidence for involvement of dopaminergic transmission in the regulation of other androgen-mediated behaviors, such as aggression in humans.[48]

These normal androgen-regulated behaviors are a subset of motivated behaviors, such as ingestive behaviors, intracranial self-stimulation, active avoidance, and maternal behaviors. Each of these behaviors involves forward locomotion to obtain a goal (e.g., food, electrical stimulation of the brain, avoidance of painful peripheral stimulation, retrieval of pups). Each of these behaviors is also dependent on dopamine transmission.[32,49]

Normal circulating androgen levels appear to facilitate dopaminergic transmission, especially in the mesolimbic and mesocortical systems. Castration of male rats has been reported to decrease concentrations of dopamine and its metabolites in the nucleus accumbens, but not in the caudate-putamen.[50,51] This effect could be prevented by administration of testosterone, dihydrotestosterone, or estradiol. Mitchell and Stewart[51] provided evidence that the decrease in dopamine and metabolite levels in the nucleus accumbens correlated with a decrease in ''female-directed behaviors,'' i.e., motivated behaviors that are directed toward sexual activity. These results support our hypothesis that androgenic-anabolic steroids increase dopamine turnover in mesolimbic/mesocortical pathways, at least in rats. Comparable studies are lacking for primates.

The distribution of neurons with androgen receptors provides a clue as to possible mechanisms of androgen-induced increased dopaminergic activity. Androgen receptors have been detected in the central nervous system by binding studies,[52,53] by immunocytochemistry,[54] and by detection of mRNA for the androgen receptor.[55] While the majority of studies have dealt with rodents, several studies have indicated the presence of androgen receptors in the brains of primates, including humans.[56-65] Due to the much larger brain size of primates compared to rodents, studies on the localization of neurons with androgen receptors in primate brains have generally focused on specific regions of the brain, rather than on the entire central nervous system. Several studies have reported the presence of neurons with androgen receptors in the primate hypothalamic-preoptic area and amygdala.[56-58,60,63] Another study has demonstrated androgen receptors in neurons of the brain stem of the rhesus monkey, especially in cranial nerve nuclei and sensory processing nuclei.[62] Of importance for the present thesis, is the existence of scattered androgen-concentrating neurons in the substantia nigra, but not in the ventral tegmental area. Other studies have reported neurons with androgen receptors in the primate cerebral cortex.[64,65] However, studies to date have indicated an absence of neurons that concentrate androgens in the nucleus accumbens. It should be emphasized, however, that steroid autoradiography is not always the most sensitive method for detection of neurons that express steroid hormone receptors; *in situ* hybridization to detect mRNA, or immunocytochemistry to detect the receptor protein, is often a more sensitive approach. To date, no studies have been published on the detection of neurons with androgen receptors in the primate brain using these more sensitive techniques.

In the rat, androgen receptor mRNA has been detected in several, but not all, of the projection sites of the mesocortical system, i.e., very high levels in several portions of the amygdala, with lower densities of labeled neurons in the anterior olfactory nucleus, piriform cortex, and entorhinal cortex;[55] labeled neurons were not detected in the caudate-putamen or nucleus accumbens. Neurons with androgen receptor mRNA were also reported in the substantia nigra and ventral tegmental area, providing an anatomical basis for postulated effects of androgenic-anabolic steroids on dopaminergic transmission in ascending systems. However, a previous study, using steroid autoradiography combined with catecholamine fluorescence, reported no androgen concentration by dopaminergic cells in the ventral midbrain.[66] The difference between these two sets of results may be due to differing sensitivities of the two techniques. The existence of neurons with androgen receptor mRNA in the cerebral cortex of rat brain[55] is consistent with reports of androgen receptors in the cerebral cortex of primates, including humans.[65]

However, one problem with the hypothesis that androgenic-anabolic steroids, taken in dosages typical of abuse, act through the androgen receptor, is that normal circulating androgen levels are sufficient to saturate androgen receptors, at least in muscle.[4] If this is the case in brain, as well as muscle, then it is difficult to ascribe the behavioral and psychological effects of androgenic-anabolic steroids to interactions with androgen receptors. Two other hypotheses will be discussed, i.e., that androgenic-anabolic steroids, or their metabolites, act through estrogen or through glucocorticoid receptors.

B. ANDROGENIC-ANABOLIC STEROIDS MAY ACT THROUGH ESTROGEN RECEPTOR SYSTEMS IN THE BRAIN

As described above, circulating estrogen levels are quite high in athletes taking large doses of androgenic-anabolic steroids,[39] presumably due to aromatization of the androgens to estrogens. Thus, it is possible that some of the behavioral and psychological effects of androgenic-anabolic steroids may be due to the increased circulating levels of estrogens.

Estrogens facilitate a variety of behaviors that include a motivational component or that are mediated by ascending dopaminergic systems in the brain (nigrostriatal or mesocortical). For instance, following estrogen treatment, female rats will traverse numerous barriers to gain access to male rates,[67] indicating that estrogens can increase certain motivated behaviors. Estrogen also alters a variety of dopamine-facilitated behaviors, but in a very complex fashion (reviewed in Reference 68). In rats, estrogen can facilitate or inhibit amphetamine-induced stereotypic behavior (continuous sniffing, licking, gnawing, and locomotion), depending on the dose and interval after estrogen administration. Rates of intracranial self-stimulation vary with the stage of the estrous cycle in rats,[69,70] suggesting that estrogen may modulate this dopamine-dependent[49] behavior. Roberts et al.[71] found that dopaminergic antagonists increase the rate of self-administration of cocaine in rats; this facilitative effect of dopaminergic antagonists varied with the stage of the estrous cycle, and decreased following ovariectomy. However, the effect of ovariectomy could not be overcome by administration of estrogen. Baseline levels of cocaine self-administration were not affected by ovariectomy. These results suggest that estrogen or some other ovarian hormone may interact with dopaminergic transmission in brain regions involved in cocaine self-administration. Another assessment of dopaminergic activity in extrapyramidal pathways is the amount of rotation induced by one of the following treatments: amphetamine administration, electrical stimulation of the nigrostriatal system, lesioning of the contralateral nigrostriatal system, or intrastriatal injection of dopaminergic agonists. The rotational responses to these challenges are altered by estrogen treatment or vary during the estrous cycle (reviewed in Reference 68). Estrogen can also modulate catalepsy induced by dopaminergic agonists (reviewed in Reference 68) and can increase the ''dorsal immobility response'' by a direct action on the striatum.[72,73]

In primates, estrogen has also been reported to modify behaviors dependent on dopamine transmission in the nigrostriatal system. For instance, in the monkey, estrogen reduces the effect of the dopamine receptor agonist apomorphine on lingual dyskinesia induced by lesioning the substantia nigra,[74] suggesting an inhibitory effect of estrogen on dopaminergic transmission in the nigrostriatal system. This inhibitory effect fits with several clinical studies that have indicated an antidopaminergic (neuroleptic) effect of estrogen on parkinsonian symptoms, tardive dyskinesia, and choreatic disorders.[68,75]

In addition to these extrapyramidal behaviors modulated by estrogen, a substantial literature exists indicating that estrogen can modulate dopaminergic activity in the basal ganglia (see ref. 68 for a review). A large number of reports, many contradictory to each other, indicate that estrogen can alter dopaminergic transmission in the nigrostriatal system. Both pre- and postsynaptic mechanisms have been invoked. The situation is complicated due to the variety of different paradigms investigated, to the metabolism of estrogen to catecholestrogen (2-hydroxyestrogen), and to estrogen-induced increases in circulating prolactin levels, which can, in turn, induce alterations in dopamine receptors. In addition, estrogen may have differential effects in the nigrostriatal and mesocortical systems. Van Hartesveldt and Joyce[68] summarized the literature, and suggested that estrogen inhibits dopaminergic activity in the nigrostriatal system, but facilitates activity in the mesocortical system.

The overall inhibitory effect of estrogen on dopaminergic transmission in the nigrostriatal system is the result of a complex interrelationship between pre- and postsynaptic events. Estrogen treatment reportedly decreases firing rates of neurons in the substantia nigra,[76,77] thus decreasing dopamine turnover in the striatum in a complex time-dependent manner.[68] Equally complex postsynaptic alterations in dopaminergic sensitivity occur following estrogen treatment. Several studies have indicated that estrogen increases the density of dopaminergic receptors (both D1 and D2) in the striatum.[68,78-80] The effect occurs in both male and female rats. However, the effect can be blocked by hypophysectomy[81] and can be mimicked by prolactin treatment,[82] suggesting that the effect is due to estrogen-induced prolactin release.

Despite this extensive literature on estrogen-induced alterations in behaviors mediated by dopamine and on alterations in dopaminergic activity, possible mechanisms have been difficult to determine. As indicated above, a complicating feature is estrogen-induced prolactin secretion, which can alter dopamine receptors in the striatum. The anatomical basis for possible direct actions of estrogens on the brain has been established by localization of neurons with estrogen receptors. A recent report by Simerly et al.[55] indicates that a subset of the brain regions which contain neurons expressing the androgen receptor gene also contain neurons with estrogen receptor mRNA, although generally fewer neurons express estrogen receptor mRNA than androgen receptor mRNA. Notably, neurons expressing estrogen receptor mRNA were not detected in the substantia nigra or ventral tegmental area. These results based on *in situ* hybridization largely confirmed previous studies based on steroid autoradiography,[66,83] with a notable exception in the dorsomedial portion of the nucleus accumbens, where estrogen-concentrating neurons were found[83] but no neurons with estrogen receptor mRNA were found.[55] Simerly et al.[55] raise the interesting possibility of the existence of an estrogen-binding protein whose mRNA does not cross-hybridize with an estrogen receptor probe.

In summary, estrogen has been shown in a variety of situations to interact with dopaminergic transmission in extrapyramidal and motivational pathways, at least in rodents. Thus, it is possible that the high levels of circulating estrogen found in athletes self-administering large doses of androgenic-anabolic steroids may mediate some of the behavioral and psychological effects of these steroids. However, in primates, including humans, estrogen appears to be generally inhibitory to dopaminergic transmission, at least in the nigrostriatal system. Possible effects of estrogen on dopaminergic transmission in mesolimbic and mesocortical systems in primates have not been examined to our knowledge. Since we postulate that androgenic-anabolic steroids exert behavioral and psychological effects by increasing dopaminergic activity in mesolimbic and mesocortical systems, we will also propose an additional possible mechanism of androgenic-anabolic steroid action, i.e., that the behavioral and psychological effects of these steroids may lie in interactions with glucocorticoid receptors.

C. ANDROGENIC-ANABOLIC STEROIDS MAY ACT THROUGH GLUCOCORTICOID RECEPTOR SYSTEMS IN THE BRAIN

The decreased circulating levels of ACTH in athletes self-administering large doses of androgenic-anabolic steroids[39] suggest that these steroids, or their metabolites, may act as glucocorticoid receptor agonists. In muscle, testosterone and other androgenic-anabolic steroids can displace dexamethasone or cortisol from the glucocorticoid receptor.[84] Moreover, expression of the glucocorticoid receptor in the levator ani muscle increases after castration in the rat,[85] suggesting that androgens normally downregulate expression of the glucocorticoid receptor, as do glucocorticoids.[86] Thus, it is possible that high doses of androgenic-anabolic steroids act as agonists, or possibly partial agonists, at glucocorticoid receptors in the brain, thus exerting behavioral and psychological effects. Indeed, we have obtained some preliminary evidence that large doses of androgenic-anabolic steroids can have an effect similar to corticosterone on inducing translocation of glucocorticoid receptors from the cytoplasm to the nucleus in the brains of adrenalectomized rats.[86a] Moreover, corticosteroids have long been recognized to exert psychological effects, some of which are similar to symptoms seen with androgen abuse, e.g., euphoria, depression, and psychotic reactions.[87-89] Moreover, a clinical syndrome of withdrawal has been described,[90,91] suggesting dependence on corticosteroids.

Two glucocorticoid receptors may be described in the nervous system, a Type I or classical mineralocorticoid receptor, and a Type II or classical glucocorticoid receptor (reviewed in Reference 92). The Type I receptor, with a high affinity for both aldosterone and corticosterone, is thought to mediate the tonic effects of corticosteroids within the diurnal rhythm of steroid levels, and perhaps the central effects of aldosterone on salt intake. The Type II receptor, with a lower affinity for aldosterone and corticosterone, is thought to mediate neural responses to stress levels of glucocorticoids. Both receptors are widely distributed throughout the nervous system, as determined by immunocytochemistry[43,93] or *in situ* hybridization.[94-96]

A considerable amount of evidence indicates that stress and glucocorticoids can activate dopaminergic systems in the rat brain (reviewed in Reference 92). For instance, repeated administration of amphetamine induces a sensitization effect, resulting in increasing levels of locomotor activity following each treatment; this sensitization effect is dependent on increased dopaminergic turnover, especially in the nucleus accumbens.[97] This sensitization effect is abolished by adrenalectomy and restored by the synthetic glucocorticoid dexamethasone, which binds selectively to the Type II receptor, but not by deoxycorticosterone, which binds selectively to the Type I receptor.[98] Recent evidence indicates that corticosterone can increase the amount of self-administration of amphetamine in rats, suggesting that corticosterone facilitates the reinforcing action of dopamine.[99] Such effects of glucocorticoids may be mediated by direct action on dopaminergic neurons in the ventral tegmental area, since these neurons have Type II receptors.[100] Alternatively, presynaptic mechanisms may also be invoked, since glucocorticoids have been reported to inhibit synaptic uptake of dopamine,[101] an effect that should increase the amount of dopamine in the synaptic cleft. Postsynaptic mechanisms of glucocorticoid action may also be invoked, since these steroids can alter the sensitivity of adenylate cyclase to catecholamines.[102]

However, the clinical syndromes associated with chronic overexposure to glucocorticoids do not match precisely those associated with abuse of androgenic-anabolic steroids; overexposure to glucocorticoids leads more often to depression than to mania.[89] Thus, it is unlikely that behavioral and psychological effects of androgenic-anabolic steroids can be attributed entirely to interactions with glucocorticoid receptor systems. Rather, it is likely that the effects of androgenic-anabolic steroids are mediated in part through androgen receptors, in part through estrogen receptors, and in part through glucocorticoid receptors. Activation of each of these steroid receptor systems can result in increased dopaminergic activity in the mesolimbic/mesocortical systems. Whether there is synergism between the effects of the different steroids on dopaminergic activity has not been studied.

V. CONSEQUENCES OF INCREASED DOPAMINE TURNOVER IN THE MESOCORTICAL AND MESOLIMBIC SYSTEMS

If our hypothesis that androgenic-anabolic steroids increase dopamine turnover in the mesocortical and mesolimbic systems is correct, what effects might this increased turnover have on expression of neuropeptide genes and on activity of other neurotransmitter systems? Several studies have indicated that expression of the proenkephalin gene is inhibited by dopamine transmission in the nigrostriatal system.[103,104] Dopamine can also increase expression of the gene encoding substance P in the striatum.[105,106] Whether such effects also occur in the nucleus accumbens has not been determined.

In addition to alterations in neuropeptide gene expression produced by steroid-induced changes in dopaminergic transmission, steroid hormones can have direct effects on expression of neuropeptide genes (reviewed in Reference 45). Estrogen has been reported to increase expression of the proenkephalin gene in the ventromedial nucleus of the hypothalamus,[107] while estrogen reportedly decreases pro-opiomelanocortin (POMC) gene expression in the hypothalamus.[108] More recent work has shown that testosterone can also inhibit POMC gene expression,[109,110] although opposite results have also been published.[111] Testosterone has also recently been reported to increase expression of the genes which code for both growth hormone-releasing factor[112] and somatostatin, the growth hormone-inhibiting factor.[113] While these reports may provide some clues as to actions of androgenic-anabolic steroids on neuropeptide gene expression in the brain, it should be emphasized that no studies have been reported to date on effects of the massive doses and combinations of steroids typically abused by athletes.

It should be emphasized that the hypothesis that we propose, i.e., that androgenic-anabolic steroids exert behavioral and psychological or psychiatric effects by increasing dopaminergic transmission in mesolimbic and mesocortical systems, is by no means proved. Very little research has been published on the effects of massive doses of androgenic-anabolic steroids on the brain. We present this hypothesis to stimulate research into this area, which is of increasing concern to society.

REFERENCES

1. **Tennant, F., Black, D. L., and Voy, R. O.,** Anabolic steroid dependence with opioid-type features, *N. Engl. J. Med.*, 319, 578, 1988.

2. **Haupt, H. A. and Rovere, G. D.,** Anabolic steroids: a review of the literature, *Am. J. Sports Med.,* 12, 469, 1984.
3. **Mooradian, A. D., Morley, J. E., and Korenman, S. G.,** Biological actions of androgens, *Endocr. Rev.,* 8, 1, 1987.
4. **Wilson, J. D.,** Androgen abuse by athletes, *Endocr. Rev.,* 9, 181, 1988.
5. American Medical Association, Council on Scientific Affairs. Drug abuse in athletes: anabolic steroids and human growth hormone, *J. Am. Med. Assoc.,* 259, 1703, 1988.
6. **Windsor, R. E. and Dimitru, D.,** Anabolic steroid use by athletes: how serious are the health hazards? *Postgrad. Med.,* 84, 37, 1988.
7. **Pope, H. G., Jr. and Katz, D. L.,** Affective and psychotic symptoms associated with anabolic steroid use, *Am. J. Psychiatry,* 145, 487, 1988.
8. **Jaffe, J. H.,** Drug addiction and drug abuse. in: *Goodman and Gilman's The Pharmacological Basis of Therapeutics,* 8th ed., Goodman, A. G., Rall, T. W., Nies, A. S., and Taylor, P., Eds., Pergamon Press, New York, 1990, 522.
9. **Cranshaw, J. P.,** Recognizing anabolic steroid abuse, *Patient Care,* 15 Aug. 28, 1985.
10. **Strauss, R. H., Wright, J. E., Finerman, G. A., and Catlin, D. H.,** Side effects of anabolic steroids in weight-trained men, *Physician Sports Med.,* 11, 87, 1983.
11. **Uzych, L.,** Steroids and the controlled substances act, *Biol. Psychiatry,* 27, 561, 1990.
12. **Pope, H. G., Jr., Katz, D. L., and Champoux, R.,** Anabolic-androgenic steroid use among 1,010 college men, *Physician Sports Med.,* 16, 75, 1988.
13. **Johnson, M. D., Jay, M. S., Shoup, B., and Rickert, V. I.,** Anabolic steroid use by male adolescents, *Pediatrics,* 83, 921, 1989.
14. **Buckley, W. E., Yesalis, C. E., Friedl, K. E., Anderson, W. A., Streit, A. L. and Wright, J. E.,** Estimated prevalence of anabolic steroid use among male high school seniors, *J. Am. Med. Assoc.,* 260, 3441, 1988.
15. **Nuzzo, N. A. and Waller, D. P.,** Drug abuse in athletes. in *Drugs, Athletes and Physical Performance,* Thomas, J. A., Ed., Plenum Press, New York, 1988, 141.
16. **Johnson, M. D.,** Anabolic steroid use in adolescent athletes, *Sports Med.,* 37, 1111, 1990.
17. **Miller, N. E., Hubert, G., and Hamilton, J. B.,** Mental and behavioral changes following male hormone treatment of adult castration, hypogonadism, and psychic impotence, *Proc. Soc. Exp. Biol. Med.,* 38, 538, 1938.
18. **Altschule, M. D. and Tillotson, K. J.,** The use of testosterone in the treatment of depressions, *N. Eng. J. Med.,* 239, 1036, 1948.
19. **Vogel, W., Klaiber, E. L., and Broverman, D. M.,** A comparison of the antidepressant effects of a synthetic androgen (mesterolone) and amitriptyline in depressed men, *J. Clin. Psychiatry,* 46, 6, 1985.
20. **Wilson, I. C., Prange, A. J., and Lara, P. P.,** Methyltestosterone with imipramine in men: conversion of depression to paranoid reaction, *Am. J. Psychiatry,* 131, 21, 1974.
21. **Annitto, W. J. and Layman, W. A.,** Anabolic steroids and acute schizophrenic episode, *J. Clin. Psychiatry,* 41, 143, 1980.
22. **Freinhar, J. P. and Alvarez, W.,** Androgen-induced hypomania, *J. Clin. Psychiatry,* 46, 354, 1985.
23. **Pope, H. G., Jr. and Katz, D. L.,** Body-builder's psychosis, *Lancet,* 11, Ap., 863, 1987.
24. **Brower, K. J., Blow, F. C., Beresford, T. P., and Fuelling, C.,** Anabolic-androgenic steroid dependence, *J. Clin. Psychiatry,* 50, 31, 1989.
25. **Sloviter, R., Valiquette, G., Abrams, G., Ronk, E., Sallas, A., Paul, L., and Neubort, S.,** Selective loss of hippocampal granule cells in the mature brain after adrenalectomy, *Science,* 243, 535, 1989.
26. **Sapolsky, R., Krey, L., and McEwen, B.,** Prolonged glucocorticoid exposure reduces hippocampal neuron number: implications for aging, *J. Neurosci.,* 5, 1221, 1985.

27. **Armanini, M. P., Hutchins, C., Stein, B. A., and Sapolsky, R. M.,** Glucocorticoid endangerment of hippocampal neurons is NMDA-receptor dependent, *Brain Res.,* 532, 7, 1990.
28. **Goodwin, F. K. and Jamison, K. R.,** *Manic-Depressive Illness,* Oxford University Press, New York, 1990.
29. **Swerdlow, N. R. and Koob, G. F.,** Dopamine, schizophrenia, mania, and depression: toward a unified hypothesis of cortico-striao-pallido-thalamic function, *Behav. Brain Sci.,* 10, 197, 1987.
30. **Depue, R. A. and Iacono, W. G.,** Neurobehavioral aspects of affective disorders, *Annu. Rev. Psychol.,* 40, 457, 1989.
31. **Moore, R. Y. and Bloom, F. E.,** Central catecholamine neuron systems: anatomy and physiology of the dopamine systems, *Annu. Rev. Neurosci.,* 1, 129, 1978.
32. **Fibiger, H. C. and Phillips, A. G.,** Reward, motivation, cognition: psychology of mesotelencephalic dopamine systems, in *Handbook of Physiology. The Nervous System,* Vol. IV, Mountcastle, V. B., Bloom, F. E., and Geiger, S. R., Eds., 1986, 647.
33. **Berger, B., Gaspar, P., and Verney, C.,** Dopaminergic innervation of the cerebral cortex: unexpected differences between rodents and primates, *Trends Neurosci.,* 14, 21, 1991.
34. **Mogenson, G. J., Jones, D. L., and Yim, C. Y.,** From motivation to action: functional interface between the limbic system and the motor system, *Prog. Neurobiol.,* 14, 69, 1980.
35. American Psychiatric Association. *Diagnostic and Statistical Manual of Mental Disorders,* 3rd ed., rev., Washington, D.C., APA, 1987.
36. **Kashkin, K. B. and Kleber, H. D.,** Hooked on hormones? An anabolic steroid addiction hypothesis, *J. Am. Med. Assoc.,* 262, 3166, 1989.
37. **Koob, G. F. and Bloom, F. E.,** Cellular and molecular mechanism of drug dependence, *Science,* 242, 715, 1988.
38. **Ritz, M. C., Lamb, R. J., Goldberg, S. R., and Kuhar, M. J.,** Cocaine receptors on dopamine transporters are related to self-administration of cocaine, *Science,* 237, 1219, 1987.
39. **Alen, M., Reinla, M., and Vihko, R.,** Response of serum hormones to androgen administration in power athletes, *Med. Sci. Sports Exerc.,* 17, 354, 1985.
40. **Alen, M. Rahkila, P., Reinla, M., and Vikho, R.,** Androgenic-anabolic steroid effects on serum thyroid, pituitary, and steroid hormones in athletes, *Am. J. Sports Med.,* 15, 357, 1987.
41. **Chicca, M. G., Sardeno, G., and Mariani, G.,** Effect of anabolic treatment on the serum levels of gonadotropins, testosterone, prolactin, thyroid hormones and myoglobin of male athletes under physical training, *J. Nucl. Med. Allied Sci.,* 25, 79, 1981.
42. **Shepard, R. J., Killinger, D., and Fried, T.,** Responses to sustained use of anabolic steroid, *Br. J. Sports Med.,* 11, 170, 1977.
43. **Ahima, R. S. and Harlan, R. E.,** Charting of type II glucocorticoid receptor-like immunoreactivity in the rat central nervous system, *Neuroscience,* 39, 579, 1990.
44. **McEwen, B. S.,** Steroid hormone interactions with the brain: cellular and molecular aspects, *Rev. Neurosci.,* 4, 1, 1979.
45. **Harlan, R. E.,** Regulation of neuropeptide gene expression by steroid hormones, *Mol. Neurobiol.,* 2, 183, 1988.
46. **Bitran, D. and Hull, E. M.,** Pharmacological analysis of male rat sexual behavior, *Neurosci. Biobehav. Rev.,* 11, 365, 1987.
47. **Hull, E. M., Bitran, D., Pehek, E. A., Warner, R. K., Band, L. C., and Holms, G. M.,** Dopaminergic control of male sex behavior in rats: effects of an intracerebrally infused agonist, *Brain Res.,* 370, 73, 1986.
48. **Cherek, D. R., Steinberg, J. L., Kelly, T. H., and Robinson, D. E.,** Effects of d-amphetamine on human aggressive behavior. *Psychopharmacology,* 88, 381, 1986.

49. **Wise, R. A. and Rompere, P.-P.,** Brain dopamine and reward, *Annu. Rev. Psychol.,* 40, 191, 1989.
50. **Alderson, L. M. and Baum, M. J.,** Differential effects of gonadal steroids on dopamine metabolism in mesolimbic and nigrostriatal pathways of male rat brain, *Brain Res.,* 218, 189, 1981.
51. **Mitchell, J. B. and Stewart, J.,** Effects of castration, steroid replacement, and sexual experience on mesolimbic dopamine and sexual behaviors in the male rat, *Brain Res.,* 491, 116, 1989.
52. **Sar, M. and Stumpf, W. E.,** Distribution of androgen target cells in rat forebrain and pituitary after [^{3}H] dihydrotestosterone administration, *J. Steroid Biochem.,* 8, 1131, 1977.
53. **Sar, M. and Stumpf, W. E.,** Androgen concentration in motor neurons of cranial nerves and spinal cord, *Science,* 197, 77, 1977.
54. **Sar, M., Lubahn, D. B., French, F. S., and Wilson, E. M.,** Immunohistochemical localization of the androgen receptor in rat and human tissues, *Endocrinology,* 127, 3180, 1990.
55. **Simerly, R. B., Chang, C., Muramatsu, M., and Swanson, L. W.,** Distribution of androgen and estrogen receptor in mRNA-containing cells in the rat brain: an *in situ* hybridization study, *J. Comp. Neurol.,* 294, 76, 1990.
56. **Handa, R. J., Connolly, P. B., and Resko, J. A.,** Ontogeny of cytosolic androgen receptors in the brain of the fetal rhesus monkey, *Endocrinology,* 122, 1890, 1988.
57. **Handa, R. J., Roselli, C. E., and Resko, J. A.,** Distribution of androgen receptor in microdissected brain areas of the female baboon *(Papio cynocephalus), Brain Res.,* 445, 111, 1988.
58. **Michael, R. P. and Rees, H. D.,** Autoradiographic localization of [^{3}H]dihydrotestosterone in the preoptic area, hypothalamus, and amygdala of a male rhesus monkey, *Life Sci.,* 30, 2087, 1982.
59. **Pomerantz, S. M., Fox, T. O., Sholl, S. A., Vito, C. C., and Goy, R. W.,** Androgen receptors in the fetal monkey brain, *Endocrinology,* 116, 83, 1985.
60. **Rees, H. D. and Michael, R. P.,** Brain cells of the male rhesus monkey accumulate ^{3}H-testosterone or its metabolites, *J. Comp. Neurol.,* 206, 273, 1982.
61. **Rees, H. D. and Michael, R. P.,** Autoradiographic localization of [^{3}H] dihydrotestosterone in the thalamus and brain stem of a male rhesus monkey, *Neuroendocrinol. Lett.,* 5, 55, 1983.
62. **Sheridan, P. J. and Weaker, F. J.,** Androgen receptor systems in the brainstem of the primate, *Brain Res.,* 235, 235, 1982.
63. **Sheridan, P. J., Hagino, N., and Weaker, F. J.,** Androgen-concentrating cells in the periventricular brain of the female rhesus monkey, *J. Comp. Neurol.,* 207, 93, 1982.
64. **Sholl, S. A. and Pomerantz, S. M.,** Androgen receptors in the cerebral cortex of fetal female rhesus monkeys, *Endocrinology,* 119, 1625, 1986.
65. **Sarrieau, A., Mitchell, J. B., Lal, S., Olivier, A., Quirion, R., and Meaney, M. J.,** Androgen binding sites in human temporal cortex, *Neuroendocrinology,* 51, 713, 1990.
66. **Heritage, A. S., Stumpf, W. E., Sar, M., and Grant, L. D.,** Brainstem catecholamine neurons are target sites for sex steroid hormones, *Science,* 207, 1377, 1980.
67. **Beach, F. A.,** Sexual attractivity, proceptivity and receptivity in female mammals, *Horm. Behav.,* 7, 105, 1976.
68. **Van Hartesveldt, C. and Joyce, J. N.,** Effects of estrogen on the basal ganglia, *Neurosci. Biobehav. Rev.,* 10, 1, 1986.
69. **Prescott, R. G. W.,** Estrous cycle in the rat: effects on self-stimulation behavior, *Science,* 152, 796, 1966.
70. **Steiner, M., Katz, R. J., and Carroll, B. J.,** Detailed analysis of estrous-related changes in wheel running and self-stimulation, *Physiol. Behav.,* 28, 201, 1982.

71. **Roberts, D. C. S., Dalton, J. C. H., and Vickers, G. J.,** Increased self-administration of cocaine following haloperidol: effect of ovariectomy, estrogen replacement, and estrous cycle, *Pharmacol. Biochem. Behav.*, 26, 37, 1987.
72. **Van Hartesveldt, C., Cottrell, G. A., and Meyer, M. E.,** Effects of intracerebral estradiol on the dorsal immobility response in the rat, *Pharmacol. Biochem. Behav.*, 33, 321, 1989.
73. **Van Hartesveldt, C., Cottrell, G. A., and Meyer, M. E.,** The effects of intrastriatal hormones on the dorsal immobility response in gonadectomized male and female rats, *Pharmacol. Biochem. Behav.*, 34, 459, 1989.
74. **Bedard, P. J., Malouin, F., Di Paola, T., and Labrie, F.,** Estradiol, TRH and striatal dopaminergic mechanisms, *Prog. Neuropsychopharmacol. Biol. Psychiatry*, 6, 555, 1982.
75. **Bedard, P. J., Langelier, P., Dankova, J., Villeneuve, A., Di Paolo, T., Barden, N., Labrie, F., Boissier, J. R., and Euvard, C.,** Estrogens, progesterone, and the extrapyramidal system, *Adv. Neurol.*, 24, 411, 1979.
76. **Chiodo, L. A. and Caggiula, A. R.,** Alterations in basal firing rate and autoreceptor sensitivity of dopamine neurons in the substantia nigra following acute and extended exposure to estrogen, *Eur. J. Pharmacol.*, 67, 165, 1980.
77. **Chiodo, L. A. and Caggiula, A. R.,** Substantia nigra dopamine neurons: alterations in basal discharge rates and autoreceptor sensitivity induced by estrogen, *Neuropharmacology*, 22, 593, 1983.
78. **Hruska, R. E.,** Elevation of striatal dopamine receptors by estrogen: dose and time studies, *J. Neurochem.*, 47, 1908, 1986.
79. **Hruska, R. E. and Nowak, M. W.,** Estrogen treatment increases the density of D1 dopamine receptors in the rat striatum, *Brain Res.*, 442, 349, 1988.
80. **Falardeau, P. and Di Paolo, T.,** Regional effect of estradiol on rat caudate-putamen dopamine receptors: lateral-medial differences, *Neurosci. Lett.*, 74, 43, 1987.
81. **Hruska, R. E., Ludmer, L. M., and Silbergeld, E. K.,** Characterization of the striatal dopamine receptor supersensitivity produced by estrogen treatment of male rats, *Neuropharmacology*, 19, 923, 1980.
82. **Hruska, R. E., Pitman, K. T., Silbergold, E. K., and Ludmer, L. M.,** Prolactin increases the density of striatal dopamine receptors in normal and hypophysectomized male rats, *Life Sci.*, 30, 547, 1982.
83. **Pfaff, D. W. and Keiner, M.,** Atlas of estradiol-concentrating cells in the central nervous system of the female rat, *J. Comp. Neurol.*, 151, 121, 1973.
84. **Mayer, M. and Rosen, F.,** Interaction of anabolic steroids with glucocorticoid receptor sites in rat muscle cytosol, *Am. J. Physiol.*, 229, 1381, 1975.
85. **DuBois, D. C. and Almon, R. R.,** Perineal muscles: possible androgen regulation of glucocorticoid receptor sites in the rat, *J. Endocrinol.*, 102, 225, 1984.
86. **Rosewicz, S., McDonald, A. R., Maddux, B. A., Goldfine, I. D., Miesfeld, R. L., and Logsdon, C. D.,** Mechanism of glucocorticoid receptor down regulation by glucocorticoids, *J. Biol. Chem.*, 263, 2581, 1988.

86a. **Ahima, R. S. and Harlan, R. E.,** unpublished.

87. **Ling, M. H. M., Perry, P. J., and Tsuang, M. T.,** Side effects of corticosteroid therapy, *Arch. Gen. Psychiatry*, 38, 471, 1981.
88. **D'Orban, P. T.,** Steroid-induced psychosis, *Lancet*, 16 Sept., 694, 1989.
89. **Gold, P. W., Goodwin, F. K., and Chrousos, G. P.,** Clinical and biochemical manifestations of depression: relation to the neurobiology of stress, *N. Engl. J. Med.*, 319, 413, 1988.
90. **Amatruda, T. T., Jr., Hurst, M. M., and D'Esopo, N. D.,** Certain endocrine and metabolic facets of the steroid withdrawal syndrome, *J. Clin. Endocrinol.*, 25, 1207, 1965.
91. **Dixon, R. B. and Christy, N. P.,** On the various forms of corticosteroid withdrawal syndrome, *Am. J. Med.*, 68, 224, 1980.

92. **McEwen, B. S., DeKloet, E. R., and Rostene, W.,** Adrenal steroid receptors and actions in the nervous system, *Physiol. Rev.*, 66, 1121, 1986.
93. **Ahima, R. S., Krozowski, Z., and Harlan, R. E.,** Type I corticosteroid receptor-like immunoreactivity in the rat CNS: distribution and regulation by corticosteroids, *J. Comp. Neurol.*, submitted.
94. **Sousa, R. J., Tannery, N. H., and Lafer, E. M.,** *In situ* hybridization mapping of glucocorticoid receptor messenger ribonucleic acid in rat brain, *Mol. Endocrinol.*, 3, 481, 1989.
95. **Arronson, M., Fuxe, K., Dong, Y., Luigi, F., Agnati, F., Okret, S., and Gustaffson, J. A.,** Localization of glucocorticoid receptor mRNA in the male rat brain by *in situ* hybridization, *Proc. Natl. Acad. Sci. U.S.A.*, 85, 9331, 1988.
96. **Arriza, J., Simerly, R., Swanson, L., and Evans, R.,** The neuronal mineralocorticoid receptor as a mediator of glucocorticoid response, *Neuron,* 1, 887, 1988.
97. **Robinson, T. E., Jurson, P. A., Bennet, J. A., and Bentgen, K. M.,** Persistent sensitization of dopamine neurotransmission in ventral striatum (nucleus accumbens) produced by prior experience with (+)-amphetamine: a microdialysis study in freely moving rats, *Brain Res.*, 462, 211, 1988.
98. **Rivet, J.-M., Stinus, L., LeMoal, M., and Mormede, P.,** Behavioral sensitization to amphetamine is dependent on corticosteroid receptor activation, *Brain Res.*, 498, 149, 1989.
99. **Piazza, P. V., Maccari, S., Deminiere, J.-M., Le Moal, M., Mormede, P., and Simon, H.,** Corticosterone levels determine individual vulnerability to amphetamine self-administration, *Proc. Natl. Acad. Sci. U.S.A.*, 88, 2088, 1991.
100. **Harfstrand, A., Fuxe, K., Cintra, A., Agnati, L. F., Zini, I., Wikstrom, A. C., Okret, S., Zhao-Ying, Y., Goldstein, M., Steinbush, H., Verhofstad, A., and Gustafsson, J. A.,** Glucocorticoid receptor immunoreactivity in monoaminergic neurons in rat brain, *Proc. Natl. Acad. Sci. U.S.A.*, 83, 9779, 1986.
101. **Gilad, G. M., Rabey, J. M., and Gilad, V. H.,** Presynaptic effects of glucocorticoids on dopaminergic and cholinergic synaptosomes. Implications for rapid endocrine-neural interactions in stress, *Life Sci.*, 40, 2401, 1987.
102. **Harrelson, A. L., Rostene, W., and McEwen, B. S.,** Adrenocortical steroids modify neurotransmitter-stimulated cyclic AMP accumulation in the hippocampus and limbic brain of the rat, *J. Neurochem.*, 48, 1648, 1987.
103. **Tang, F., Costa, E., and Schwartz, J. P.,** Increase of proenkephalin mRNA and enkephalin content of rat striatum after daily injection of haloperidol for two to three weeks, *Proc. Natl. Acad. Sci. U.S.A.*, 80, 3841, 1983.
104. **Romano, G. J., Shivers, B. D., Harlan, R. E., Howells, R. D., and Pfaff, D. W.,** Haloperidol increases proenkephalin mRNA levels in the caudate-putamen of the rat: a quantitative study at the cellular level using *in situ* hybridization, *Mol. Brain Res.*, 2, 33, 1987.
105. **Young, W. S., III, Bonner, T. I., and Brann, M. R.,** Mesencephalic dopamine neurons regulate the expression of neuropeptide mRNAs in the rat forebrain, *Proc. Natl. Acad. Sci. U.S.A.*, 83, 9827, 1986.
106. **Bannon, M. J., Lee, J.-M., Giraud, P., Young, A., Affolter, H.-U., and Bonner, T. I.,** Dopamine antagonist haloperidol decreases substance P, substance K and preprotachykinin mRNAs in rat striatonigral neurons, *J. Biol. Chem.*, 261, 6640, 1986.
107. **Romano, G. J., Harlan, R. E., Shivers, B. D., Howells, R. D., and Pfaff, D. W.,** Estrogen increases proenkephalin messenger ribonucleic acid levels in the ventromedial hypothalamus of the rat, *Mol. Endocrinol.*, 2, 1320, 1988.
108. **Wilcox, J. N. and Roberts, J. L.,** Estrogen decreases rat hypothalamic pro-opiomelanocortin messenger ribonucleic acid levels, *Endocrinology,* 117, 2392, 1985.
109. **Blum, M., Roberts, J. L., and Wardlaw, S. L.,** Androgen regulation of pro-opiomelanocortin gene expression and peptide content in the basal hypothalamus, *Endocrinology,* 124, 2283, 1989.

110. **Wardlaw, S. L. and Blum, M.,** Biphasic effect of orchiectomy on pro-opiomelanocortin gene expression in the hypothalamus, *Neuroendocrinology,* 52, 521, 1990.
111. **Clifton, D. K. and Steiner, R. A.,** Testosterone regulation of proopiomelanocortin messenger ribonucleic acid in the arcuate nucleus of the male rat, *Endocrinology,* 124, 1697, 1989.
112. **Zeitler, P., Argente, J., Chowen-Breed, J. A., Clifton, D. K., and Steiner, R. A.,** Growth hormone-releasing hormone messenger ribonucleic acid in the hypothalamus of the adult male rat is increased by testosterone, *Endocrinology,* 127, 1362, 1990.
113. **Argente, J., Chowen-Breed, J. A., Steiner, R. A., and Clifton, D. K.,** Somatostatin messenger RNA in hypothalamic neurons is increased by testosterone through activation of androgen receptors and not by aromatization to estradiol, *Neuroendocrinology,* 52, 342, 1990.

Chapter 11

INFLUENCES OF THE HORMONAL MILIEU ON ACUTE AND CHRONIC BENZODIAZEPINE RESPONSES IN RATS

Marlene A. Wilson

TABLE OF CONTENTS

ABSTRACT

Benzodiazepine agonists, such as diazepam, are a widely prescribed class of compounds noted for their anticonvulsant, anxiolytic, sedative/hypnotic and muscle relaxant properties. The clinical utility of the benzodiazepines, however, is hampered by the development of tolerance to the pharmacological effects of these drugs following their chronic use. The mechanism of benzodiazepine (BZ) agonist action involves potentiation of the inhibitory actions of the neurotransmitter GABA through interaction at the GABA/benzodiazepine/chloride ionophore receptor complex. This chapter reviews recent evidence that gonadal steroid hormones or their derivatives also alter the GABA/benzodiazepine/ionophore complex and GABA-mediated responses. *In vivo* studies have indicated neuron- and region-specific effects of gonadal hormones on GABA/BZ receptors, while *in vitro* analyses have demonstrated direct steroid effects on the GABA receptor complex and GABAergic actions. Further, nonreproductive effects of gonadal steroid hormones on convulsability, anxiety, and sedation markedly resemble the pharmacological profile of the benzodiazepines. Such evidence would suggest that through their ability to modulate GABAergic systems, gonadal steroids might be capable of influencing benzodiazepine responses. This report specifically examines whether the steroid hormone milieu in animals can modulate responses in rats following acute and chronic benzodiazepine exposure. Recent results indicate that *in vivo* modifications in the hormonal status of rats do not affect neural benzodiazepine receptors and do not significantly modulate the acute anticonvulsant effects of the benzodiazepine agonist diazepam. Electrophysiological studies indicate that neuronal responses to acute benzodiazepine administration may be altered in specific brain areas by the hormonal milieu of an animal. Further, it appears that the presence of gonadal steroids may play a permissive role in the development of tolerance to the anticonvulsant effects of the benzodiazepines in rats. Future studies should provide a better understanding of the influences of gonadal hormone-related factors on both acute benzodiazepine responsiveness and the development of tolerance to the pharmacological actions of these drugs.

I. INTRODUCTION

Benzodiazepine (BZ) agonists, such as diazepam, are a widely prescribed class of compounds noted for their anticonvulsant, anxiolytic, sedative/hypnotic and muscle relaxant properties. The clinical utility of the benzodiazepines, however, is hampered by the development of tolerance to the pharmacological effects of these drugs following their chronic use. Investigations of neural changes associated with both acute and prolonged exposure to the benzodiazepines have helped elucidate the mechanisms through which these compounds exert their pharmacological effects and extend our understanding of the development of tolerance to these drugs.

The mechanism of benzodiazepine agonist action is known to involve potentiation of the inhibitory actions of the amino acid neurotransmitter γ-aminobutyric acid (GABA).[1,2] Enhancement of GABAergic responses by BZ agonists is mediated through binding to a specific neural BZ receptor site that forms part of a supramolecular receptor complex for GABA.[1,2] The binding of various compounds to this receptor site is correlated with their potency in several tests of their benzodiazepine-like properties, suggesting that this site mediates the actions of these drugs.[3,4] This $GABA_A$/BZ receptor complex is a ligand-gated channel receptor that includes a GABA-activated ionophore for chloride. Both electrophysiological techniques and neurochemical assays of ^{36}Cl influx into brain vesicle preparations have demonstrated that the benzodiazepines alter GABA-stimulated chloride flux, apparently by allosterically increasing the number of channel openings.[5-7] This allosteric coupling is further suggested by the ability of GABA to increase the affinity of the BZ recognition site in *in vitro* binding assays.[8] GABA's influences on BZ binding and on neurophysiological responses appear to be similarly mediated by the low affinity state of the $GABA_A$ receptor.[8,9] Since BZ ligands lack activity in the absence of GABA, understanding the mechanisms' underlying responses to the benzodiazepines cannot easily be dissociated from understanding their interactions with the GABAergic system.

Recently, several studies have demonstrated that gonadal steroid hormones or their derivatives can alter the GABA/BZ/ionophore complex and GABA-mediated responses.[10-15] Clinical and animal studies have also shown that gonadal steroids can be anticonvulsant or proconvulsant, anxiolytic or anesthetic.[10,16-19] The similarity of these steroid effects with benzodiazepine actions not only supports the idea that gonadal steroids can modulate GABAergic systems, but suggests that these steroids might be capable of influencing BZ responses. This chapter reviews some of the evidence concerning the influences of gonadal steroid hormones on GABAergic processes as it relates to acute and/or chronic responses to the benzodiazepines. Further, it presents some results of studies examining whether the steroid hormone milieu in animals can modulate BZ responses.

II. HORMONAL INFLUENCES ON γ-AMINOBUTYRIC ACID AND BENZODIAZEPINE RECEPTOR SITES

Animal studies have shown that various aspects of GABAergic neurotransmission can be altered by gonadal steroid treatments, including various components of the GABA/BZ/chloride ionophore complex. Estrogen-induced increases[20-25] and decreases[26-28] in $GABA_A$ receptor levels have been demonstrated in gonadectomized rats or hamsters. Estrogen treatments also generally increase levels of BZ recognition sites[23,29,30] and channel-related sites (*t*-butylbicyclophosphorothionate binding)[23] in gonadectomized rats, although decreases in BZ receptors are observed in some brain areas.[30] While estrogen modulates GABA/BZ receptor sites in a regionally specific manner, identical changes are not consistently reported for any single brain area.

Progesterone treatments have more consistently been shown to increase levels of binding sites within the GABA receptor complex in animal studies. Progesterone regimens given to rats and hamsters led to increased levels of GABA receptors in various brain areas.[20,24,25,31] Increased BZ receptor sites were similarly observed in very specific regions of amygdala and hippocampus following progesterone treatments.[30] Prolonged progesterone exposure (10 to 14 d) enhanced cortical levels of BZ binding sites, although hippocampal levels were unaltered.[32] Administration of progesterone or its metabolites also alters physiological responses to GABA (see below).[11-15]

These results suggest that exogenous administration of gonadal steroid hormones can modulate various sites within the GABA/BZ/ionophore receptor complex in a regionally specific manner. Alternatively, relatively few studies have examined whether physiological fluctuations in gonadal steroid levels can alter various aspects of the GABA receptor complex. $GABA_A$ receptor levels do not vary over the estrous cycle in rats,[26] but are increased in late gestational periods and decreased postpartum.[33] Sex differences have also been demonstrated in high-affinity $GABA_A$ sites.[34,35] Males show higher levels,[34,35] but decreased affinity,[35] of GABA sites in hypothalamic areas than do female rats.

Studies in this laboratory have examined the influences of gender, estrous cycle, and diminution of female hormones on BZ receptor parameters of several brain areas in rats. We also determined whether there were correlations between circulating levels of estrogen and progesterone and BZ receptors in cycling females. Benzodiazepine receptors in cortex, hypothalamus-preoptic area, and hippocampus were not significantly different between males and females and were unaffected by removal of gonadal steroids by ovariectomy (Table 1). Thus, our studies do not support the sex differences in BZ binding observed in Roman rat strains.[36] In agreement with a previous report,[26] BZ receptors did not vary over the rat estrous cycle (Table 1). Further, no significant correlations were seen between circulating levels of estradiol or progesterone and BZ binding parameters in cycling female rats (data not shown). The BZ binding assay is not altered by the direct *in vitro* addition of up to 1 μM 17β-estradiol, progesterone, or testosterone (data not shown).[26,30,32] Thus, while prior studies have indicated that exogenous hormone administration can modulate both GABA and BZ receptor sites, physiological cyclic variations in such hormones do not appear to have substantial effects on BZ receptors. Such discrepancies may be related to a variety of factors, some of which are discussed below.

TABLE 1
Benzodiazepine Receptor Parameters in Various Hormone Groups

	Cortex		Hypothalamus		Hippocampus	
	B_{max}	K_d	B_{max}	K_d	B_{max}	K_d
Intact Females						
Diestrous 1	1346 ± 99	7.3 ± 0.2	998 ± 35	7.7 ± 0.6	1153 ± 70	7.2 ± 0.2
Diestrous 2	1388 ± 124	6.8 ± 0.4	1076 ± 81	7.6 ± 1.0	1004 ± 115	6.7 ± 0.2
Proestrous	1470 ± 98	7.0 ± 0.3	973 ± 53	7.1 ± 0.4	1078 ± 67	7.1 ± 0.2
Estrous	1299 ± 96	7.2 ± 0.3	1018 ± 146	7.1 ± 1.0	1104 ± 107	6.7 ± 0.5
Males	1498 ± 88	5.7 ± 0.1*	867 ± 98	8.0 ± 1.5	892 ± 73	6.3 ± 0.4
Ovariectomized females	1483 ± 130	6.6 ± 0.3	1017 ± 84	7.6 ± 0.9	1199 ± 89	7.0 ± 0.3

Note: The concentration (B_{max}; fmol/mg protein) and affinity (K_d, n*M*) of BZ receptors were determined by saturation analysis of [^{3}H]-flunitrazepam binding (see refs. 87,88) in cerebral cortex, hippocampus, and hypothalamus-preoptic area of cycling female, ovariectomized female (OVX), and male rats. Neither the concentration nor affinity of BZ receptors varied significantly over the estrous cycle. Female (all stages combined), male, and OVX groups also showed comparable BZ receptor numbers and K_d's in hypothalamus-preoptic area and hippocampus. In cortex, all groups had similar concentrations of BZ receptors, but males showed a greater receptor affinity (lower K_d) than the female or OVX groups. *, indicates $p<0.05$ using ANOVA for female-OVX-male comparison. Rats ($N = 5$ to 12 per group) were sacrificed at late night 3 to 4 weeks after gonadectomy or sham surgery. Estrous cyclicity was determined with vaginal smears and confirmed with uterine weights and serum estradiol and progesterone levels (measured by radioimmunoassay). Rats were handled and habituated to the decapitation procedure daily in order to minimize the effects of stress.

III. HORMONAL EFFECTS ON COUPLING BETWEEN γ-AMINOBUTYRIC ACID/BENZODIAZEPINE SITES

The ability of GABA to stimulate BZ binding is suggestive of the allosteric coupling between the GABA and BZ sites within the receptor complex.[8] In gonadectomized animals, GABA modulation of BZ binding in the brain regions examined was unaffected by prior estrogen treatments.[23] This suggests a lack of estrogen-related effects on coupling between the GABA and BZ sites in gonadectomized animals.[23] In contrast, intact rats showed an estrogen-induced decrease in the effects of GABA on BZ binding.[26]

We have compared GABA shifts in benzodiazepine binding in groups of intact males, intact females, and ovariectomized rats. A sex-specific difference was observed in GABAergic stimulation of BZ sites in hippocampal, but not cortical, hypothalamic, or cerebellar membranes (see Table 2). Further, the nature of this difference was related to whether the animals were habituated to the sacrifice procedures used to obtain neural tissue. Divergent responses in naive and habituated animals suggest that stress may play some role in determining the effects of gonadal hormones on GABA/BZ receptors and/or

TABLE 2
GABA Enhancement of Benzodiazepine Binding in Various Hormone Groups

	Unhabituated (%)	Habituated (%)
Intact female	58 ± 7	55 ± 1
Male	52 ± 5	67 ± 3*
Ovariectomized female	64 ± 5	55 ± 6

Note: Percentage increases in hippocampal BZ binding observed with the addition of GABA in unhabituated and habituated groups of cycling female, male, and ovariectomized rats. Unhabituated animals were naive while habituated animals were familiarized daily for at least 2 weeks with the sacrifice procedure. GABA enhancement in BZ binding at 1 n*M* [^{3}H]-flunitrazepam was assessed by adding GABA concentrations of 10^{-7} to 10^{-4} to the BZ receptor assay (see Reference 8). Increases at 100 μM GABA are shown. Analysis of variance with repeated measures (GABA concentration) was used to determine group differences. GABA significantly increased BZ binding in both unhabituated and habituated groups (F = 146 and 446, respectively, $p<0.0001$). No difference between hormone groups was observed in unhabituated rats ($F=0.5$, $p>0.05$). In habituated rats, males showed a significantly greater GABA shift than the female groups ($F=5.2, p<0.02$). *, Indicates male group differs significantly ($p<0.05$) from other groups by ANOVA. Values are mean ± SEM percentage increases over binding in the absence of GABA with $N=3$ per group.

responses. In naive animals (unhabituated), males showed a diminished GABA-related shift in hippocampal BZ binding compared to ovariectomized females (Table 2). When animals were habituated prior to sacrifice, males showed a significantly greater increase in BZ binding due to GABA than either female or OVX groups (Table 2). Although these results are far from conclusive, they suggest that stress may have an impact on the nature of effects associated with gonadal hormone status. Several studies have indicated that stressors can modulate the GABA/BZ/ionophore complex.[37-39] Further, certain stress-related responses are altered by gender and/or the gonadal hormone milieu.[40-43] Therefore, some alterations in GABA/BZ receptors related to gonadal status may actually be mediated by hormone- or sex-related modifications in stress responses, which subsequently modulate GABAergic systems.

Thus, receptor binding studies indicate that various *in vivo* regimens of estrogen or progesterone and/or certain aspects of *in vivo* hormonal status can modify the GABA/BZ receptor complex in a region-specific manner. The discrepancies between reports, however, confuse the exact nature and mechanisms of such hormonal influences. The inconsistent effects of various *in vivo* hormone manipulations may be related to a variety of factors, including differences in the hormone regimens, brain areas analyzed, and/or methods used to measure GABA/BZ receptors. For example, autoradiographic methodologies have demonstrated hormone-induced GABA receptor changes specific to CA_1, but not CA_3 or CA_4, neurons within the hippocampus.[24] When receptors were measured in the hippocampal region as a whole, this localized change was masked and an overall lack of steroid effects was seen.[24] Our results examining GABA/BZ coupling in habituated and unhabituated rats indicate that a variety of factors associated with stress may further complicate the effects of hormones observed in any particular study. Disparate effects of hormones might be attributable to subtle methodological differences between studies, such as whether subjects are gonadectomized or gonadectomized *plus* adrenalectomized and, therefore, have divergent stress-related responses. Further, steroid influences may also become more substantial during periods of dramatic fluctuations in the hormonal milieu, such as during gestation or postpartum.[33]

IV. MECHANISM OF STEROID HORMONE EFFECTS ON γ-AMINOBUTYRIC ACID/BENZODIAZEPINE RECEPTORS

Steroid hormones bind to specific intracellular receptor proteins through which they are capable of modifying protein synthesis. Since this influence on protein synthesis utilizes genomic mechanisms, such effects require from hours to days for their actions to be seen.[44] Many of the estrogen-induced changes in GABA/BZ receptors appear to be mediated primarily through such receptor-dependent genomic alterations.[21,30] Altered numbers of GABA receptors following estrogen administration are observed in brain areas considered to be estrogen-sensitive "target areas" such as the preoptic and hypothalamic regions,[22-28] although "nontarget" regions such as cortex, striatum, hippocampus, and cerebellum are also affected.[20-25,27] Autoradiographic analysis has indicated that localization of estrogen-induced decreases in GABA sites does not match the pattern of estrogen receptor distribution in the hypothalamus or other estrogen-sensitive areas.[28] These results suggest that some steroid effects, including those of estrogens, may be mediated through mechanisms independent of steroid receptors.

Steroid hormones have been shown to exert direct (i.e., nonsteroid receptor-mediated) effects on various neural responses and on receptors for neurotransmitters and drugs.[44,45] Of particular relevance is the recent evidence suggesting that steroid hormone derivatives are capable of specifically modulating the GABA receptor and its responses. Several derivatives have consistently been shown to enhance GABA function, including the reduced progesterone metabolite 5α-pregnane-3α-ol-20-one (3α-OH-DHP; 3α-hydroxy-

5α-dihydroprogesterone) and 3α,5α-tetrahydrodeoxycorticosterone (THDOC; 5α-pregnane-3α,21-diol-20-one). Steroid derivatives enhance BZ binding,[11,12,30,46] inhibit channel-site (TBPS) binding[11,12,47] and modulate GABA-activated chloride influx into synaptosomes[11,12,48] when added directly to the *in vitro* assays for these parameters. Direct, local applications of steroids also affect various physiological indices of GABA function with a very rapid time course.[11,12,14,15] Steroid metabolites can potentiate GABA actions on neuronal[11,12] or adrenal[14] cells in culture, on cuneate slices *in vitro*,[15,49] and on cells transfected with various $GABA_A$ receptor subunit combinations.[50] Androgen-derived steroids can similarly enhance GABA complex binding and potentiate physiological responses to GABA agonists.[15,51] Several derivatives (e.g., 3α-OH-DHP), exert effects within a physiologically relevent concentration range,[12,14] particularly in the presence of GABA.[47] Furthermore, progesterone administration *in vivo* enhances GABAergic responses of cerebellar Purkinje cells, with a time lag consistent with the conversion of progesterone into an active metabolite.[13] "Neurosteroids," such as pregnenolone sulfate, attenuate GABAergic effects,[52,53] although the physiological relevance of such an effect has been questioned.[54]

The rapid effects of steroids on physiological responses combined with their *in vitro* modulation of receptor sites has suggested that these derivatives interact directly with the GABA/BZ/ionophore complex. The structural specificity required for producing these direct GABAergic effects has led to the suggestion that these compounds may act through a distinct "steroid" binding site in the complex.[47,48,51] For example, compounds with a 3α-OH configuration exert substantially greater effects on GABAergic functions than 3β-OH steroids, suggesting a stereoselective requirement for activity.[12,15,46,49] Interestingly, membrane binding sites for progesterone have recently been demonstrated in brain.[55] Thus, while steroid influences on general membrane properties (e.g., fluidity) could mediate some of their rapid effects, evidence has supported the notion that steroid derivatives specifically and directly modulate the GABA receptor complex.

These direct steroid-GABA/BZ receptor interactions could be capable of modifying BZ responses, since BZs rely on a similar interaction with the GABA receptor complex for pharmacological activity. Furthermore, *in vivo* modulation of BZ responses could result from a combination of receptor-mediated, time-dependent, genomic steroid influences as well as direct, rapid steroid effects on the GABA receptor complex and its responses. Steroid derivatives have, in fact, been shown to modify BZ enhancement of GABA-activated chloride flux *in vitro*,[48] but relatively few studies have examined the ability of the steroid milieu to modulate *in vivo* BZ responses.

This laboratory has been examining whether *in vivo* modifications in the hormonal status of an animal can alter BZ responses. We have begun to determine the influences of gender and castration on BZ effects by electrophysiologically analyzing drug responses of single neurons in rats. At the other end of the spectrum, the anticonvulsant effects of the benzodiazepines have been used as a "behavioral" index of altered BZ responses due to hormone-dependent factors. The role of gonadal hormones in the development of tolerance to these drugs was also assessed in rats following chronic BZ exposure.

V. HORMONAL INFLUENCES ON BENZODIAZEPINE RESPONSES OF SUBSTANTIA NIGRA NEURONS

Several studies have indicated that influences of gonadal steroids on GABA/BZ/ionophore receptor sites are not only regionally specific, but can be confined to single populations of neurons within a given brain area.[24,31] Therefore, we have begun to compare BZ responses at the level of the single neuron in groups of rats with distinct gonadal hormone milieus. The substantia nigra parts reticulata (SNr) was chosen for these studies because it receives a dense GABAergic innervation which forms part of the striatonigral feedback loop. This input from striatum tonically exposes neurons of this region to GABA,[56,57] permitting BZ agonists to decrease neuronal activity.[58-61] The effects of iontophoretically applied GABA are similarly modified by administration of various benzodiazepine ligands.[60,61] Thus, the SNr provides an extremely versatile system in which to examine BZ effects on responses to both endogenous (tonic input) and exogenous (iontophoretic) GABA.

We have tested various hormonally distinct groups of rats for substantia nigra pars reticulata (SNr) responses to systemically and locally administered benzodiazepines. SNr activity was assessed in anesthetized rats using extracellular single-unit recording techniques concomitant with microiontophoretic applications of GABA and the BZ agonist midazolam (see References 60 and 61). The ability of local applications of midazolam to potentiate SNr responses to GABA was analyzed in intact male, intact female, orchidectomized (ORCH) male, and ovariectomized (OVX) female rats. Local BZ enhancement of GABAergic responses was not significantly different between these hormonally distinct groups. Iontophoretically applied midazolam led to increases of 31% ± 5, 36% ± 4, 24% ± 8, and 40% ± 3 in GABA sensitivity of male, female, ORCH, and OVX groups, respectively ($N = 5$ per group; $F = 1.6$, $p = 0.23$ using Analysis of Variance). In contrast, early results indicate that hormonal status may modulate reticulata responses to systemic diazepam administration (Figure 1). As seen in Figure 1, female rats demonstrated an attenuated inhibition in firing rate due to the systemic administration of diazepam when compared to males. These results suggest that the *in vivo* hormonal state of the animal can apparently modulate systemic, but not local, BZ responses of SNr neurons. This could indicate that these hormone-related influences on BZ responses are not intrinsic to this region, but involve modifications in some afferent pathway. Altered SNr responsiveness to systemic BZ administration could reflect modified GABAergic control from striatum

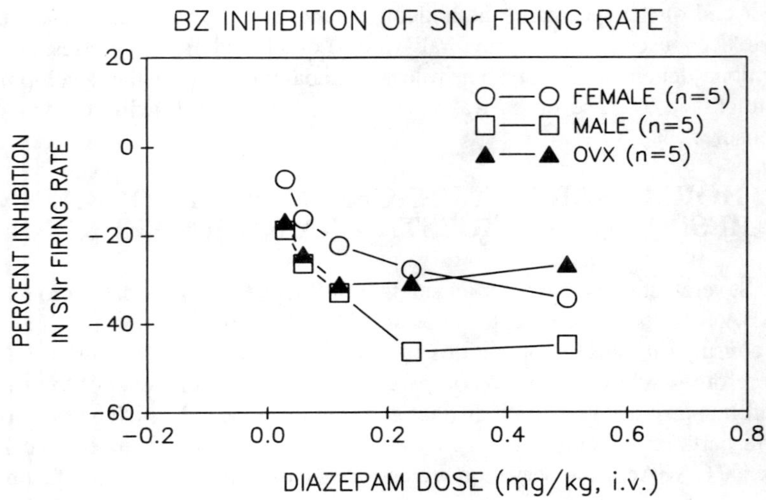

FIGURE 1. Hormonal status altered responses of single substantia nigra pars reticulata (SNr) neurons to systemic BZ administration. Decreases in SNr activity resulting from systemic administration of the BZ agonist diazepam were compared in groups of intact female, intact male, and ovariectomized (OVX) female rats. SNr activity was assessed in chloral hydrate anesthetized rats using extracellular single-unit recording techniques (see References 60,61). Results are shown as the mean percentage decreases in SNr firing rate induced by diazepam administered intravenously at 2-min intervals in increasing cumulative doses from 0.03 to 0.5 mg/kg. Analysis of Variance with repeated measures (dose of diazepam) indicated significant effects of diazepam ($F = 3.3$, $p < 0.02$) and hormone group ($F = 3.3$, $p < 0.05$), but no interaction between hormone group and diazepam response.

or some other brain region, which results in altered endogenous GABA release. Estrogen administration has been shown to decrease levels of the GABA-synthetic enzyme glutamic acid decarboxylase in substantia nigra,[62] and several studies have documented the ability of gonadal hormones to modulate striatal functions.[63-65]

These studies in SNr indicate that neuronal responses to systemic administration of BZ agonists may be altered by the hormonal milieu of an animal. Further, they demonstrate that hormonal status may modulate many aspects of GABAergic neurotransmission, any or all of which are capable of influencing BZ responsiveness.

VI. HORMONAL INFLUENCES ON ACUTE ANTICONVULSANT EFFECTS OF BENZODIAZEPINES

Clinical evidence suggesting gonadal steroids can influence GABAergic and/or BZ responses is indirectly based on the similarity in some of the pharmacological effects of steroids and the benzodiazepines. Both clinical and animal studies have shown that gonadal hormones can influence seizure activity.[10,16-18] In female epileptic patients, the increased seizure activity observed about the time of menstruation (Catamenial epilepsy) has been linked to decreased progesterone levels, increased estrogen levels, and a heightened estrogen:progesterone ratio.[10,16-18,66-68] Animal studies have generally supported the view that estrogens enhance seizure susceptibility.[23,66,69-72] Interestingly, steroid hormone treatments and antiestrogen administration have been shown in a few cases to be antiepileptic (see References 16 and 17) and both estrogen and testosterone treatments have been shown to protect against chemically induced seizures in rats.[73] Progesterone treatments can be anxiolytic[19] or anticonvulsant[17,69-75] and certain progesterone metabolites[46,68] show anticonvulsant activity.

The mechanism by which gonadal steroid hormones alter seizure susceptibility or reduce anxiety is unclear, but one possibility would be through modulation of the GABAergic system. This would be consistent with (1) the ability of BZ ligands to modify convulsability and anxiogenesis through GABA-mediated mechanisms and (2) the recent evidence that steroid metabolites can modulate GABAergic responses in a manner similar to the benzodiazepines. While relatively few studies have examined the influences of gonadal hormone on responses to benzodiazepines, their results suggest that the steroid hormone milieu can, in fact, modulate BZ responsiveness. For example, sex differences are seen in the BZ-induced alterations in corticosterone release in animals.[44] Clinical studies have also demonstrated fluctuations in psychomotor effects of benzodiazepines related to the oral contraceptive cycle in women.[76]

Recent data from this laboratory suggest, however, that the hormonal status of rats does not significantly modify the acute anticonvulsant effects of the BZ agonist diazepam. We have examined the effects of gonadectomy on the ability of diazepam to protect against seizures induced by the GABA antagonist bicuculline in both male and female rats. As seen in Figure 2, similar dose-related elevations in seizure threshold values were seen in groups of diazepam-treated intact male, cycling female, ovariectomized, and orchidectomized rats. Brain levels of diazepam, as indicated by radioreceptor analysis, were comparable in all hormone groups after a single, acute diazepam injection (see Figure 4, acute DZ injection). Thus, these studies fail to show substantial hormonal modulation of *acute* BZ responses in rats. Additional studies suggest that the steroid milieu does alter the effects of *chronic* exposure to the benzodiazepines.

VII. HORMONAL INFLUENCES ON RESPONSES TO CHRONIC BENZODIAZEPINE EXPOSURE

Clinical and behavioral studies have shown that tolerance to the sedative, muscle relaxant, and anticonvulsant actions of the benzodiazepines develops during prolonged exposure.[77-84] Since the acute administration of benzodi-

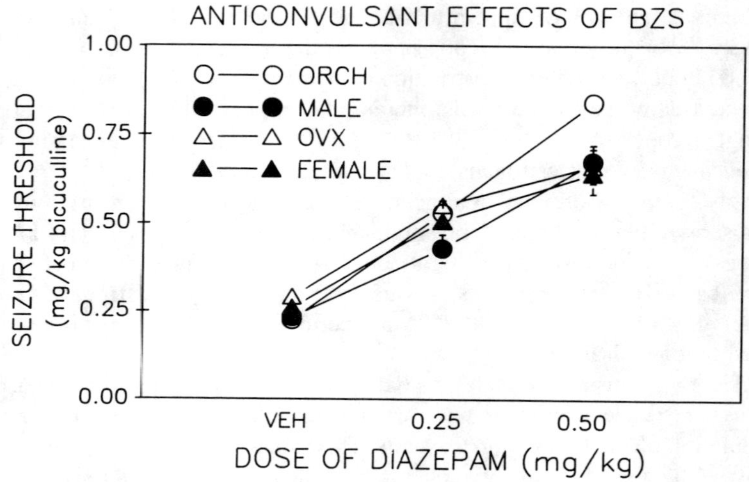

FIGURE 2. The anticonvulsant effects of the BZ agonist diazepam (DZ) were similar in groups of intact and gonadectomized male and female rats, indicating that hormonal status does not alter acute BZ responses. Mean ± SEM seizure threshold values are shown for groups of orchidectomized (ORCH), intact male, and ovariectomized (OVX) and intact female rats given either vehicle (40% propylene glycol, 10% ethanol), 0.25 mg/kg, or 0.5 mg/kg diazepam (iv) 5 min prior to testing. Seizures were induced using a constant infusion of bicuculline (0.05 mg/ml at 1.55 ml/min) through an indwelling iv catheter and seizure thresholds indicate the dose required to induce the first myoclonic twitch. Two-way Analysis of Variance with dose of diazepam and hormone group as factors revealed a significant effect of diazepam dose on seizure threshold ($F=23.7$, $p<0.0001$), no effect of hormone status on diazepam's anticonvulsant effects ($F=2.0$, $p>0.05$), and a lack of interaction between diazepam dose and hormone group. *N*'s are 8,22,18,16 for control points; 3,7,11,7 for 0.25 mg/kg points, and 3,12,11,10 for 0.5 mg/kg points from ORCH, male, OVX, and female groups, respectively.

azepines facilitates the inhibitory effects of GABA on neuronal activity,[1,2] studies examining the neural adaptations underlying the development of BZ tolerance have focused on GABAergic systems.

In male rats, prolonged exposure to benzodiazepines diminishes the ability of these compounds to potentiate GABA responses in several brain areas, i.e., tolerance develops.[60,85,86] Reduced neuronal responsiveness to benzodiazepines has been seen in cerebellar,[86] dorsal raphe,[87,88] and substantia nigra pars reticulata neurons.[60] Interestingly, microinjected benzodiazepines also have diminished behavioral effects when infused into the substantia nigra after prolonged exposure.[89] Measurements of GABA-stimulated chloride influx into membrane vesicles have similarly demonstrated a reduced enhancement of GABAergic effects with BZ agonists following chronic exposure, at least in some brain areas.[90-93]

The neural alterations associated with prolonged BZ exposure in male rats include regionally specific decreases in GABA sensitivity, alterations in levels of GABA/BZ receptor sites, and uncoupling between the GABA and BZ receptor sites. Electrophysiological studies in male animals have shown a reduced responsiveness or dorsal raphe neurons to iontophoretically applied GABA following chronic treatment with benzodiazepines.[87,88] This GABA subsensitivity observed in the dorsal raphe develops with a time course paralleling the development of tolerance to the anticonvulsant effects of diazepam[94] and appears to be intrinsic to the midbrain region.[95] Similarly, decreases in GABA activation of chloride influx are observed in cortical membrane vesicles after chronic BZ treatment in male rats.[92] GABA subsensitivity is not observed in all brain areas,[60,61,92] indicating that the decreased functional responsiveness to GABA associated with chronic exposure to the benzodiazepines occurs in a regionally specific manner.

Biochemical studies in male animals have demonstrated contradictory effects of chronic BZ exposure on the numbers of various components of the GABA/BZ/ionophore complex, including the BZ recognition site. Several laboratories have failed to observe any changes in BZ receptors following chronic exposure,[87,96-98] although decreased[80,85,89,93,99] and increased[100] levels of BZ binding have been seen by some investigators. Similarly, inconsistent changes have been observed in GABA receptor numbers following chronic exposure to BZ agonists,[87,88,97] although a shift to the high-affinity, and perhaps less functional, form of the GABA receptor following such treatments has been observed.[87,88] Decreased levels of the α_1, but not β_1, GABA receptor subunit following prolonged BZ treatment[101] further indicate that such treatments may modify the GABA receptor subunit composition. Rather than consistently changing the number of GABA or BZ receptor sites, chronic BZ treatment appears to alter the allosteric modulation between the two sites.[87,88,98,102] This altered coupling is supported by the decreased ability of GABA to "shift" BZ binding in membrane preparations from animals or cultured brain cells exposed to BZ agonists for prolonged periods.[85,87,88,98,102]

These results in male subjects suggest that the mechanism(s) controlling the development of tolerance to the benzodiazepines may involve complex interactions between neural modifications in several brain structures. We have recently begun to examine whether hormonal status can modulate the development of tolerance to the benzodiazepines and/or some of the neural changes associated with prolonged exposure. Although still preliminary, they suggest that the gonadal hormone milieu of an animal can modify the effects of *chronic* exposure to BZ agonists.

In male rats, tolerance to the anticonvulsant effects of diazepam occurs following chronic exposure to diazepam-filled silastic capsules.[88,94] These capsules are implanted subcutaneously and provide prolonged continuous exposure to low amounts of the compound.[88] Using this novel method of benzodiazepine treatment, we have examined whether tolerance to the anticonvulsant effects of diazepam develops in female, orchidectomized (ORCH), and ovariectomized (OVX) rats in a manner comparable to that seen in males.

As seen in Figure 3, the *acute* anticonvulsant effects of diazepam released from the silastic implants were similar in all hormone groups following 3 d of exposure. Following *chronic* (3-week) exposure to diazepam-filled silastic implants, both intact male and female rats showed tolerance to the anticonvulsant effects of diazepam. This is evidenced by the lack of enhancement in bicuculline seizure thresholds in chronic diazepam-treated males and females compared to vehicle implant groups (compare Figures 3A and 3B, male and female bars). In contrast, gonadectomized animals failed to develop tolerance to the anticonvulsant effects of diazepam. In both ORCH and OVX groups, similar increases in seizure thresholds were observed after 3-d and 3-week exposures to diazepam-filled implants compared to appropriate vehicle-treated control values (compare Figures 3A and 3B). Interestingly, while no tolerance to the anticonvulsant effects of diazepam released from the capsules was observed in gonadectomized animals, preliminary evidence suggests that these groups do show tolerance to additional challenge doses of diazepam. We are currently investigating the reasons for this discrepancy.

Gonadal hormones are known to modify the metabolism of many drugs, including some benzodiazepines.[103] Hormonal influences on diazepam levels do not appear to underlie the altered tolerance development in gonadectomized animals, since similar brain concentrations of drug were observed after acute (3-d) and chronic (3-week) silastic implants in all hormone groups (Figure 4). As mentioned, acute injection also produced similar brain levels of diazepam in all groups.

VIII. SUMMARY OF HORMONAL INFLUENCES ON BENZODIAZEPINE RESPONSES

Several lines of evidence suggest that gonadal hormones can modulate neural receptors for GABA and benzodiazepines, as well as modifying GABAergic responses. *In vivo* studies have indicated neuron- and region-specific effects of gonadal hormones on GABA/BZ receptors, while *in vitro* analyses have demonstrated direct steroid effects on the GABA/BZ/chloride ionophore receptor complex and GABAergic actions. Further, nonreproductive effects of steroid hormones on convulsability, anxiety, and sedation markedly resemble the pharmacological profile of the benzodiazepines. Such evidence would suggest that through their ability to modulate GABAergic systems, gonadal steroids might be capable of influencing BZ responses.

Results from this laboratory suggest that neither sex-related factors nor removal of gonadal steroids dramatically modifies the acute anticonvulsant effects of BZ agonists, at least in rats. This result is consistent with the lack of BZ receptor changes associated with altered *in vivo* hormonal status. Findings in substantia nigra support the existence of subtle, sex-related differences in neuronal responses to benzodiazepines. This may suggest that certain neuronal populations show steroid-related modulation of BZ responsiveness, but

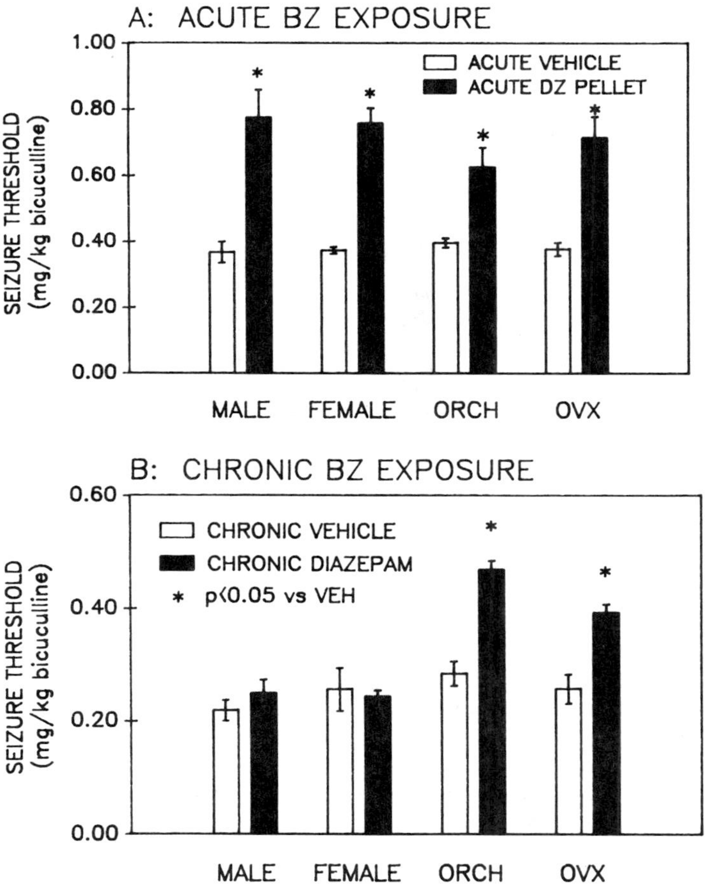

FIGURE 3. Seizure thresholds are shown in groups of intact and gonadectomized rats treated for 3 days (A: acute BZ exposure) or 3 weeks (B: chronic BZ exposure) with empty (vehicle) or diazepam-filled (diazepam; DZ) silastic capsules. Panel (A) shows that acute (3-d) exposure to DZ-filled pellets elevated bicuculline-induced seizure thresholds to a similar degree in all hormone groups. Two-way Analysis of Variance revealed a significant effect of DZ-pellet implant ($F = 40.7$, $p < 0.0001$), but no effect of hormone treatment ($F = 0.3$, $p > 0.05$) and no interaction between these two factors. Panel (B) shows that 3 weeks of exposure to DZ-filled capsules induced tolerance to the anticonvulsant effects of diazepam in intact males and females. In contrast, gonadectomized rats with DZ-pellet implants still showed elevated seizure thresholds after 3 weeks of BZ exposure. This suggests that gonadectomy prevented the development of tolerance to the anticonvulsant effects of diazepam. Two-way ANOVA revealed significant effects of pellet implant ($F = 18.2$, $p < 0.0003$), hormone group ($F = 10.7$, $p < 0.0001$), and an interaction between hormone group and DZ treatment ($F = 5.0$, $p < 0.009$). *, Indicates a significant difference between diazepam-treated and vehicle groups for each hormone treatment. Bars indicate mean ± SEM dose of bicuculline to induce a myoclonic twitch (see Figure 2 and Reference 88 for details) with N's of 3 to 6 each.

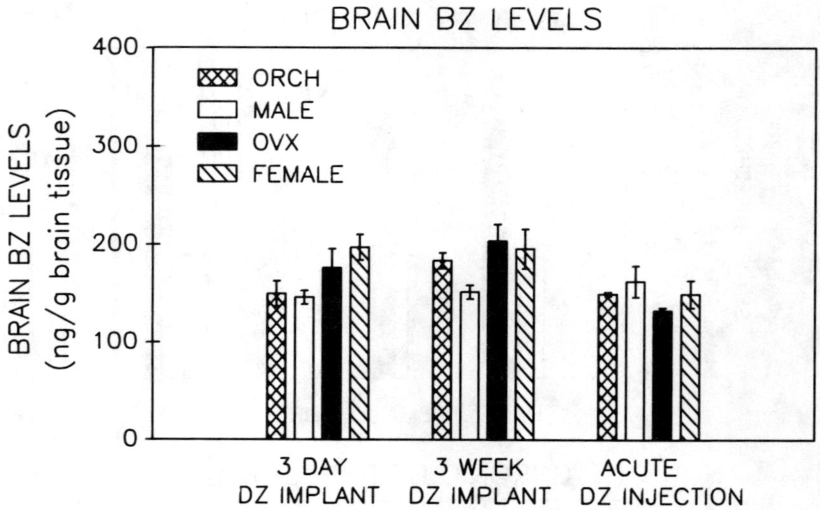

FIGURE 4. Brain levels of BZ-like activity after 3-d treatment with DZ-filled silastic implants, 3-week exposure to DZ-filled implants, and an acute injection of diazepam (0.5 mg/kg, iv) were comparable in groups of intact and gonadectomized male and female rats. Brain levels were assayed by radioreceptor procedures using [^{3}H]-flunitrazepam binding, which indicate levels of BZ-like activity (see Reference 88). ANOVAs revealed that hormone groups all had similar brain levels of diazepam following 3-d DZ implant, 3-week DZ implant, and acute DZ injection. These three diazepam treatments also produce comparable brain levels of drug. *N*'s are 3 to 6 per group.

that such localized changes are obscured during whole animal tests of anticonvulsant activity. Alternatively, perhaps sex-specific differences in BZ responses of this brain area are unimportant for anticonvulsant effects, but may serve to modify other acute pharmacological effects of the benzodiazepines. The *in vivo* hormonal milieu of an animal does seem capable of modulating the diminished anticonvulsant responses associated with chronic BZ exposure. Further, these studies do not preclude the possibility that other pharmacological actions of the benzodiazepines may be modified by hormonal status in other species.[42,76]

There are several ways in which gonadal steroid hormones could modify BZ responsiveness. Various aspects of the GABA/BZ/ionophore complex are altered by gonadal steroid hormones or their derivatives, either through genomic mechanisms or through direct modulation of the receptor complex itself. Hormones could modulate BZ (and GABA) actions by altering the subunit composition of the GABA receptor complex. Alterations in subunit structure have been shown to influence various aspects of GABA/BZ receptor function, including GABA sensitivity[104] and BZ potentiation.[105] This possibility is highly consistent with the ability of steroid hormones to regulate protein synthesis.[44] Alterations in other aspects of GABA neurotransmission could also be involved. Steroid treatments have been shown to modify GABA-synthetic enzyme levels[62] and GABA release.[106] The divergence between hormonal influences on local versus systemic BZ responses in substantia nigra is consistent with hormonal modulation of endogenous GABA inhibition unrelated to receptor changes within this region. Hormonal influences could also involve alterations in several different aspects of GABA neurotransmission concomitantly.

Alternatively, effects of gonadal steroids could be indirectly mediated through steroid-induced changes in some other compound. Likely candidates include other hormones and/or catecholestrogens. The latter are catechol-like metabolites of estrogen that can mimic certain effects of the steroid[107] and can affect activity of the GABA-synthetic enzyme glutamic acid decarboxylase.[108] Other hormones which are regulated by gonadal steroids could also mediate various *in vivo* effects. For example, prolactin levels are influenced by estrogen administration and prolactin has been shown to influence GABAergic systems.[109] Further, some effects of estrogen can be prevented by hypophysectomy and/or other treatments that prevent estrogen-induced hyperprolactinemia.[26] The demonstration of gender differences in levels of the proposed endogenous BZ receptor ligand provides an additional site at which hormones may modulate BZ responsiveness.[110]

Studies using an animal model of prolonged BZ exposure suggest that gonadal steroids serve a permissive role in the development of tolerance to the anticonvulsant effects of the benzodiazepines. The specific hormonal change responsible for the altered responses to chronic BZ treatment in gonadectomized rats is unknown. Since both castrated males and females were affected, the underlying modification may involve loss of some common steroid derivative or some other change common to the diminution of both male and female gonadal steroids. This finding resembles the ability of castration to modify neural receptor changes associated with chronic antidepressant treatments.[111] We are currently investigating whether hormone status modulates the various neural adaptations in GABAergic systems associated with prolonged BZ exposure in rats. Such studies may help elucidate the mechanism by which gonad removal modifies responses to prolonged BZ exposure. Further studies should provide a better understanding of the influences of gonadal hormone-related factors on both acute BZ responsiveness and the development of tolerance to the pharmacological actions of these drugs.

REFERENCES

1. **Haefely, W., Polc, P., Pieri, L., Schaffner, R., and Laurent, J.-P.,** Neuropharmacology of benzodiazepines: synaptic mechanisms and neural basis of action, in *The Benzodiazepines: From Molecular Biology to Clinical Practice,* Costa, E., Ed., Raven Press, New York, 1983, p. 21.
2. **Tallman, J. F. and Gallager, D. W.,** The GABAergic system: a locus of benzodiazepine action, *Annu. Rev. Neurosci.,* 8, 21, 1985.
3. **Mohler, H. and Okada, T.,** Benzodiazepine receptor: demonstration in the central nervous system, *Science,* 198, 849, 1977.
4. **Squires, R. F. and Braestrup, C.,** Benzodiazepine receptors in rat brain, *Nature (London),* 266, 732, 1977.
5. **Study, R. E. and Barker, J. L.,** Diazepam and (−)-pentobarbital: fluctuation analysis reveals different mechanisms for potentiation of γ-aminobutyric acid responses in cultured central neurons, *Proc. Natl. Acad. Sci. U.S.A.,* 78, 7180, 1981.
6. **Obata, T., Morelli, M., Concas, A., Serra, M., and Yamamura, H. I.,** Modulation of GABA-stimulated chloride influx into membrane vesicles from rat cerebral cortex by benzodiazepines and nonbenzodiazepines, in *Chloride Channels and their Modulation by Neurotransmitters and Drugs,* Biggio, G. and Costa, E., Eds., Raven Press, New York, 1988, 175.
7. **Morrow, A. L., Suzdak, P. D., and Paul, S. M.,** Benzodiazepine, barbiturate, ethanol and hypnotic steroid hormone modulation of GABA-mediated chloride ion transport in rat brain synaptoneurosomes, in *Chloride Channels and Their Modulation by Neurotransmitters and Drugs,* Biggio, G. and Costa, E., Eds., Raven Press, New York, 1988, 247.
8. **Tallman, J. F., Thomas, J. W., and Gallager, D. W.,** GABAergic modulation of benzodiazepine binding site sensitivity, *Nature (London),* 274, 383, 1978.
9. **Krespan, B., Springfield, S. A., Haas, H., and Geller, H. M.,** Electrophysiological studies on benzodiazepine antagonists, *Brain Res.,* 295, 265, 1984.
10. **Maggi, A. and Perez, J.,** Role of female gonadal hormones in the CNS: clinical and experimental aspects, *Life Sci.,* 37, 893, 1985.
11. **Majewska, M. D., Harrison, N. L., Schwartz, R. D., Barker, J. L., and Paul, S. M.,** Steroid hormone metabolites are barbiturate-like modulators of the GABA receptor, *Science,* 232, 1004, 1986.
12. **Harrison, N. L., Majewska, M. D., Harrington, J. W., and Barker, J. L.,** Structure-activity relationships for steroid interaction with the γ-aminobutyric $acid_A$ receptor complex, *J. Pharmacol. Exp. Ther.,* 241(1), 346, 1987.
13. **Smith, S. S., Waterhouse, B. D., and Woodward, D. J.,** Sex steroid effects on extrahypothalamic CNS. II. Progesterone, along and in combination with estrogen, modulates cerebellar responses to amino acid neurotransmitters, *Brain Res.,* 422, 52, 1987.
14. **Callachan, H., Cottrell, G. A., Hather, N. Y., Lambert, J. J., Nooney, J. M., and Peters, J. A.,** Modulation of the $GABA_A$ receptor by progesterone metabolites, *Proc. R. Soc. London B,* 231, 359, 1987.
15. **Simmonds, M. A. and Turner, J. P.,** Potentiators of responses to activation of γ-aminobutyric acid ($GABA_A$) receptors, *Neuropharmacology,* 26, 923, 1987.
16. **Newmark, M. E. and Penry, J. K.,** Catamenial epilepsy: a review, *Epilepsia,* 21, 281, 1980.
17. **Mattson, R. H. and Cramer, J. A.,** Epilepsy, sex hormones, and antiepileptic drugs, *Epilepsia,* 26(Suppl.1), S40, 1985.
18. **Schachter, S. C.,** Hormonal considerations in women with seizures, *Arch. Neurol.,* 45, 1267, 1988.
19. **Rodriquez-Sierra, J. F., Horward, J. L., Pollard, G. T., and Hendrickes, S. E.,** Effect of ovarian hormones on conflict behavior, *Psychopharmacology,* 9, 293, 1984.

20. **Maggi, A. and Perez, J.,** Progesterone and estrogens in rat brain: modulation of GABA (γ-aminobutyric acid) receptor activity, *Eur. J. Pharmacol.,* 103, 165, 1984.
21. **Maggi, A. and Perez, J.,** Estrogen-induced up-regulation of γ-aminobutyric acid receptors in the CNS of rodents, *J. Neurochem.,* 47(6), 1793, 1986.
22. **Perez, J., Zucchi, I., and Maggi, A.,** Sexual dimorphism in the response of the GABAergic system to estrogen administration, *J. Neurochem.,* 47(6), 1798, 1986.
23. **Perez, J., Zucchi, I., and Maggi, A.,** Estrogen modulation of the γ-aminobutyric acid receptor complex in the central nervous system of rat, *J. Pharmacol. Exp. Ther.,* 244(3), 1005, 1988.
24. **Schumacher, M., Coirini, H., and McEwen, B. S.,** Regulation of high-affinity $GABA_A$ receptors in the dorsal hippocampus by estradiol and progesterone, *Brain Res.,* 487, 178, 1989.
25. **Canonaco, M., O'Connor, L. H., Pfaff, D. W., and McEwen, B. S.,** $GABA_A$ receptor level changes in female hamster forebrain following *in vivo* estrogen progesterone and benzodiazepine treatment: a quantitative autoradiography analysis, *Exp. Brain Res.,* 75, 644, 1989.
26. **Hamon, M., Goetz, C., and Euvrard, C., et al.,** Biochemical and functional alterations of GABA receptors during chronic estradiol treatment, *Brain Res.,* 279, 141, 1983.
27. **O'Connor, L. H., Nock, B., and McEwen, B. S.,** Regional specificity of gamma-aminobutyric acid receptor regulation of estradiol, *Neuroendocrinology,* 47, 473, 1988.
28. **Schumacher, M., Coirini, H., and McEwen, B. S.,** Regulation of high-affinity $GABA_A$ receptors in specific brain regions by ovarian hormones, *Neuroendocrinology,* 50, 315, 1989.
29. **Wilkinson, M., Bhanot, R., Wilkinson, D. A., and Brawer, J. R.,** Prolonged estrogen treatment induces changes in opiate, benzodiazepine and β-adrenergic binding sites in female rat hypothalamus, *Brain Res. Bull.,* 11, 279, 1983.
30. **Canonaco, M., Valenti, A., Tavolaro, R., Bettini, E., and Maggi, A.,** Differential modulation of [^{3}H]flunitrazepam binding in female rat brain by sex steroid hormones, *Eur. J. Pharmacol.,* 170, 95, 1989.
31. **Canonaco, M., O'Connor, L. H., Pfaff, D. W., and McEwen, B. S.,** Longer term progesterone treatment induces changes of $GABA_A$ receptor levels in forebrain sites in female hamster: quantitative autoradiography study, *Exp. Brain Res.,* 77, 407, 1989.
32. **Gavish, M., Weizman, A., Youdim, M. B. H., and Okun, F.,** Regulation of central and peripheral benzodiazepine receptors in progesterone-treated rats, *Brain Res.,* 409, 386, 1987.
33. **Majewska, M. D., Ford-Rice, F., and Falkay, G.,** Pregnancy-induced alteration of $GABA_A$ receptor sensitivity in maternal brain; an antecedent of post-partum "blues"? *Brain Res.,* 482, 397, 1989.
34. **Francois-Bellan, A. M., Segu, L., and Hery, M.,** Regulation of estradiol of $GABA_A$ and $GABA_B$ binding sites in the diencephalon of the rat: an autoradiographic study, *Brain Res.,* 503, 144, 1989.
35. **Juptner, M. and Hiemke, C.,** Sex differences in $GABA_A$ receptor binding in rat brain measured by an improved *in vitro* binding assay, *Exp. Brain Res.,* 81, 297, 1990.
36. **Shephard, R. A., Nielsen, E. B., and Broadhurst, P. L.,** Sex and strain differences in benzodiazepine receptor binding in Roman rat strains, *Eur. J. Pharmacol.,* 77, 327, 1982.
37. **Soubrie, P., Thiebot, M. H., Jobert, A., Montastruc, J. L., Hery, F., and Hamon, M.,** Decreased convulsant potency of picrotoxin and pentetrazol and enhanced [^{3}H]flunitrazepam cortical binding following stressful manipulations in rats, *Brain Res.,* 189, 505, 1980.
38. **Medina, J. H., Novas, M. L., Wolfman, C. N. V., Levi de Stein, M., and DeRobertis, E.,** Benzodiazepine receptors in rat cerebral cortex and hippocampus undergo rapid and reversible changes after acute stress, *Neuroscience,* 9, 331, 1983.

39. **Havoundjian, H., Paul, S. M., and Skolnick, P.,** Acute, stress-induced changes in the benzodiazepine/γ-aminobutyric acid receptor complex are confined to the chloride ionophore, *J. Pharmacol. Exp. Ther.*, 237, 787, 1986.
40. **Riegle, G.,** Chronic stress effects on adrenocortical responsiveness in young and aged rats, *Neuroendocrinology,* 11, 1, 1973.
41. **Ramaley, J. A.,** Effects of ovariectomy on dexamethasone suppression of the adrenal axis in adult rats, *Neuroendocrinology,* 20, 260, 1976.
42. **Pericic, D., Manev, H., and Lakic, N.,** Sex differences in the response of rats to drugs affecting GABAergic transmission, *Life Sci.*, 36, 541, 1985.
43. **Pohorecky, L. A., Cotler, S., Carbone, J. J., and Roberts, P.,** Factors modifying the effect of diazepam on plasma corticosterone levels in rats, *Life Sci.*, 43, 2159, 1988.
44. **Pfaff, D. W. and McEwen, B. S.,** Actions of estrogens and progestins on nerve cells, *Science,* 219, 808, 1983.
45. **Schumacher, M.,** Rapid membrane effects of steroid hormones: an emerging concept in neuroendocrinology, TINS, 13(9), 359, 1990.
46. **Gee, K. W., Bolger, M. B., Brinton, R. E., Coirini, H., and McEwen, B. S.,** Steroid modulation of the chloride ionophore in rat brain; structure-activity requirements, regional dependence and mechanism of action, *J. Pharmacol. Exp. Ther.*, 246(2), 803, 1988.
47. **Gee, K. W., Chang, W. C., Brinton, R. E., and McEwen, B. S.,** GABA-dependent modulation of the Cl ionophore by steroids in rat brain, *Eur. J. Pharmacol.*, 136, 419, 1987.
48. **Morrow, A. L., Pace, J. R., Purdy, R. H., and Paul, S. M.,** Characterization of steroid interactions with γ-aminobutyric acid receptor-gated chloride ion channels: evidence for multiple steroid recognition sites, *Mol. Pharmacol.*, 37, 263, 1990.
49. **Turner, J. P. and Simmonds, M. A.,** Modulation of the $GABA_A$ receptor complex by steroids in slices of rat cuneate nucleus, *Br. J. Pharmacol.*, 96, 409, 1989.
50. **Puia, G., Santi, M., and Vicini, S., et al.,** Neurosteroids act on recombinant human $GABA_A$ receptors, *Neuron,* 4, 759, 1990.
51. **Turner, D. M., Ransom, R. W., Yang, J., S.-J., and Olsen, R. W.,** Steroid anesthetics and naturally occurring analogs modulate the γ-aminobutyric acid receptor complex at a site distinct from barbiturates, *J. Pharmacol. Exp. Ther.*, 248, 960, 1989.
52. **Majewska, M. D., Mienville, J. M., and Vicini, S.,** Neurosteroid pregnenolone sulfate antagonizes electrophysiological responses to GABA in neurons, *Neuroscience Lett.*, 90, 279, 1988.
53. **Majewska, M. D. and Schwartz, R. D.,** Pregnenolone-sulfate: an endogenous antagonist of the γ-aminobutyric acid receptor complex in brain? *Brain Res.*, 404, 355, 1987.
54. **Gee, K. W., Joy, D. S., and Belelli, D.,** Complex interactions between pregnenolone sulfate and the *t*-butylbicyclophosphorothionate-labeled chloride ionophore in rat brain, *Brain Res.*, 482, 169, 1989.
55. **Ke, F.-C. and Ramirez, V. D.,** Binding of progesterone to nerve cell membranes of rat brain using progesterone conjugated to ^{125}I-bovine serum albumin as a ligand, *J. Neurochem.*, 54, 467, 1990.
56. **Grace, A. A. and Bunney, B. S.,** Paradoxical GABA excitation of nigral dopaminergic cells: indirect mediation through reticulata inhibitory neurons, *Eur. J. Pharmacol.*, 59, 211, 1979.
57. **Waszczak, B. L., Bergstrom, D. A., and Walters, J. R.,** Single unit responses of substantia nigra and globus pallidus neurons to GABA agonist and antagonist drugs, in *GABA and the Basal Ganglia,* DiChiara, G., and Gessa, G. L., Eds., Raven Press, New York, 1981, 79.
58. **Ross, R. J., Waszczak, B. L., Lee, E. K., and Walters, J. R.,** Effects of benzodiazepines on single unit activity in the substantia nigra pars reticulata, *Life Sci.*, 31, 1025, 1982.

59. **Mereu, G. and Biggio, G.,** Effect of agonists, inverse agonists, and antagonists of benzodiazepine receptors on the firing rate of substantia nigra pars reticulata neurons, in *Benzodiazepine Recognition Site Ligands: Biochemistry and Pharmacology,* Biggio, G. and Costa, E., Eds., Raven Press, New York, 1983, 201.
60. **Wilson, M. A. and Gallager, D. W.,** Effects of chronic diazepam exposure on GABA sensitivity and on benzodiazepine potentiation of GABA-mediated responses of substantia nigra pars reticulata neurons of rats, *Eur. J. Pharmacol.,* 136, 333, 1987.
61. **Wilson, M. A. and Gallager, D. W.,** Responses of substantia nigra pars reticulata neurons to benzodiazepine ligands after acute and and prolonged diazepam exposure. I. Modulation of γ-aminobutyric acid sensitivity, *J. Pharmacol. Exp. Ther.,* 248, 879, 1989.
62. **Nicoletti, F., Patti, F., and Ferrara, N., et al.,** Comparative effects of estrogens and prolactin on nigral and striatal GAD activity, *Brain Res.,* 232, 238, 1982.
63. **Bedard, P. J., DiPaolo, T., Langelier, P., Poyet, P., and Labrie, F.,** Behavioral and biochemical evidence of an effect of estradiol on striatal dopamine receptors, in *Steroid Hormone Regulation of the Brain,* Fuxe, K., Gustafsson, J.-A., and Wetterberg, L., Eds., Pergamon Press, New York, 1981, 331.
64. **Clopton, J. K. and Gordon, J. H.,** The possible role of 2-hydroxyestradiol in the development of estrogen-induced striatal dopamine receptor heterogeneity, *Brain Res.,* 333, 1, 1985.
65. **Becker, J. B. and Beer, M. E.,** The influence of estrogen on nigrostriatal dopamine activity: behavioral and neurochemical evidence for both pre- and postsynaptic components, *Behav. Brain Res.,* 19, 27, 1986.
66. **Logothetis, J., Harner, R., Morrell, F., and Torres, F.,** The role of estrogens in catamenial exacerbation of epilepsy, *Neurology,* 9, 352, 1958.
67. **Backstrom, T.,** Epileptic seizures in women related to plasma estrogen and progesterone during the menstrual cycle, *Acta Neurol. Scand.,* 54, 321, 1976.
68. **Rosciszewska, D., Buntner, B., Guz, I., and Zawisza, L.,** Ovarian hormones, anticonvulsant drugs and seizures during the menstrual cycle in women with epilepsy, *J. Neurol. Neurosurg. Psychiatry,* 49, 47, 1986.
69. **Woolley, D. E. and Timiras, P. S.,** The gonad-brain relationship: effects of female sex hormones on electroshock convulsions in the rat, *Endocrinology,* 70, 196, 1962.
70. **Stitt, S. L. and Kinnard, W. J.,** The effect of certain progestins and estrogens on the threshold of electrically induced seizure patterns, *Neurology,* 18, 213, 1968.
71. **Nicoletti, F., Speciale, C., and Sortino, M. A., et al.,** Comparative effects of estradiol benzoate, the antiestrogen clomiphene citrate, and the progestin medroxyprogesterone acetate on kainic acid-induced seizures in male and female rats, *Epilepsia,* 26(3), 252, 1985.
72. **Hom, A. C. and Buterbaugh, G. G.,** Estrogen alters the acquisition of seizures kindled by repeated amygdala stimulation or pentylenetetrazol administration in ovariectomized female rats, *Epilepsia,* 27(2), 103, 1986.
73. **Schwartz-Giblin, S., Korotzer, A., and Pfaff, D. W.,** Steroid hormone effects on picrotoxin-induced seizures in female and male rats, *Brain Res.,* 476, 240, 1989.
74. **Selye, H.,** Correlations between the chemical structure and pharmacological actions of the steroids, *Endocrinology,* 30, 437, 1942.
75. **Craig, C. R.,** Anticonvulsant activity of steroids: separability of anticonvulsant from hormonal effects, *J. Pharmacol. Exp. Ther.,* 153(2), 337, 1966.
76. **Ellinwood, E. H., Easler, M. E., Linnolla, M., Molter, D. W., Heatherly, D. G., and Bjornsson, T. D.,** Effects of oral contraceptives on diazepam-induced psychomotor impairment, *Clin. Pharmacol. Ther.,* 35, 360, 1983.
77. **Goldberg, M. E., Manian, A. A., and Efron, D. H.,** A comparative study of certain pharmacologic responses following acute and chronic administrations of chlordiazepoxide, *Life Sci.,* 6, 481, 1967.

78. **Browne, T. R. and Penry, J. K.,** Benzodiazepines in the treatment of epilepsy. A review, *Epilepsia,* 14, 277, 1973.
79. **Greenblatt, D. J. and Shader, R. I.,** Dependence, tolerance and addiction to benzodiazepines: clinical and pharmacokinetic considerations, *Drug Metab. Rev.,* 8, 13, 1978.
80. **Rosenberg, H. C. and Chiu, T. H.,** Tolerance during chronic benzodiazepine treatment associated with decreased receptor binding, *Eur. J. Pharmacol.,* 70, 453, 1981.
81. **Gent, J. P. and Haigh, J. R. M.,** Development of tolerance to the anticonvulsant effects of clobazam, *Eur. J. Pharmacol.,* 94, 155, 1983.
82. **File, S. E.,** Tolerance to the behavioral actions of benzodiazepines, *Neurosci. Biobehav. Rev.,* 9, 113, 1985.
83. **Stephens, D. N. and Schneider, H. H.,** Tolerance to the benzodiazepine diazepam in an animal model of anxiolytic activity, *Psychopharmacology,* 87, 322, 1985.
84. **Lader, M. H. and File, S.,** The biological basis of benzodiazepine tolerance, *Psychol. Med.,* 17, 539, 1987.
85. **Sher, P. K., Study, R. E., Mazzetta, J., Barker, J. L., and Nelson, P. G.,** Depression of benzodiazepine binding and diazepam potentiation of GABA-mediated inhibition after chronic exposure of spinal cord cultures to diazepam, *Brain Res.,* 268, 171, 1983.
86. **Waterhouse, B. D., Moises, H. C., Yeh, H. H., Geller, H. M., and Woodward, D. J.,** Comparison of norepinephrine-and benzodiazepine-induced augmentation of Purkinje cell responses to γ-aminobutyric acid (GABA), *J. Pharmacol. Exp. Ther.,* 228, 257, 1984.
87. **Gallager, D. W., Lakoski, J. M., Gonsalves, S. F., and Rauch, S. L.,** Chronic benzodiazepine treatment decreases postsynaptic GABA sensitivity, *Nature London,* 308, 74, 1984.
88. **Gallager, D. W., Malcolm, A. B., Anderson, S. A., and Gonsalves, S. F.,** Continuous release of diazepam: electrophysiological, biochemical and behavioral consequences, *Brain Res.,* 342, 26, 1985.
89. **Tietz, E. I. and Rosenberg, H. C.,** Behavioral measurement of benzodiazepine tolerance and GABAergic subsensitivity in the substantia nigra pars reticulata, *Brain Res.,* 438, 41, 1988.
90. **Yu, O., Chiu, T. H., and Rosenberg, H. C.,** Modulation of GABA-gated chloride ion flux in rat brain by acute and chronic benzodiazepine administration, *J. Pharmacol. Exp. Ther.,* 246, 107, 1988.
91. **Miller, L. G., Greenblatt, D. J., Barnhill, J. G., and Schrader, R. I.,** Chronic benzodiazepine administration I. Tolerance is associated with benzodiazepine receptor downregulation and decreased γ-aminobutyric $acid_A$ receptor function, *J. Pharmacol. Exp. Ther.,* 246, 170, 1988.
92. **Marley, R. J. and Gallager, D. W.,** Chronic diazepam treatment produces regionally specific changes in GABA-stimulated chloride influx, *Eur. J. Pharmacol.,* 159, 217, 1989.
93. **Miller, L. G., Woolverton, S., Greenblatt, D. J., Lopez, F., Roy, R. B., and Schrader, R. I.,** Chronic benzodiazepine administration IV. Rapid development of tolerance and receptor downregulation associated with alprazolam administration, *Biochem. Pharmacol.,* 38, 3773, 1989.
94. **Gonsalves, S. F. and Gallager, D. W.,** Time course for development of anticonvulsant tolerance and GABAergic subsensitivity after chronic diazepam, *Brain Res.,* 405, 94, 1987.
95. **Wilson, M. A. and Gallager, D. W.,** GABAergic subsensitivity of dorsal raphe neurons *in vitro* after chronic benzodiazepine treatment *in vivo, Brain Res.,* 473, 198, 1988.
96. **Braestrup, C., Nielsen, M., and Squires, R. F.,** No changes in rat benzodiazepine receptors after withdrawal from continuous treatment with lorazepam and diazepam, *Life Sci.,* 24, 347, 1979.

97. **Mohler, H., Okada, T., and Enna, S. J.,** Benzodiazepine and neurotransmitter receptor binding in rat brain after chronic administration of diazepam or phenobarbital, *Brain Res.,* 156, 391, 1978.
98. **Schiller, G. D. and Farb, D. H.,** Enhancement of benzodiazepine binding by GABA is reduced rapidly during chronic exposure to flurazepam, *Ann. N.Y. Acad. Sci.,* 463, 221, 1986.
99. **Crawley, J. N., Marangos, P. J., Stivers, J., and Goodwin, F. K.,** Chronic clonazepam administration induces benzodiazepine receptor subsensitivity, *Neuropharmacology,* 21, 85, 1982.
100. **DiStefano, P., Case, K. R., Colello, G. D., and Bosmann, H. B.,** Increased specific binding of [^{3}H]diazepam in rat brain following chronic diazepam administration, *Cell Bio. Int. Rep.,* 3, 163, 1979.
101. **Heninger, C., Saito, N., Tallman, J. F., Garrett, K. M., Vitek, M. P., Duman, R. S., and Gallager, D. W.,** Effects of continuous diazepam administration on $GABA_A$ subunit mRNA in rat brain, *J. Mol. Neurosci.,* 2, 101, 1990.
102. **Tietz, E. I., Chiu, T. H., and Rosenberg, H. C.,** Regional GABA/benzodiazepine receptor/chloride channel coupling after acute and chronic benzodiazepine treatment, *Eur. J. Pharmacol.,* 167, 57, 1989.
103. **Ochs, H. R., Greenblatt, D. J., Divoll, M., Abernethy, D. R., Feyerabend, H., and Dengler, H. J.,** Diazepam kinetics in relation to age and sex, *Pharmacology,* 23, 24, 1981.
104. **Levitan, E. S., Schofield, P. R., Burt, D. R., Rhee, L. M., Wisden, W., Kohler, M., Fujita, N., Rodriquez, H. F., Stephenson, A., Darlison, M. G., Barnard, E. A., and Seeburg, P. H.,** Structural and functional basis for $GABA_A$ receptor heterogeneity, *Nature (London),* 335, 76, 1988.
105. **Pritchett, B. D., Sontheimer, H., Shivers, B. D., Ymer, S., Kettenmann, H., Schofield, P. R., and Seeburg, P. H.,** Importance of a novel $GABA_A$ receptor subunit for benzodiazepine pharmacology, *Nature (London),* 338, 582, 1989.
106. **Fleischmann, A., Makman, M. H., and Etgen, A. M.,** Ovarian steroids increase veratridine-induced release of amino acid neurotransmitters in preoptic area synaptosomes, *Brain Res.,* 507, 161, 1990.
107. **MacLusky, N. J., Krey, L. C., Parsons, B., Merriam, G. R., Loriaux, D. L., Pfeiffer, D. G., and Naftolin, F.,** Are catechol oestrogens obligatory mediators of oestrogen action in the central nervous system? II. Potencies of natural and synthetic oestrogens for induction of gonadotrophin release and female sexual behavior in the rat, *J. Endocrinol.,* 110, 499, 1986.
108. **Nicoletti, F., Speciale, C., Sortino, M. A., Panetta, M. S., DiGiorgio, R. M., and Canonico, P. L.,** Estrogen effects on nigral glutamic acid decarboxylase activity: a possible role for catecholestrogen, *Eur. J. Pharmacol.,* 115, 297, 1985.
109. **Duvilanski, B. H., Seilicovish, A., del Carmen Diaz, M., Lasage, M. and Debeljuk, L.,** GABA-prolactin interactions, in *GABA and Endocrine Function,* Racagni, G. and Donoso, A. O., Eds., Raven Press, New York, 1986, 119.
110. **Barbaccia, M. L., Guarneri, P., Berkovish, A., Wambebe, C., Guidotti, A., and Costa, E.,** Studies on the endogenous modulator of $GABA_A$ receptors in human brain and CSF, in *Allosteric Modulation of Amino Acid Receptors: Therapeutic Implications,* Barnard, E. A. and Costa, E., Eds., Raven Press, New York, 1989, 125.
111. **Kendall, D. A., Stancel, G. M., and Enna, S. J.,** Imipramine: effect of ovarian steroids on modifications in serotonin receptor binding, *Science,* 211, 1183, 1981.

Chapter 12

IMMUNOCHEMICAL STUDIES OF PHENCYCLIDINE BINDING SITES

S. Michael Owens and Mark A. Zorbas

TABLE OF CONTENTS

I. INTRODUCTION

Biochemical and pharmacological studies suggest that phencyclidine (PCP) and other PCP receptor ligands produce a noncompetitive antagonism of an ion channel associated with the *N*-methyl-D-aspartate (NMDA) receptor subtype.[1-4] This antagonism of NMDA receptor ligands (e.g., glutamate) indicates the PCP receptor is somehow functionally associated with the ion channel modulated by the NMDA receptor. For example, in response to brain ischemia there is excessive neuronal discharge of excitatory amino acid neurotransmitters, such as glutamate, which can cause neuronal cell death.[5] PCP receptor ligands such as PCP and MK-801 can produce some neuroprotection against this ischemic brain damage, if administered prior to or shortly after ischemia.[6-9]

In addition to interaction with the NMDA receptor complex, some of the pharmacological properties of PCP, such as psychotomimetic effects, could result from its interaction with the dopamine uptake carrier.[10] Although both the PCP receptor and the dopamine uptake carrier site can bind arylcyclohexylamines (ACH), their molecular selectivities for most other compounds are different.[11,12] For example, MK-801 and (+)SKF 10,047, which bind with high affinity to the PCP receptor, bind with very low affinity to the PCP site associated with the dopamine uptake carrier.[11] Thus, a better understanding of the molecular criteria for ACH binding to the PCP receptor and dopamine uptake site could provide clues for development of compounds which possess neuroprotective properties with little to none of the psychotomimetic effects characteristic of many current PCP receptor ligands.

Current methods employed to discern the molecular features needed for binding to the PCP receptor include the Dreiding modeling[13] and computer-assisted molecular modeling.[14-16] While both methods provide useful information about the overlapping molecular similarities of PCP-like receptor ligands, they fall short of considering how small drugs like PCP interact with specific amino acid groups or sequences within the PCP receptor. An experimental approach which considers both specific molecular features and conformational characteristics of the drug and the receptor is the generation of receptorlike (antiligand) antibodies.[17-20] The complementary determining regions (i.e., ligand binding site) of these receptorlike antibodies appear to possess drug class-specific molecular contacts which are similar to the molecular contacts within the receptor itself.

These antidrug (or antiprotein or antipeptide) antibodies have then been used to generate a second set of antibodies against the complementarity determining regions of the primary antibodies. These antibodies, known as anti-idiotypes, have been shown, in some cases, to bind to the original receptor in a fashion similar to the original ligand, and in some cases they have been shown to elicit a signal or pharmacological response. Table 1 lists some of the immunogens and biologically active proteins, peptides, and drugs which

TABLE 1
Examples of Anti-Idiotypic Antibodies which Recognize Receptors

Immunogen	Receptor	Antibody-Induced Signal or Pharmacological Response
Insulin	Insulin	Oxidation of glucose[21]
Retinol	Retinol binding protein	Retinol uptake[21]
Substance P	Substance P	Stimulate turnover of membrane phosphatidylinositol[22]
Catecholamine	β-Adrenergic	cAMP activation[23]
Thyrotropin	Thyrotropin	Stimulate adenylate cyclase activity[24]
BisQ	Nicotinic acetylcholine	Muscle weakness[25]

have been successfully modeled by this immunological idiotype-anti-idiotype approach.

Using this immunological approach, we have generated antibodies against seven unique epitopes of PCP-like ligands. The characterization of five of the seven have been previously reported.[26,27] This chapter discusses the general strategy for developing antidrug and anti-idiotype antibodies based on the knowledge gained from these experiments. These studies are an attempt to improve our understanding of the fundamental rules for generating antibodies with ACH binding specificity similar to the PCP receptor.

II. SYNTHESIS OF ARYLCYCLOHEXYLAMINE ANTIGENS

Each ACH hapten was covalently bound to bovine serum albumin (BSA) using a carbodiimide reagent.[26] This procedure creates a peptide bond between the carboxylic acid group on the ACH derivative and the ε-amino group on the lysine moieties of BSA.[28] A two-step carbodiimide procedure was used for synthesizing the antigen because this technique significantly reduces cross-linking of BSA molecules.[29] Figure 1 shows the chemical structures of PCP, PCHAP (as an example of an ACH derivative used as a hapten to generate anti-ACH antibodies), and the PCHAP-BSA conjugate used as the immunogen. Note that the site of antibody recognition would be distal to the site of BSA conjugation (i.e., the aromatic and cyclohexyl rings). In previous studies in our laboratory[26,27] with five different PCP-like antigens, it was found that carbodiimide coupling of ACH derivatives to BSA using 1-ethyl-3-(3-dimethylaminopropyl)carbodiimide results in 14 to 30 mol of hapten per mole of BSA.

Figure 2 shows a summary of all seven ACH derivatives used for generation of the antibodies. After carefully considering the appropriate types of haptens for generation of the antibodies, two types of molecular templates

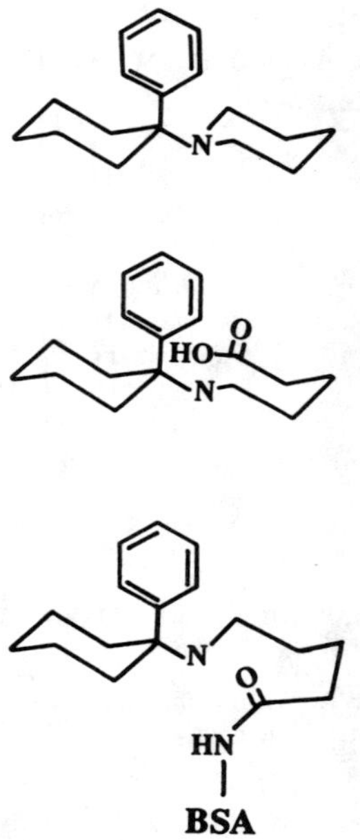

PCP

PCHAP (hapten)

PCHAP-BSA (antigen)

FIGURE 1. The chemical structures of PCP, PCHAP (a hapten used for generating one set of anti-ACH antibodies), and PCHAP-BSA (the ACH-antigen conjugate used for immunization). The structure of PCP is shown as the possible receptor-active configuration as suggested by Kamenka[45] and Carroll[16] based on physicochemical evidence. Based on these data and the immunological evidence in this study, the receptor-active conformation and orientation of arylcyclohexylamines would be approximated by the PCHAP moiety of PCHAP-BSA.

were chosen. The first series of epitopes had three intact ring structures (i.e., an aromatic ring, a cyclohexyl ring, and a nitrogen-containing ring) and were designed to be PCP-like in their presentation to the immune system as antigens. The other series was designed to be PCE-like (PCE, *N*-ethyl-1-phenylcyclohexylamine). A PCE-like template was chosen because PCE is known to be one of the most potent ACHs in both behavioral studies[30-32] and PCP receptor binding studies.[33,34] These haptens had an intact aromatic ring (either a benzene (PCHAP) or thienyl (TCHAP) ring structure) and an intact cyclohexyl ring, but unlike the PCP-like haptens, the nitrogen molecule was not incorporated into a ring structure. This presumably provided a more flexible molecule when attached to the antigen.

PCP-like Antigens

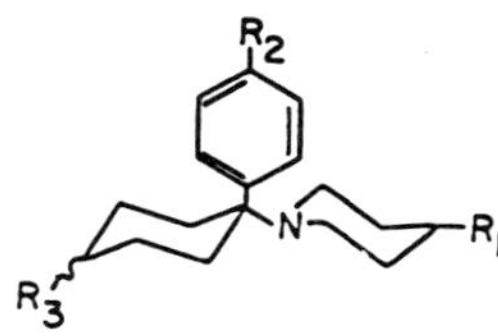

PCHP, R_1 = OCO (CH_2) 2CONH-BSA
R_2 = R_3 = H
PCHBA, R_2 = CONH-BSA
R_1 = R_3 = H
cPPC, R_3 = OCO (CH_2) 2CONH-BSA
R_1 = R_2 = H
tPPC, R_3 = OCO (CH_2) 2CONH-BSA
R_1 = R_2 = H

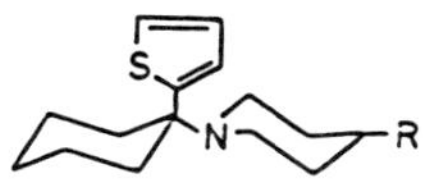

TCHAP, R = OCO (CH_2) 2CONH-BSA

PCE-like Antigens

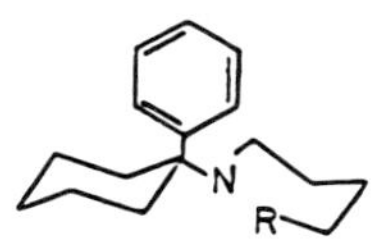

PCHAP, R = CONH-BSA

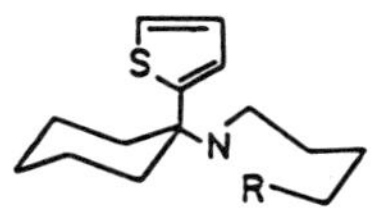

TCHAP, R = CONH-BSA

FIGURE 2. The chemical structures of the seven arylcyclohexylamine-BSA conjugates used for immunizing rabbits.

III. PRODUCTION OF ANTI-ARYLCYCLOHEXYLAMINE ANTIBODIES

In these studies, rabbits were used for production of antibodies since it was thought that the specificity of monoclonal antibodies would be too refined for choosing the appropriate hapten during these initial screening studies. In addition, the production and characterization of large numbers of individual monoclonal antibodies was considered too difficult and unnecessary. However, as will be discussed later, once the appropriate hapten was chosen, monoclonal antibodies were produced against the single best hapten.

Two to three rabbits were used for production of antibodies against each of the seven antigens. The antiserum from each rabbit was titered in a $[^3H]$-PCP radioimmunoassay (RIA). All rabbits used for antibody production generated high-affinity antibodies. For instance, the TCHAP antigen generated

TABLE 2
Chemical Names for Drugs Used in the RIA Relative Potency Studies

Drug	Chemical name
	Arylcyclohexylamines
PCP	1-(1-Phenylcyclohexyl)piperidine
PCE	*N*-Ethyl-1-phenylcyclohexylamine
TCP	1-[1-(2-Thienyl)cyclohexyl]piperidine
PHP	1-(1-Phenylcyclohexyl)pyrrolidine
THP	1-[1-(2-Thienyl)cyclohexyl]pyrrolidine
PCDMA	*N,N*-Dimethyl-1-phenylcyclohexylamine
PCHP	1-(1-Phenylcyclohexyl)-4-hydroxypiperidine
4MePCP	1-(1-Phenylcyclohexyl)-4-methylpiperidine
PCM	1-(1-Phenylcyclohexyl)morpholine
TCM	1-[1-(2-Thienyl)cyclohexyl]morpholine
*c/t*PPC	*cis/trans*-1-(1-Phenyl-4-hydroxycyclohexyl)piperidine
KETAMINE	2-(*o*-Chlorophenyl)-2-(methylamino)cyclohexanone
	Non-arylcyclohexylamines
(+) SKF 10,047	(+)-*N*-Allyl-*N*-normetazocine
(+) MK-801	(+)-5-Methyl-10,11-dihydro-5*H*-dibenzo[a,d]-cyclohepten-5,10-imine
Dexoxadrol	*a*-(+)-Dioxadrol
DTG	1,3-Di(2-tolyl)guanidine
Morphine	7,8-Didehydro-4,5-epoxy-17-methylmorphinan-3,6-diol

antibodies with K_ds for PCP ranging from 4.3 to 5.2 n*M*. In most cases, the antiserum from one rabbit from each of the antigens (Figure 2) was chosen for an extensive characterization. A single rabbit antiserum from each of the seven antigens was chosen on the basis of the rabbit producing the highest titer in the [^{3}H]-PCP RIA. These antibody titers for the seven antisera ranged from about 1:1500 to 1:8000.

IV. CHARACTERIZATION OF THE ANTI-ARYLCYCLOHEXYLAMINE ANTISERUM

An antiserum against each of the seven antigens was carefully characterized for their ability to bind a large series of ligands (see Table 2). As an example of the data from these RIA relative potency studies, Table 3 shows the relative potencies of the various ACH and other test ligands in a [^{3}H]-PCP RIA using TCHAP antiserum from three different rabbits. Figure 3 shows a typical RIA dose-response curve generated for the antiserum from one of the anti-PCHAP antiserum.

Table 4 summarizes the relative potency data for some of these same ligands in four different pharmacological studies conducted in two species (rat and pigeon). The RIA relative potency data were then correlated with

TABLE 3
Representative Data from RIA Relative Potencies of Various Arylcyclohexylamines in the ^{3}H-Labeled PCP Anti-TCHAP RIA[a]

Drug[b]	Anti-TCHAP Serum 1	Anti-TCHAP Serum 2	Anti-TCHAP Serum 3
PCP[c]	1.0 (6.5)	1.0 (5.5)	1.0 (5.7)
PCE	0.66	0.72	0.38
TCP	0.62	1.53	1.47
PHP	0.62	0.45	0.37
THP	0.27	0.35	0.35
PCDMA	0.06	0.07	0.04
PCHP	0.57	0.46	0.49
4MePCP	0.93	0.85	0.94
PCM	0.08	0.20	0.17
TCM	0.02	0.09	0.07
PPC	0.003	0.014	0.002
KETAMINE	0.00005	0.00002	0.00008
(+) SKF 10,047	0.00002	0.00001	0.00001
(+) MK-801	0.00003	0.00002	0.00002
Dexoxadrol	0.0004	0.0003	0.0003
DTG[d]	<0.00001	<0.00003	<0.00003
Morphine	0.00001	0.00001	0.00001

[a] Each number represents the relative potency to PCP calculated from an RIA IC_{50} value (RIA IC_{50} value for PCP/RIA IC_{50} value for the test ligand).
[b] See Table 1 for chemical names.
[c] Numbers in parentheses indicate the IC_{50} value for PCP (n*M*).
[d] This drug did not inhibit binding at 840 m*M*, the highest concentration tested. Thus, a conservative minimum estimate of cross-reactivity is given.

these pharmacological relative potency data to determine if there was a relationship between antibody and pharmacological specificities.

Least-squares linear regression analysis was used to correlate the relative potency data from each antiserum (e.g., the data from rabbit serum 2 in Table 3) with the relative potency data from each of the pharmacological assays in Table 4. Note that data for all of the ligands tested in the RIA studies (e.g., Table 3) were not available for all of the pharmacological studies (Table 4).

Figure 4 summarizes the coefficients of determination (r^2) for each linear regression model and the statistical significance of each model. The results from the antiserum generated against the PCE-like haptens (PCHAP or TCHAP) showed the best correlation with the relative potency data from the pharmacological studies. However, the anti-PCHAP serum was clearly the best, since relative potency data from this antiserum showed statistically significant correlations ($p<0.05$) with three out of the four comparisons. The only pharmacological study that did not correlate with these data was from a drug discrimination study in the pigeon.[32]

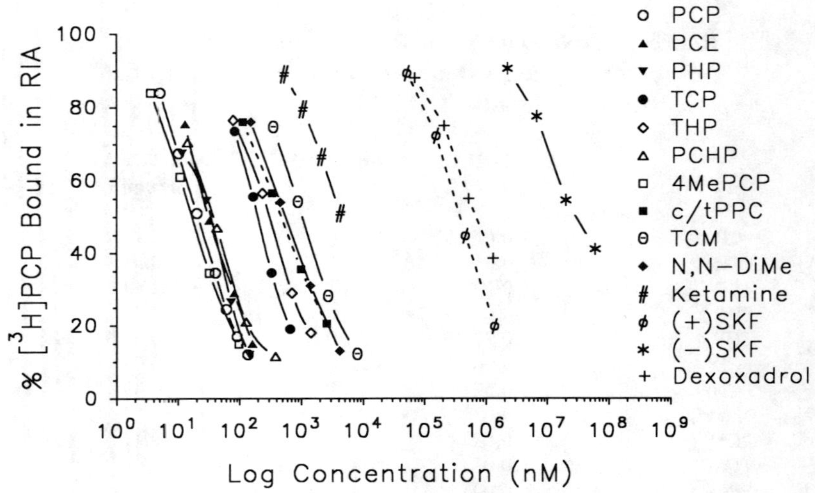

FIGURE 3. A representative RIA dose-response curve for antibody relative potency data using antibodies against the PCHAP hapten. These data show the percentage [^{3}H]-PCP bound vs. the log concentration for the average data from duplicate RIA standard curves. This antibody was tested against 14 different PCP receptor ligands.

TABLE 4
Relative Potencies of Arylcyclohexylamines in Various Pharmacological Assays[a]

	Drug Discrimination		Receptor Binding	
Drug	**Shannon[b]**	**McMillan[c]**	**Quirion[d]**	**Zukin[e]**
PCP	1.0	1.0	1.0	1.0
PCE	5.79	1.45	6.0	1.64
TCP	1.31	2.05	1.7	1.44
PHP	0.97	0.43	1.4	1.15
THP	0.87	1.13	0.80	0.77
PCDMA	0.5	0.38	0.30	0.45
PCHP	0.37	0.09		
4MePCP	0.10	0.88		
PCM	0.10	0.51		
TCM	0.07	0.16	0.05	0.14
PPC	0.03	0.00		
Ketamine	0.10	0.11	0.10	0.23

a Blank spaces indicate the drug was not tested in this assay.
b Data from Shannon.[30,31]
c Data from McMillan et al.[32]
d Data from Quirion et al.[34]
e Data from Zukin and Zukin.[33]

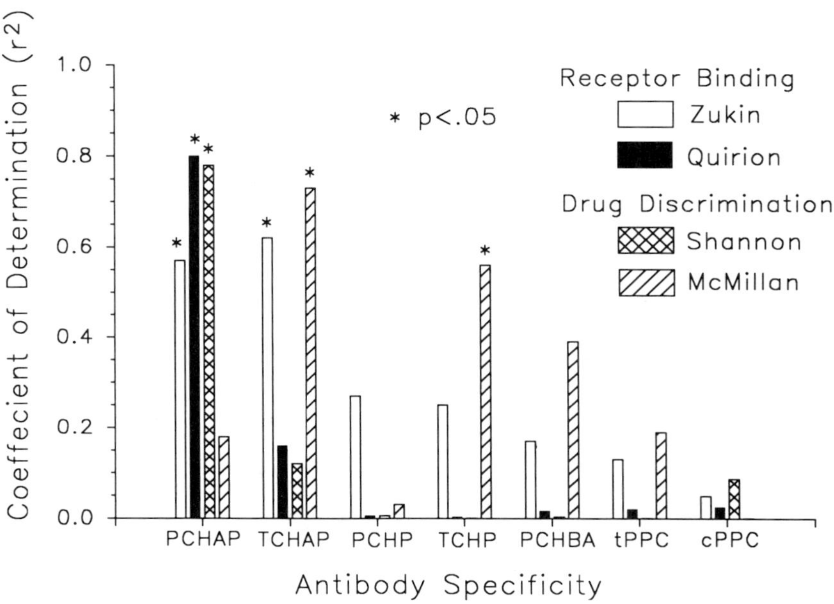

FIGURE 4. Summary of linear regression coefficients of determination (r^2) from the correlation of [^{3}H]-PCP antibody RIA relative potency data from each of the seven anti-PCP antibodies with relative potency data from each of the four pharmacological assays (see Table 4). RIA data for the PCHAP, PCHP, TCHP, PCHBA, and tPPC antiserum are from Owens et al.[26] Data for the TCHAP and cPPC antiserum were produced for this study.

We also compared the relationship between the antibody potency data and pharmacological potency data by their rank order in each type of assay. Nonparametric analysis using Spearman's rank order correlation[35] showed statistical significance ($p<0.05$) between antibody-binding potency for the PCHAP and TCHAP antibodies and pharmacological potency in all cases. In other words, the most potent compounds in the RIA were also the most potent compounds in the pharmacological assays. The other antiserum did not produce these significant results.

V. CONCLUSIONS

The major purpose of this research was to determine if protein-bound ACH analogs could stimulate production of rabbit antibodies which could recognize a series of ACH ligands with a potency order similar to the PCP receptor. All of the rabbits generated high-affinity antisera. The affinities of the seven antiserum for [^{3}H]-PCP (K_ds of about 2 to 52 nM[26]) were, in most cases, greater than the affinity of the PCP receptor for [^{3}H]-PCP (e.g., 150 nM[33]; 48 nM[34]), but they were similar to the K_d of [^{3}H]-TCP for the PCP receptor (7.4 nM;[36]). However, the molecular specificity of the antibodies (or pattern of ligand recognition) was considered to be a much more important characteristic, as it has been for similar immunological receptor modeling studies.[20,37,38] This is because the underlying premise was as follows. If antibody specificity is similar to pharmacological specificity (i.e., relative potency), then the antibody binding site is probably recognizing at least some of the critical molecular features of arylcyclohexylamines which are necessary for binding to the PCP binding site.[17,25,39]

The antibodies selectively stimulated by immunization with the various ACH-protein conjugates should recognize parts of the hapten distal to the site of protein conjugation.[40] However, because of the flexibility of the various haptens, a single stable conformation of the hapten would not be expected. Instead, a preferred conformation in a given environment is a more realistic expectation. Regardless of the exact nature of the conformation, only two of the haptens stimulated antibodies capable of detecting the pharmacologically active features in structurally dissimilar molecules such as PCP and PCE (see Table 2 for chemical names). These haptens were PCHAP and TCHAP which are quite similar in structure (see Figure 2). The only difference in these molecules is that TCHAP has a thienyl ring structure in place of the benzene ring structure in PCHAP. The aromatic nature of these two ring structures is known to be quite similar.

Although six structurally distinct classes of compounds bind to the PCP receptor[41] (e.g., arylcyclohexylamines, benzomorphans, and dioxolanes), the non-ACH PCP receptor ligands (e.g., MK-801, SKF 10,047, and dexoxadrol) did not cross-react with any of the antibodies (see Figure 3 as an example). This large difference in recognition between the antibody and receptor could be explained on the basis of the nature of the antibody binding site which fits PCHAP or TCHAP and possibly a small portion of the carrier protein used for immunization. By comparison, the ion channel which is the PCP receptor could accommodate more molecular features, such as those contained in other classes of compounds (e.g., benzomorphan or dioxolones). Similar results are used to explain the refined specificity of anti-fMet-Leu-Phe antibodies.[42] In this study, antibodies against the potent neutrophil chemoattractant, fMet-Leu-Phe, do not have as broad a specificity as the actual chemotactic receptor. Nevertheless, the anti-idiotype antibodies against the binding site of the anti-fMet-Leu-Phe antibody can still bind to the neutrophil chemotactic receptor. Other immunological modeling studies using antimorphine monoclonal antibodies[38] also find high-affinity binding for morphine but little or no antibody recognition for potent mu receptor ligands like opiate peptides and the antagonist, naloxone. However, anti-idiotype antibodies generated against these antimorphine monoclonal antibodies[43] can be inhibited by the full spectrum of mu receptor analogs and antagonists (i.e., morphine, naloxone, and enkephalins) in a rat morphine receptor binding assay. This concept is also in agreement with pharmacological evidence from Childers et al.,[44] which evaluates the differential ability of various opiate receptor ligands to displace [^{3}H]-enkephalin and [^{3}H]-opiate binding. To account for this differential effect in a homogeneous population of opiate binding sites, they suggest that some classes of opiate receptor ligands (e.g., enkephalins) might compete for all or most of the ligand attachment points within the receptor, while other ligands like morphine might combine with only some of the attachment points within the receptor.

The current anti-TCHAP data, considered with previous studies of anti-PCHAP data from our laboratory,[26,27] could provide important molecular clues concerning the preferred three-dimensional conformation of ACH in the PCP receptor. Considering the statistically significant association between antibody binding specificity and PCP receptor specificity along with immunochemical principles, these data suggest that ACH interaction with the PCP binding site could consist of a deeply embedded hydrophobic pocket containing hydrophobic amino acid contact points with the cyclohexane and aromatic ring. The nitrogen atom, with its lone pair of electrons, could project toward the more polar aqueous environment surrounding the membrane or ion channel and/or associate with complementary charged or polar amino acids. This suggested orientation and preferred conformation is approximated by the structure of the PCHAP moiety on PCHAP-BSA as shown in Figure 1. This preferred molecular configuration, represented by this molecular model, is consistent with molecular configuration studies of PCP and TCP using physicochemical measurements of PCP and TCP in aqueous and lipid environments.[16,45] However, the antibody data also suggest molecular orientation in the receptor (as discussed above) may be a critical factor.

Although not determined in the current studies, our antibody binding site may also mimic at least some of the molecular features necessary for ACH binding to the dopamine uptake site. Recent studies show a good correlation between the inhibition of [^{3}H]-dopamine uptake and [^{3}H]-PCP binding for ACH with an unmodified phenyl group.[11,12] These findings suggest that the dopamine uptake site and the PCP receptor could share some common molecular features. Although some common features could exist, there are also important molecular differences between the two PCP binding sites. For example, potent PCP receptor compounds such as ketamine (an ACH), MK-801, and SKF 10,047, bind very poorly to the dopamine uptake site.[11] These same ligands bind very poorly to our antibodies. The full range of compounds necessary to correlate our antibodies with the dopamine uptake binding site has not yet been tested.

Because monoclonal antibodies are homogeneous proteins, they could provide more precise and more easily interpreted information about the nature of the molecular conformation and orientation of ACH in a PCP binding site. For this purpose we have generated monoclonal antibodies against the PCHAP hapten.[46] In addition, if antiidiotype antibodies generated against a (PCHAP or TCHAP) monoclonal antibody could bind to a PCP binding site, this could ultimately confirm that the anti-PCHAP or -TCHAP antibody binding site mimics some aspects of the ''internal image'' of the PCP receptor or the dopamine uptake binding site or both.

In summary, an extensive library of high-affinity rabbit antibodies has been generated against ACH analogs. Based on structure-activity studies, only antibodies against a PCE-like derivative, either PCHAP or TCHAP, were able to distinguish between a series of pharmacologically active ACH. Parametric and nonparametric statistical analysis showed statistically significant correlations between antibody relative potency data and pharmacological relative potency data. These data indicated PCHAP-[26,27] and TCHAP-BSA were able to stimulate the production of antibodies whose binding site recognized at least some of the important molecular features needed for binding of ACH to the PCP receptor. However, based on comparison of statistical correlations between antibody binding data and pharmacological data, antibodies generated against the PCHAP hapten[26,27] should be better for these purposes than antibodies generated against a TCHAP hapten (Figure 4). Although isolated cases of statistically significant correlation exist for antibodies against some other haptens (e.g., TCHP; see Figure 4), PCE-like amino pentanoic acid analogs of PCP (PCHAP and TCHAP) are clearly superior to a wide variety of other haptens in their ability to stimulate production of antibodies with ACH binding specificity similar to the PCP receptor.

ACKNOWLEDGMENTS

This work was supported by Grant DA 04136 and a Research Scientist Development Award (K02 DA 00110) to S. M. Owens from the National Institute on Drug Abuse. The authors thank Drs. Richard Hawks and George Brine for providing TCHAP and Melinda Gunnell for excellent technical assistance. We also thank the Upjohn Co. for supplying dexoxadrol and the National Institute on Drug Abuse for supplying various other drugs.

REFERENCES

1. **Lodge, D., Anis, N. A., Berry, S. C., and Burton, N. R.,** Arylcyclohexylamines selectively reduce excitation of mammalian neurones by aspartate-like amino acids, in *Arylcyclohexylamines: Present and Future Applications,* Kamenka, J. M., Domino, E. F., and Geneste, P., Eds., NPP Books, Ann Arbor, MI, 1983, 595.
2. **Anis, N. A., Berry, S. C., Burton, N. R., and Lodge, D.,** The dissociative anaesthetics, ketamine and phencyclidine, selectively reduce excitation of central mammalian neurons by *N*-methyl-aspartate, *Br. J. Pharmacol.,* 79, 565, 1983.
3. **Harrison, N. L. and Simmonds, M. A.,** Quantitative studies on some antagonists of *N*-methyl-D-aspartate in slices of rat cerebral cortex, *Br. J. Pharmacol.,* 84, 381, 1985.
4. **Honey, C. R., Miljkovic, Z., and MacDonald, J. F.,** Ketamine and phencyclidine cause a voltage-dependent block of responses to L-aspartic acid, *Neurosci. Lett.,* 61, 135, 1985.
5. **Kirino, T., Tamura, A., and Sano, K.,** Selective vulnerability of the hippocampus to ischaemia-reversible and irreversible types of ischaemic cell damage, *Prog. Brain Res.,* 63, 39, 1985.

6. **Wong, E. H. F., Kemp, J. A., Priestly, T., Knight, A. R., Woodruff, G. N., and Iverson, L. L.,** The anti-convulsant MK-801 is a potent *N*-methyl-D-aspartate antagonist, *Proc. Natl. Acad. Sci. U.S.A.,* 83, 7104, 1986.
7. **Foster, A. C., Gill, R., Kemp, J. A., and Woodruff, G. N.,** Systemic administration of MK-801 prevents *N*-methyl-D-aspartate-induced neuronal degeneration in rat brain, *Neurosci. Lett.,* 76, 307, 1987.
8. **Foster, A. C., Gill, R., Iversen, L. L., and Woodruff, G. N.,** Systemic administration of MK-801 protects against ischaemia-induced hippocampal neurodegeneration in the gerbil, *Br. J. Pharmacol.,* (Suppl), 9P, 1987.
9. **Leach, M. J., Hollox, K. J., O'Donnell, R. A., and Miller, A. A.,** Hippocampal NMDA/phencyclidine receptor binding sites are reduced following forebrain ischaemia in the gerbil, *Eur. J. Pharmacol.,* 152, 189, 1988.
10. **Vignon, J. and Lazdunski, M.,** Structure-function relationship in the inhibition of synaptosomal dopamine uptake by phencyclidine and its analogues: potential correlation with binding sites, *Biochem. Pharmacol.,* 33, 700, 1984.
11. **Vignon, J., Pinet, V., Cerruti, C., Kamenka, J.-M., and Chicheportiche, R.,** [^{3}H]*N*-[1-(2-benzo(*b*)thienyl)cyclohexyl]piperidine ([^{3}H]BTCP): a new phencyclidine analog selective for the dopamine uptake complex, *Eur. J. Pharmacol.,* 148, 427, 1988.
12. **Chaudieu, I., Vignon, J., Chicheportiche, M., Kamenka, J.-M., Trouiller, G., and Chicheportiche, R.,** Role of the aromatic group in the inhibition of phencyclidine binding and dopamine uptake by PCP analogs, *Pharmacol. Biochem. Behav.,* 32, 699, 1989.
13. **Cone, E. J., McQuinn, R. L., and Shannon, H. E.,** Structure-activity relationship studies of phencyclidine derivatives in rats, *J. Pharmacol. Exp. Ther.,* 228, 147, 1984.
14. **Jacobson, A. E., Harrison, E. A., Jr., Mattson, M. V., Rafferty, M. F., Rice, K. C., Woods, J. H., Winger, G., Solomon, R. E., Lessor, R. A., and Silverton, J. V.,** Enantiomeric and diastereomeric dioxodrols: behavioral, biochemical and chemical determination of the configuration necessary for phencyclidine-like properties, *J. Pharmacol. Exp. Ther.,* 243, 110, 1987.
15. **Thurkauf, A., Zenk, P. C., Balster, R. L., May, E. L., George, C., Carroll, F. I., Mascarella, S. W., Rice, K. C., Jacobson, A. E., and Mattson, M. V.,** Synthesis, absolute configuration, and molecular modeling study of etoxadrol, a potent phencyclidine-like agonist, *J. Med. Chem.,* 31, 2257, 1988.
16. **Carroll, F. I., Brine, G. A., Boldt, K. G., Mascarella, S. W., Moreland, C. G., Summer, S. J., Burgess, J. P., and Stejskal, E. O.,** Solid state conformation of phencyclidine and phencyclidine analogs, in *Sigma and Phencyclidine-Like Compounds as Molecular Probes in Biology,* Domino, E. F. and Kamenka, J.-M., Eds., NPP Books, Ann Arbor, MI, 1988, 91.
17. **Rockson, S. G., Homcy, C. J., and Haber, E.,** Anti-alprenolol antibodies in the rabbit: a new probe for the study of β-adrenergic receptor interaction, *Circ. Res.,* 46, 808, 1980.
18. **Marullo, S., Hoebeke, J., Guillet, J.-G., and Strosberg, A. D.,** Structural analysis of the epitope recognized by a monoclonal antibody directed against tricyclic antidepressants, *J. Immunol.,* 135, 471, 1985.
19. **Couraud, P.-O.,** Structural analysis of the epitopes recognized by monoclonal antibodies to angiotensin II, *J. Immunol.,* 136, 3365, 1986.
20. **Sherman, M. A., Linthicum, D. S., and Bolger, M. B.,** Haloperidol binding to monoclonal antibodies: conformational analysis and relationships to D-2 receptor binding, *Mol. Pharmacol.,* 29, 589, 1986.
21. **Sege, K. and Peterson, P. A.,** Use of anti-idiotypic antibodies as cell surface receptors, *Proc. Natl. Acad. Sci. U.S.A.,* 75, 2432, 1978.
22. **Couraud, J. Y., Escher, E., Regoli, D., Imhoff, V., Rossignol, B., and Pradelles, D.,** Anti-substance P anti-idiotypic antibodies. Characteristics and biological activities, *J. Biol. Chem.,* 260, 9461, 1985.
23. **Schreiber, A. B., Cauraud, P. O., Andre, C., Vray, B., and Strosberg, A. D.,** Anti-alprenolol anti-idiotypic antibodies bind to β-adrenergic receptors and modulate catecholamine-sensitive adenylate cyclase, *Proc. Natl. Acad. Sci. U.S.A.,* 77, 7385, 1980.

24. **Farid, N. R., Pepper, B., Urbina-Briones, R., and Islam, N. R.,** Biological activity of anti-thyrotropin anti-idiotypic antibodies, *J. Biol. Chem.*, 19, 305, 1982.
25. **Wasserman, N. H., Penn, A. S., Freimuth, P. I., Treptoro, N., Wentzel, S., Cleveland, W. L., and Erlanger, B. F.,** Anti-idiotypic route to anti-acetylcholine receptor antibodies and experimental myasthenia gravis, *Proc. Natl. Acad. Sci. U.S.A.*, 79, 4810, 1982.
26. **Owens, S. M., Zorbas, M., Lattin, D. L., Gunnell, M., and Polk, M.,** Antibodies against arylcyclohexylamines and their similarities in binding specificity with the phencyclidine receptor, *J. Pharmacol. Exp. Ther.*, 246, 472, 1988.
27. **Owens, S. M., Zorbas, M., Gunnell, M., Polk, M., and Lattin, D. L.,** Molecular requirements for an immunological model of the phencyclidine receptor, in *Sigma and Phencyclidine-Like Compounds as Molecular Probes in Biology,* Domino, E. F. and Kamenka, J.-M., Eds., NPP Books, Ann Arbor, MI, 1988, 663.
28. **Erlanger, B. F.,** Principles and methods for the preparation of drug protein conjugates for immunological studies, *Pharmacol. Rev.*, 25, 271, 1973.
29. **Tam,-M., Davis, B., and Preston, J. F.,** A simple modified carbodiimide method for conjugation of small-molecular weight compounds to immunoglobulin G with minimal protein crosslinking, *Anal. Biochem.*, 116, 402, 1981.
30. **Shannon, H. E.,** Evaluation of phencyclidine analogs on the basis of their discriminative stimulus properties in the rat, *J. Pharmacol. Exp. Ther.*, 216, 543, 1981.
31. **Shannon, H. E.,** Discriminative stimulus effects of phencyclidine: structure-activity relationships, in *Phencyclidine and Related Arylcyclohexylamine: Present and Future Applications,* Kamenka, J.-M., Domino, E. F., and Geneste, P., Eds., NPP Books, Ann Arbor, MI, 1983, 311.
32. **McMillan, D. E., Evans, E. B., Wessinger, W. D., and Owens, S. M.,** Structure-activity relationships of arylcyclohexylamines as discriminative stimuli in pigeons, *J. Pharmacol. Exp. Ther.*, 247, 1086, 1988.
33. **Zukin, S. R. and Zukin, R. S.,** Specific ^{3}H phencyclidine binding in rat central nervous system, *Proc. Natl. Acad. Sci. U.S.A.*, 76, 5372, 1979.
34. **Quirion, R., O'Donohue, T. L., Everist, H., Pert, A., and Pert, C. B.,** Phencyclidine receptors and possible existence of an endogenous ligand, in *Phencyclidine and Related Arylcyclohexylamines: Present and Future Applications,* Kamenka, J.-M., Domino, E. F., and Geneste, P., Eds., NPP Books, Ann Arbor, MI, 1983, 667.
35. **Seigel, S.,** *Nonparametric Statistics for the Behavioral Sciences,* Mcgraw-Hill, New York, 1956.
36. **Vignon, J., Chicheportiche, R., Chicheportiche, M., Kamenka, J. M., Geneste, P., and Lazdunski, M.,** [^{3}H]TCP: a new tool with high affinity for the PCP receptor in rat brain, *Brain Res.*, 280, 194, 1983.
37. **Sherman, M. A., Deans, R. J., and Bolger, M. B.,** Haloperidol binding to monoclonal antibodies: hypervariable region amino acid sequence determination, *J. Biol. Chem.*, 263, 4059, 1988.
38. **Glasel, J. A., Bradbury, W. M., and Venn, R. F.,** Properties of murine anti-morphine antibodies, *Mol. Immunol.*, 20, 1419, 1983.
39. **Strosberg, A. D.,** Anti-idiotypic antibodies as immunological internal images of hormones, in *Idiotypy in Biology and Medicine,* Kohler, H., Urbain, J., and Cazenane, P.-A., Eds., Academic Press, San Diego, 1984, 365.
40. **Landsteiner, N. K.,** *The Specificity of Serological Reactions,* Dover Press, New York, 1962.
41. **Thurkauf, A., Mattson, M. V., Jacobson, A. E., and Rice, K. C.,** Synthesis of a molecular hybrid of phencyclidine and dexoxadrol, in *Sigma and Phencyclidine-Like Compounds as Molecular Probes in Biology,* Domino, E. F. and Kamenka, J. M., Eds., NPP Books, Ann Arbor, MI, 1988, 47.
42. **Marasco, W. A., Showell, H. J., Freer, R. J., and Becker, E. L.,** Anti-fMet-Leu-Phe: similarities in fine specificity with the formyl peptide chemotaxis receptor of the neutrophil, *J. Immunol.*, 128, 956, 1982.

43. **Glasel, J. A.,** Opiate receptors and molecular shapes, in *Anti-Idiotypes, Receptors, and Molecular Mimicry,* Linthicum, D. S. and Farid, N. R., Springer-Verlag, New York, 1988, 135.
44. **Childers, S. R., Creese, I., Snowman, A. M., and Snyder, S. H.,** Opiate receptor binding affected differentially by opiates and opiate peptides, *Eur. J. Pharmacol.,* 55, 11, 1979.
45. **Kamenka, J.-M. and Chicheportiche, R.,** The conformational adaptation of the phencyclidine molecular pattern to the lipophilicity of its surroundings, in *Sigma and Phencyclidine-Like Compounds as Molecular Probes in Biology,* Domino, E. F. and Kamenka, J.-M., Eds., NPP Books, Ann Arbor, MI, 1988, 1.
46. **Owens, S. M., Arnold, L. W., Sasser, D., and Gunnell, M.,** Anti-PCHAP monoclonal antibodies as models of phencyclidine receptor binding to arylcyclohexylamines, *Pharmacologist,* 30, 135, 1988.

Chapter 13

BARBITURATE DEPENDENCE: NEUROCHEMICAL AND METHODOLOGICAL CONSIDERATIONS

Gerald J. Yutrzenka and Graham A. Patrick

TABLE OF CONTENTS

I. INTRODUCTION

Use of pharmacologic agents that produce a generalized depression of the central nervous system (CNS) has been a prevalent type of substance abuse throughout the history of that phenomenon. The abuse of CNS depressants began with the excessive consumption of alcoholic beverages in man's prehistory, grew to include anesthetic gases and vapors along with chloral and bromide salts in the 19th century, and has expanded to include more chemically complex synthetic agents, such as the barbiturates, methaqualone, and the benzodiazepines during the current century. All of these agents seem to share the ability to produce a qualitatively similar subjective "high" resulting, apparently, from disinhibitory effects in the CNS. It is this subjective effect that accounts for the initial abuse liability of the CNS depressants, and later development of psychological and physical dependence helps to perpetuate their use.

Although CNS depressants have always comprised a major category of abused substances, the incidence of usage of various members of the class has varied over time. The most precise data on usage of abused drugs in the U.S. has been generated by the "National High School Senior Drug Abuse Survey" which has been conducted annually by the National Institute on Drug Abuse (NIDA) since 1975. Some of these data, which have more recently included additional information on drug usage among college students and young adults, are described below.[1,2] It should be borne in mind that the population described by these data is a more promiscuous drug-taking population than are older Americans. That is to say that, while the older population probably has a lower incidence of abuse of drugs, a relatively greater fraction of their abuse would likely entail CNS depressants.

Among high school seniors, those reporting use of any illicit drug during their lifetime varied from 55.2% in 1975 to high values of around 65% from 1979 to 1982. This was followed by gradual decline to the 1989 figure of 50.9%. Corresponding percentages for those reporting current (last 30 d) use of an illicit drug were 30.7 in 1975, rising to peak values of 38.9 in 1978 and 1979, and subsequently declining to 19.4 in 1989.

Reported lifetime illicit use of sedatives (barbiturates and methaqualone) was 18.2% in 1975 and has declined rather steadily to an estimated 6.5% in 1989. Over the same period, use of "tranquilizers" has diminished from the reported 17 to 18% in the mid-to-late 1970s to 7.6% in 1989. Current (last 30 d) use of barbiturates and tranquilizers is at rates of 1.4 and 1.3%, respectively. Use of alchol during life has remained between 90.4 and 93.2% throughout the period, while current use has declined from around 70% in the early 1980s to a low of 60% in 1989. Inhalant drug abuse was reported at a lifetime rate of 10.3% in 1975, but the rate rose to around 18% in 1979, with the addition of volatile nitrites to the category, and has remained near that level since. When young adults, age 19 to 30 years, are included in the polling, the lifetime usage rates among young adults are lower for inhalants (0.5%) and barbiturates (0.7%), the same for tranquilizers, and slightly higher for alcohol (around 70%) when compared with high school seniors. Usage rates among college students closely resembled those among young adults, but lifetime and current rates were slightly higher for young adults, 19 to 22, who were not in college.

While the use of barbiturates and tranquilizers among high school seniors has declined by more than half over the last 15 years, the numbers of students introduced to the use of these agents in grades 8 and 9 has remained relatively constant at around 5 to 6%. The apparent decline in use among high school seniors is due, in part, to a lesser number of older students beginning use or continuing to use these drugs.

Among seniors in the class of 1988, approximately 70% of recent barbiturate users achieved some degree of "high" with the drug, but only around 15% described the effect as "very high." Corresponding percentages were obtained from recent users of alcohol (approximately 82 and 7), tranquilizers (45 and 7), and methaqualone (88 and 15).

In terms of perceived risk in drug usage, 27 to 30% of seniors saw "great risk" in trying barbiturates once or twice or in taking one or two drinks nearly every day. About 70% considered either regular barbiturate use or taking four or five drinks nearly every day as involving great risk. In comparison, about 80% of young adults aged 23 to 26 years perceived great risk in the same activities.

Only 1.4% of high school seniors reported being around people taking barbiturates "often" in the past year, while 87.6% were around such activity "not at all." Data for exposure to use of tranquilizers were similar (2.2 and 81.8%, respectively), while those for exposure to use of alcohol were quite different (56.4 and 6.9%, respectively). Around 50% of the seniors and young adults felt that barbiturates and tranquilizers would be "fairly easy" or "very easy" to obtain.[1,2]

Another source of statistical information on medical problems associated with substance abuse is the Drug Abuse Warning Network (DAWN).[3] In visits to hospital emergency rooms in the U.S., barbiturate sedatives were identified as a precipitating factor in 1915 cases. Specific drugs mentioned included phenobarbital (1067 cases), secobarbital (115), and pentobarbital (39), with the agent unspecified in 635 cases. In addition, there were 5746 cases involving nonbarbiturate sedatives, including over-the-counter sedatives (1766 cases), flurazepam (724), and unspecified agents (2623). There were also 703 cases involving butalbital combination products. Approximately one-half of the barbiturate-related cases were reported from the five largest U.S. cities and the District of Columbia.

In about 70% of the reports concerning barbiturates the drug was taken in combination with other agents. The route of administration was known to be oral in 82% and was unknown in an additional 15.5%. Users were predominantly in the age ranges of 20 to 29 years (28.5%) and 30 to 39 years (35.8%), with 61% of the users being Caucasian and 24% Black. The reported users were evenly divided between the sexes. In only 7% of cases was the barbiturate known to be procured illegally, with 44% of the cases involving prescribed medications, while 49% involved unknown sources of the drug. The motivation for drug taking was described as a suicide attempt or gesture in 54.2% of cases, as dependence in 15.4%, and as psychic effects in 10.4%. New Orleans was the only metropolitan area in which dependence was the most frequent motivation (51.9%).

In cases involving nonbarbiturate sedatives, the modal age of patients was lower than that observed for barbiturates (33.6% were 20 to 29 years old) and the population was predominantly female (61%). Suicidal intent/gesture was the motivating force in 75.5% of cases and dependence in only 5%. In addition, tranquilizer usage was reported in 13,335 cases. Approximately 36% of the subject population was in the age range of 30 to 39 years and 61% were female. Principal factors motivating usage were suicidal intent/gesture in 62.5% of cases and dependence in 15% of cases.

Data from medical examiners' offices[3] indicate that barbiturates were involved in 431 deaths with phenobarbital (173 cases), secobarbital (89 cases), and pentobarbital (22 cases) as the principal agents. The drug was used along in 10.8% of cases and was part of a combined ingestion in 89.1% of cases. The nonbarbiturate sedatives were implicated in 189 deaths, with 96.2% of the cases involving multiple drugs. Tranquilizers were associated with 825 deaths with multiple drug involvement occurring in 97.2% of cases.

The foregoing data indicate that use and abuse of barbiturates have declined substantially since the middle 1970s. Such use continues at a relatively low, but stable and significant rate, and it still contributes significantly to drug abuse and associated morbidity and mortality. The nonbarbiturate sedatives and tranquilizers have supplanted much barbiturate use in the past two to three decades and they also make an important contribution to the overall problem of substance abuse. Ethanol continues to occupy its preeminent place in the problems of drug abuse, dependence, and toxicity.

The extent and the danger of abuse of CNS depressant agents give impetus to the desire and need to gain a greater understanding of the nature of CNS depressant abuse and the elucidation of processes involved in the development of dependence on these drugs.

II. ABUSE POTENTIAL AND DEPENDENCE LIABILITY

With the foregoing background in mind, some experimental findings related to barbiturate dependence will be described. The scope of the studies discussed will be limited to the barbiturates, and will, in fact, focus on prototypical members of the three subclasses of barbiturates. Barbital will be considered as representative of the long-acting barbiturates, pentobarbital will represent the intermediate-acting barbiturates, and methohexital the ultrashort-acting barbiturates. Other members of each class will be mentioned when utilized in the same studies.

A. BARBITAL

Dependence upon barbital was first documented by Seevers and Tatum[4] using the dog as a model. The drug was administered by stomach tube in a dose of 100 mg/kg/d from 4.5 months to 3 years. Physical signs of dependence such as tremor, irritability, and convulsive seizures were observed following withdrawal of the barbital after 2 to 6 months of exposure. These observations were confirmed and extended by Fraser and Isbell,[5] who treated dogs with barbital (from 217 to 337 d) and other barbiturates for extended periods. In addition to observing the signs described earlier, they reported a type of withdrawal, "canine delirium," which was characterized by bizarre behavior as well as the signs of hyperirritability.

With the characterization of the barbital withdrawal syndrome in dogs, Deneau and Weiss[6] developed a substitution method for evaluating other drugs for liability to produce barbiturate-like physiological dependence. With barbital as the drug of primary dependence, given in doses of 100 mg/kg/d orally, substitution and single-dose suppression studies were performed to assess the ability of other drugs to prevent or alleviate signs of barbital abstinence. A number of other barbiturates (pentobarbital, phenobarbital), nonbarbiturate sedatives (e.g., ethyl alcohol, chloral, methaqualone), and antianxiety agents (meprobamate, chlordiazepoxide) were found to be effective in suppressing barbital abstinence signs. In addition, Deneau[7] reported that signs of barbital withdrawal increased both in incidence and severity with increasing duration of treatment and that maximum dependence, characterized by seizures and behavioral delirium in all animals, occurred only after more than 12 months of treatment.

Withdrawal seizures following long-term treatment with barbital have also been described in the cat.[8] The incidence of seizures correlated well with the final dose administered. The same investigators described the occurrence of paroxysmal EEG abnormalities during withdrawal from barbital following 100 to 260 days of treatment.

Okamoto and her colleagues[9] produced physical dependence more rapidly in cats using a "maximally tolerable" intragastric dosing regimen, which allowed for increasing dosage in a stepwise fashion over a period of 35 d. Withdrawal symptoms observed included motor, autonomic, and behavioral signs of moderate severity (as reflected by abstinence scores of about 8 to 11 out of 21 possible) and around 10% incidence of seizures. Characteristics of dependence upon barbital have been compared with dependence upon pentobarbital using a "chronically equivalent" dosing method in the cat.[10-13] Signs of barbital abstinence were noted to appear later and with less severity than were signs of pentobarbital abstinence. The observed differences in the abstinence syndromes of the two drugs appeared to be due to pharmacokinetic factors, i.e., the difference in rates of elimination of the drugs from tissues accounted for the difference in appearance of signs. This view was substantiated by the fact that withdrawal from barbital could be made to mimic that from pentobarbital by hastening the rate of elimination of barbital to a degree comparable to elimination of pentobarbital.

Barbital dependence has also been reported in the rhesus monkey[14] following oral administration of barbital in doses of 75 mg/kg, twice daily, for 3 months. Signs of withdrawal were similar to those in cats and dogs, and included: hyperirritability, tremor, weight loss, convulsions, and delirium and hyperthermia, in severe cases. The monkey has been used by the same investigators as a model for evaluating the barbiturate like dependence liability of other sedative-hypnotic agents in a drug-substitution study similar to that described earlier for the dog.

For several years it was generally believed that the rat would not become dependent upon the barbiturates. However, Crossland and Leonard[15] demonstrated that the rat could be made addicted to barbital by the forced consumption of barbital in the drinking fluid. Evidence of physical dependence on barbital was provided by the occurrence of spontaneous seizures and "convulsive activity in response to auditory stimulation" after 4 to 5 weeks of exposure. Essig[16] extended this study by administering barbital to rats in their drinking water for 111 to 159 d. He observed withdrawal seizures and significant loss of body weight (12.5%), both reaching maxima on the third day of withdrawal. Another sign of abstinence observed in the rat following lesser exposure included an increased sensitivity to the cortical spiking effect of pentylenetetrazol.[17] More recent studies of barbital dependence have suggested that antimuscarinic (atropine) treatment may diminish the withdrawal seizures[18] and that withdrawal from barbital results in increased utilization of catecholamines in the brain.[19]

Another method of promoting barbital dependence in the rat is by chronic dosing using a drug-admixed food (DAF) technique.[20] Exposure to high doses of drug mixed into the food for 35 to 40 d led to physical dependence as manifested by such withdrawal signs as twitching, irritability, hyperthermia, convulsions, and a weight loss which occurred, most prominently, between 24 to 40 h of the abstinence period. The DAF method has also been used as a model in conducting substitution studies for assessing cross-dependence of other sedative agents with barbital.[21]

Only recently has the mouse been utilized as a model of barbital dependence. To this point, the periods of treatment have been quite short and the degree of dependence rather mild in comparison with the studies described in other species. In general, mice have been exposed to barbital using either a subcutaneously implanted osmotic minipump for 1 week[22] or from subcutaneously implanted drug-laden pellets for 3 to 6 d.[23,24] In neither case has seizure activity occurred spontaneously nor has there been any consistent loss of body weight. However, the barbital-treated mice have been shown to be more susceptible to seizures induced by handling, pentylenetetrazol, or electroshock.[22,23,25]

B. PENTOBARBITAL

1. Studies in Animals

In an extension of earlier studies in the dog, Fraser and Isbell[5] made a number of dogs dependent upon pentobarbital by oral administration of 5 doses/d, eventually reaching a maintenance dose of approximately 100 mg/kg/d. After about 6 months of exposure, withdrawal signs were noted to be relatively mild with only some tremor and weight loss being evident. The investigators concluded that this dosage regimen was inadequate to produce severe dependence.

Some years later, Jones and colleagues[26,27] successfully produced a quantifiable pentobarbital withdrawal syndrome following cessation of chronic pentobarbital treatment in dogs. In this method, pentobarbital was automatically infused, periodically, to provide a total daily dose of 200 mg/kg. During withdrawal, signs of exaggerated startle response, twitches, tremor, and weight loss were prevalent and occasional seizure activity was observed. This model has been used for assessing dependence liability of other depressant drugs (e.g., thiopental, secobarbital, and methaqualone) in substitution studies. The substituted drugs did suppress the signs of abstinence in a dose-related manner. Interestingly, if the automatic infusions of pentobarbital were discontinued for 6 h during the day and the dogs were allowed to self-inject the drugs during that period, the animals self-administered each of the drugs in a dose approximately equivalent to the deleted dosage. In addition, it was noted that the incidence of withdrawal convulsion declined as the period of drug treatment at the maintenance dosage (200 mg/kg/d) increased, probably reflecting tolerance to the chronic depressant effect of pentobarbital.

Deneau[7] did not employ pentobarbital as a primary dependence-producing agent, but did attempt substitution in barbital-dependent dogs. Pentobarbital, in a dose of 15 mg/kg every 6 h, effectively suppressed signs of abstinence in dogs that had been maintained on barbital at doses of 100 mg/kg/d. Dependence on pentobarbital in the cat was initially described by Jaffe and Sharpless,[28] who treated cats by iv injection 3 or 4 times daily with sufficient quantities to produce a state of surgical anesthesia. The period of treatment lasted from 26 h to 21 d. Abstinence (or degree of dependence) was measured by determining the lowering of the threshold for pentylenetetrazole-induced seizures following cessation of barbiturate administration. After as little as 26 h of exposure to pentobarbital, the seizure threshold was found to be significantly reduced at 20 h of abstinence. The maximum reduction in seizure threshold (about 30% of control value) was found to occur after 2 weeks of chronic pentobarbital treatment. In addition, some animals displayed spontaneous seizure activity after 2 to 3 weeks exposure.

Okamoto and co-workers[9] described a more severe, spontaneously occurring abstinence syndrome in the cat following 5 weeks of "maximally tolerable" dosing with pentobarbital. Numerous signs of hyperirritability and disturbances of autonomic and motor function were observed. In fact, frank convulsions occurred in all cats so treated and death following seizures occurred in 41% of the animals. In a related study, phenobarbital was shown to be quite effective in reducing withdrawal seizures, while phenytoin was found to be ineffective.[29]

Further studies by Boisse and Okamoto,[10-13] compared the physical dependence consequent to pentobarbital to that resulting from barbital using a "chronically equivalent dosing" method in which doses of the two drugs were adjusted to provide equal maxima and minima of central depression for each drug. Some important differences between the two dependencies did emerge. For example, the tolerance to pentobarbital appeared to be both dispositional and functional while barbital seemed to elicit only functional tolerance. The withdrawal syndrome for pentobarbital appeared more quickly and was shorter in duration and it was also much more severe in terms of incidence of seizures and death. However, as noted above, pharmacokinetic factors seemed quite important in contributing to the differences in the withdrawal syndromes, because the two syndromes became quite similar when the elimination rate of each drug was artificially altered to mimic the rate of the other.

In rhesus monkeys, pentobarbital was found to be readily self-administered by iv injection.[30] In fact, at a dose of 5 mg/kg/injection, monkeys would infuse themselves up to 85 times per day.[31] This high rate of self-administration led to the development of marked physical dependence after 2 weeks of exposure. Upon cessation of pentobarbital administration, the abstinence signs of convulsion, delirium, and hyperthermia were considered "severe." Winger and colleagues[32] also observed self-administration of pentobarbital, amobarbital, methohexital, thiopental, and barbital in rhesus monkeys. An inverse relationship between dosage per injection and response rate was observed, such that total drug dosage per session was maintained. Yanagita and Takahashi[14] also found that pentobarbital was self-administered by rhesus monkeys via the intragastric route, albeit the rate of self-administration was more modest than by iv administration. In the same series of experiments, pentobarbital, given in an iv dose of 25 mg/kg, suppressed signs of withdrawal in monkeys made dependent upon barbital by twice daily oral administration in doses of 75 mg/kg.

Pentobarbital self-administration has been demonstrated in the baboon, with a response rate of administration comparable to that maintained by cocaine.[33] Interestingly, self-administration of pentobarbital in this study did not affect the subsequent administration rate of benzodiazepines. In addition, baboons can be trained to discriminate pentobarbital from vehicle and will generalize to lorazepam. However, not all lorazepam-trained baboons will generalize to pentobarbital.

Although methods for assessing abuse and dependence liability of depressant drugs in mammalian species seem reliable and well established, as described above, the expense and difficulty of obtaining large animals has prompted some laboratories to seek methods utilizing rodents. It was reported some years ago that the rat did not become dependent upon barbiturates,[34] but that suggestion was refuted by the demonstration of withdrawal from barbital.[16] Recently, Martin and colleagues[35] have reported the development of physical dependence on pentobarbital in rats following chronic oral administration of the drug in food. Dosage of pentobarbital was increased to a maintenance level of almost 700 mg/kg/d in female rats. When the pentobarbital was withdrawn after 2 to 3 months of chronic treatment, a rapidly progressing (10 to 16 h) abstinence syndrome including grand mal, clonic, and atypical convulsive movements was observed. Yutrzenka and co-workers[36] utilized pentobarbital as a primary dependence-producing agent in the rat using a chronic intraperitoneal infusion technique. Following 12 d of infusion in escalating dosages up to 900 to 1000 mg/kg/d, a withdrawal syndrome including hyperactivity, hyperirritability, weight loss (10 to 12%), and occasional convulsions (15 to 20% incidence) was observed. Substitution studies have also been performed with this technique as a means of assessing dependence liability of such agents as diazepam, bromazepam, temazepam, and methaqualone.[37,38]

2. Studies in Humans

In early clinical reports, Isbell and co-workers[39] and Fraser and colleagues[40] noted that cessation of barbiturate ingestion after prolonged oral administration at a high dose lead to signs of anxiety, nervousness, tremors, weakness, anorexia, nausea and vomiting, and insomnia. They observed that these signs could intensify to the point of seizures, delirium, hyperpyrexia, and death. These observations were followed by studies in which human subjects were given large doses of either secobarbital or pentobarbital chronically, and were then withdrawn from drug.[41,42] Evidence of paroxysmal changes in the EEG of some patients occurred following administration of a dosage of 0.4 g/d for up to 3 months. After doses of 0.6 g/d for 1 to 2 months, 50% of patients exhibited some EEG abnormality and minor physical signs of abstinence (tremor, hyperactivity). Withdrawal from high doses (0.9 to 2.2 g/d) given for several months resulted in seizures and delirium, as well as all of the minor physical signs.

Recent studies have been more concerned with the subjective effects of the barbiturates that might contribute to their abuse potential. In this regard, pentobarbital given in a single im injection of 150 to 250 mg was observed to produce pleasurable subjective effects, comparable to im morphine in reinforcing properties.[43,44] In addition, pentobarbital and secobarbital, given either im or orally, produced dose-related increases in the "liking" and euphoria scales, as well as increases in the PCAG (Pentobarbital, Chlorpromazine, Alcohol Group) Scale, in human subjects being studied in the Addiction Research Center.[45]

In drug self-administration studies in humans, pentobarbital is found to be self-administered. In fact, in a double-blind choice procedure, pentobarbital is self-administered in preference to either placebo or to diazepam over a wide range of doses.[46,47] Pentobarbital also produces dose-related increases in responding on a progressive-ratio schedule, indicating reinforcing power in that the subject is required to perform more work in order to obtain reinforcement.[48]

As another indicator of abuse potential in humans, data from DAWN indicate a higher incidence of abuse of pentobarbital as compared to either the licit opiates (except for propoxyphene) or other sedative-hypnotic agents (except diazepam and secobarbital). The "relative abuse index" (incidence of abuse relative to incidence of licit use) was higher for pentobarbital than for diazepam but was lower than for secobarbital.[27]

C. METHOHEXITAL

A few recent studies have addressed the abuse potential of methohexital through self-administration procedures.[32,49] Winger and co-workers[32] described iv self-administration of methohexital and thiopental in rhesus monkeys. Ator and Griffiths[50] reported that methohexital could be administered to baboons in drinking water, after drinking of the solution had been instigated by a schedule-induced polydipsia procedure. Once the drug-taking had been established, it continued at a high level even after the food inducement had been removed.

It seems that little or nothing has been done with regard to attempting induction of physical dependence on methohexital. Its very potent depressant effect and very short duration would seem to make it difficult to maintain a prolonged tissue level that would be high enough to ensure dependence without depressing function to the point of death. Whether the paucity of data relating to this matter reflects a lack of attempts at establishing dependence or a lack of success in those attempts is unclear.

The World Health Organization Expert Committee on Drug Dependence[51] cited no experimental or clinical evidence of physical dependence upon methohexital or thiopental. That committee recommended against scheduling of those drugs for that reason.

III. NEUROCHEMICAL MECHANISMS INVOLVED IN PHYSICAL DEPENDENCE ON BARBITURATES

A. INTRODUCTION

The various neurochemical changes which occur as the result of chronic administration of barbiturates are of importance in understanding the mechanisms that may underlie the establishment of both tolerance and physical dependence. It is often held that the processes which are involved in the establishment of functional (pharmacodynamic) tolerance may be similar to those processes which account for the establishment of physical dependence.[52,53]

It has been postulated that the chronic administration of a drug, resulting in the sustained influence of the drug at the site of action, may alter the normal functions and processes at the site of drug action and, eventually, may induce the targeted system to give rise to adaptive changes in the function of the various processes subserved at this site of drug action. Such adaptation may be thought of as a response to the drug-induced alteration of the homeostasis of the targeted system. The adaptive changes, in turn, serve to reestablish the "normal" level of functioning of the drug-targeted system even in the continued presence of the drug. These drug-induced adaptations in the function of the system are thought to lead to the development of functional tolerance.

Various theories have been proposed in attempts to explain how the development of physical dependence on a drug may be an extension of the various adaptive processes that underlie the development of functional tolerance to that drug.[53] The withdrawal syndrome, the occurrence of which signifies that physical dependence has indeed been established, is postulated to be a consequence of those adaptive changes which have occurred in the targeted system. The withdrawal syndrome can be manifested in the physically dependent individual upon abrupt removal of the drug either by discontinuation of drug administration or by inhibition of the action of the drug with the use of drug-specific antagonists. The adaptive changes that have occurred in order to allow the targeted system to function at a "normal" level in the presence of the drug are now being expressed in a situation in which the drug no longer has maximal influence over the targeted system. Thus, the homeostasis of the adapted system has again been altered. Over time the adapted system will again revert back to the state prior to the drug exposure and the withdrawal syndrome will wane and, eventually, end. The time course for the establishment of tolerance to the drug's action on the system as well as the time course of the withdrawal syndrome, itself, suggests that various alterations in the functioning of the drug-targeted system are occurring as the affected system attempts to adapt to either the sustained presence of the drug or the subsequent removal of the drug from the site of action.

General theories have been advanced to explain the process which may underlie the development of both functional tolerance and physical dependence. One theory suggests that the alterations occurring as a consequence of sustained exposure to the drug are developed at some point beyond the initial site of drug-receptor interaction and may include alteration in a variety of biochemical or neurochemical processes.[54-56] Alternatively, it has been proposed that the sustained presence of the drug leads to alteration in the response of the specific receptor which is involved in the mediation of the action of the drug.[53] The changes induced by the chronic presence of the drug may take the form of an alteration in the number of drug-specific receptors and/or an alteration in the affinity of the receptor for this specific drug.

The drug-induced alteration in receptor function has been extensively investigated for opiates[57-64] as well as for CNS depressants such as barbiturates and benzodiazepines.[52,65-73] The processes by which the alteration of receptor function subserves the development of physical dependence is not entirely clear but, with respect to barbiturates, this adaptive process appears to involve the $GABA_A$ receptor, which is composed of at least three binding sites which show specificity for GABA, benzodiazepines, and barbiturates as well as a transmembrane chloride ion channel. The specific binding sites on the $GABA_A$ receptor may allosterically influence each other and have as a final common pathway a modulatory role in the regulation of chloride ion flux through the chloride ion channels in the membrane.[59,65,69,74-78]

B. INVOLVEMENT OF GABA IN BARBITURATE PHYSICAL DEPENDENCE

The involvement of the GABAergic neurotransmitter system in the mediation of the action of the barbiturates has been extensively investigated. Essig,[79,80] first reported that the GABA transaminase (GABA-T) inhibitor, aminooxyacetic acid (AOAA), was effective in preventing the occurrence of spontaneous convulsive activity in barbital-dependent dogs and rats following abrupt discontinuation of barbital. It was suggested that the administration of AOAA may have reversed a deficiency in GABA activity which may have occurred as a result of either the chronic administration of barbital or as a result of the withdrawal process itself.

Subsequent investigations yielded somewhat inconsistent results with regard to the effect of barbiturate administration on the GABAergic system. Crossland and Turnbull[81] found that there was no significant change in rat whole brain GABA levels during either 5 weeks of barbital treatment or at 48 h into the barbital abstinence period. Audiogenic seizures were found to peak at 48 h of abstinence and AOAA administration, leading to more than a doubling of brain GABA levels, was unable to provide significant protection from these seizures. Sutton and Simonds[82] also reported that neither acute nor chronic administration of pentobarbital sodium had a significant effect on whole brain GABA content or on the activity of either glutamic acid decarboxylase (GAD) or GABA-T. This was in contrast to the results of previous experiments in which it was demonstrated that chronic ethanol administration led to a significant increase in whole brain GABA content as well as an increase in GABA-T activity.[83]

In further studies, Blagoeva and colleagues[84] reported that chronic barbital administration to rats lead to EEG synchronization during the initial 2 weeks of the administration period. There then followed a second stage during which EEG records showed increased signs of excitation which progressed, during the barbital withdrawal period, to EEG manifestations consistent with convulsive activity. Interestingly, during withdrawal from barbital, brain GABA levels were found to be increased. In subsequent investigations, Lennox and co-workers[85] noted that in a number of brain regions there was a significant tendency for GABA levels to be decreased during chronic barbital administration. However, by 36 h into the withdrawal period, brain GABA levels tended to be similar to that seen in control rats.

Tzeng and Ho[86] and Ho[24] reported that chronic pentobarbital administration to mice led to a decrease in levels of both GABA and glutamate as well as to a decrease in brain GAD activity. The observed decreases were noted to extend to at least 6 h into the abstinence period. Dependent mice that convulsed following pentylentetrazole administration were also found to have both lower GABA levels and lower GAD activity as compared to dependent mice which did not convulse.[87] In addition, it was noted that chronic pentobarbital administration led to a decrease in the spontaneous release of [^{3}H]-GABA from the striatum and the cerebellum.[88] The spontaneous release of [^{3}H]-GABA was also found to be significantly lower during the subsequent pentobarbital abstinence period. In support of these findings, other investigators have reported that at 72 h of abstinence there was observed both a significant decrease in GABA levels and a decrease in the rate of GABA turnover in the striatum of barbital-dependent rats.[89] In addition, it was demonstrated that these rats were more susceptible to convulsions induced either by sound or by chemical convulsants and that the incidence of audiogenic seizures was reduced following administration of either baclofen or 4,5,6,7-tetrahydroisoxazolo[5,4-c]pyridin-3-ol (THIP).

Using a continuous infusion method to produce pentobarbital-dependent rats, Yutrzenka and colleagues[90] reported that there was a 30 to 50% increase in GABA levels in cortex, cerebellum, striatum, hippocampus, medulla/pons and midbrain regions by the sixth day of continuous administration. By the tenth day of drug administration, the GABA levels in all brain regions were noted to be declining toward levels exhibited by control rats. By 8 h of the pentobarbital abstinence period, at a time when the rats were exhibiting maximal behavioral signs of withdrawal, the GABA levels in all brain regions were noted to be similar to levels observed in concurrently assayed control rats.

Thus, the investigations noted above have provided somewhat contradictory results as to the role of GABA in barbiturate dependence. There is a suggestion that the GABAergic system is involved in the mediation of the actions of barbiturates. However, these attempts to establish the involvement of the GABAergic system by monitoring changes in either regional or whole brain GABA levels or in the activity of enzymes responsible for synthesis or turnover of GABA have yielded inconsistent results and have failed to clearly establish the role of GABA in the action of the barbiturates.

C. ROLE OF THE $GABA_A$ RECEPTOR

More recent investigations of the role of GABA in the mediation of the action of CNS depressants has led to the characterization of a GABA/benzodiazepine/barbiturate-chloride ion channel macromolecular binding site complex which is located on the postsynaptic neuronal membrane.[65,69,74,75,91-93] This binding site complex, also termed the $GABA_A$ receptor, functions to mediate the activity of the chloride ion channel. Binding of GABA to this site results in opening of the chloride ion channel with a subsequent increase in inward-directed chloride ion flux leading to a hyperpolarization of the neuronal membrane[69,91,94] and the generation of inhibitory postsynaptic potentials.[94-97]

Besides the $GABA_A$ receptor there has also been described a $GABA_B$ receptor subtype.[96,98-100] The $GABA_B$ receptor appears to mediate both an increase in K^+ conductance[101,102] as well as an inhibition of Ca^{2+} conductance.[103,104] The $GABA_B$ receptor binds GABA and baclofen but is insensitive to the action of the GABA antagonist, bicuculline. $GABA_B$ receptors play a role in the function of neurons in the CNS[105,106] as well as in the modulation of norepinephrine release from sympathetic neurons in the autonomic nervous system.[99,100] In addition, it has recently been demonstrated that GABA-induced generation of inhibitory postsynaptic potentials (IPSP) can be resolved into an early component which is mediated by the $GABA_A$ receptor, and a late component mediated by the $GABA_B$ receptor.[97] The $GABA_B$ receptor does not appear to contain binding sites for either the barbiturates or benzodiazepines and is not thought to be involved in the mediation of the action of these drugs.[101,107]

The $GABA_A$ receptor from rat, bovine, and human brain has recently been purified.[108-111] Complimentary DNA has been obtained and the DNA sequence and predicted primary structure of the receptor has been described.[112-115] The $GABA_A$ receptor complex is characterized as being a heterologous tetramer composed of 2α- and 2β-subunits. The α-subunit has an apparent molecular mass (M_r) of 53,000 while the β-subunit has a M_r of 57,000. Through the use of photoaffinity labeling, it has been determined that the α-subunit contains the benzodiazepine binding site.[109,116-118] The β-subunit appears to contain the site for the binding of GABA ligands[113] as well as a site which binds the convulsant, picrotoxinin.[66,69,91,111] While it is generally held that barbiturates bind to the same site as picrotoxinin, some investigators have also proposed that the barbiturates may, in fact, bind to other distinct areas on the $GABA_A$ receptor complex.[119-122]

In the proposed structure of the $GABA_A$ receptor it appears that each subunit has four regions of hydrophobic sequences which function to span the width of the membrane.[115] It is suggested that these membrane-spanning regions of the individual subunits are intimately involved in the formation of the chloride ion channel. Both the C-terminal and N-terminal ends of the subunits are located extracellularly and there are numerous points which allow for glycosylation to occur. The GABA and benzodiazepine binding sites are located extracellularly. Studies have demonstrated that the deglycosylation of the $GABA_A$ receptor complex alters the ability of benzodiazepine ligands to bind to the specific benzodiazepine binding site. Intracellularly, there is a site on the β-subunit for cAMP-dependent serine phosphorylation.

The membrane-spanning hydrophobic regions of each subunit are believed to form the chloride ion selective channel.[115] The chloride ion channel has a 5.6 Å diameter lumen and there is a clustering of positively charged side chains at both ends of the channel which effectively render the channel selective for anions. It is speculated that the binding of GABA to the β-subunit results in a conformational change somewhere in the membrane-spanning domain which, in turn, "opens" the channel and allows the chloride ions to flow through the channel. Several investigations have provided evidence that the benzodiazepines and barbiturates allosterically modulate the conformational change induced by GABA.[67,68,75,123,124]

The investigations to date provide evidence which suggests that there is a good deal of homology in the DNA sequence of GABA receptors, nicotinic acetylcholine receptors,[125,126] and glycine receptors[127] and leads to the suggestion that there may be a common, superfamily, of chemically gated ion channel receptors.[115,128] Recent investigations have also made it apparent that the molecular arrangement of the $GABA_A$ receptor is more heterogeneous than originally determined. For example, there has been reported the isolation of γ-, δ-, and ϵ-subunits of the $GABA_A$ receptor[129-134] as well as subtypes of the individual subunits.[131-133,135,136]

The importance of these additional subunits and subtypes of individual subunits has been noted in several experiments. Ymer and colleagues[132,133] report that either β_1-, β_2-, or β_3-subunits, when coexpressed with an α subunit, may form functional $GABA_A$ receptors. Coexpression of an α_4- with a β_1-subunit in *Xenopus* oocytes yields functional GABA-gated chloride channels. The activity of this channel, however, was not found to be potentiated by the addition of benzodiazepines.[132,133] Pritchett and colleagues[129] have reported a γ_2-subunit, which can be coexpressed with both α_1- and β_1-subunits producing functional $GABA_A$ receptors which display high-affinity binding for benzodiazepines. This activity was found to be lacking when only α_1- and β_1-subunits were coexpressed.

Verdoorn and colleagues[134] demonstrated that $GABA_A$ receptors formed from α_1- and β_2-subunits formed chloride channels with a shorter mean open time than did $GABA_A$ receptors made up of α_1- and γ_2-subunits. Addition of the γ_2-subunit yielded $GABA_A$ receptors which had a greater chloride conductance. It has also been demonstrated that the expression of β_1 RNA, alone, in *Xenopus* oocytes may result in the formation of a homomeric anion selective channel which has affinity for picrotoxinin but which is not gated by GABA.[111] These studies point out the major role that these subunits play in providing for the full expression of the $GABA_A$ receptor and serve to underscore the complexity of the interaction of these subunits.

The localization of the $GABA_A$ receptor complex within the CNS has been determined using both autoradiography and monoclonal antibodies directed at the specific subunits of this binding site complex. It was reported that [^{3}H]-muscimol bound to receptor sites located on the plasma membranes of both presynaptic and postsynaptic neurons in the cerebellum.[137] Additional investigations brought forth indications of a high density of $GABA_A$ receptors in cerebral cortex, cerebellum, hippocampus, medial-basal hypothalamus, anterior hypothalamus, substantia nigra, and superior colliculus with lower densities of GABA-chloride ion channel complexes being found in regions such as striatum, medulla, and pons.[138-141] Monoclonal antibodies directed against each of the specific components of the $GABA_A$ receptor have provided evidence for the co-localization of GABA, benzodiazepine, and barbiturate binding sites within the same brain regions and has shown that the distribution of the $GABA_A$ receptor within the CNS is similar to that observed using autoradiographic techniques.[112,142-144]

The functional activity of the GABA/benzodiazepine/barbiturate-chloride ion channel complex has been extensively studied using radioligands specific for each of the binding sites. The GABA binding site has been characterized using [^{3}H]-labeled agonists such as GABA or muscimol[69,145-148] while diazepam or flunitrazepam have been used to define the characteristics of the benzodiazepine binding site.[149-153] Finally, the barbiturate binding site has been extensively investigated using a variety of ligands such as [^{3}H]-dihydropicrotoxinin ([^{3}H]-DHP) or [^{35}S]-*t*-butylbicyclophosphorothionate ([^{35}S]-TBPS).[66,76,77,91,154-158] More recently, $^{36}Cl^-$ flux studies have been used to more fully characterize the functional interactions of the individual components of the binding site complex as it relates to the regulation of chloride ion flux across the membrane.[78,159-162]

Various investigators have attempted to characterize the functional interactions between the individual binding sites of the $GABA_A$ receptor complex. It has been demonstrated that benzodiazepines, barbiturates, and a variety of convulsant compounds may modulate the binding of agonists and antagonists to the GABA binding site and thus alter the ability of GABA to regulate Cl^- conductance.[67,68,75,123,124,163] It has been demonstrated using *in vitro* preparations that a variety of benzodiazepine agonists have the ability to enhance GABA agonist binding to the GABA binding site.[123,163-165] In addition, there is also evidence that GABA agonists enhance the binding of benzodiazepines to the benzodiazepine site and that bicuculline blocks this GABA enhancement of benzodiazepine binding in *in vitro* membrane preparations,[166-179] in highly washed and detergent-treated membrane preparations,[152,179,181] and in *in vivo* preparations.[173] The ability of the GABA agonists to enhance the binding of benzodiazepines correlates well with the potency of the GABA agonists to bind to the GABA receptor.[69,91,182,183]

There is a good deal of evidence to suggest that the barbiturates may also influence the ability of both GABA and benzodiazepine agonists to bind to the respective binding sites on the $GABA_A$ receptor. The barbiturates appear to promote an increase in the number of high-affinity binding sites and there may also be an increase in the affinity of the low-affinity sites for GABA and benzodiazepines.[75,92,184,185] Barbiturates appear to bind to a site on the receptor which also binds a variety of convulsant compounds including picrotoxin and TBPS.[69,76,154,155,186,187] A number of depressant barbiturates can inhibit the binding of either [^{3}H]-dihydropicrotoxinin or [^{35}S]-TBPS at these sites and, in addition, have been demonstrated to enhance the binding of GABA agonist to the GABA receptor.[67,69,76,91,120,121,124,184,187-193] It has also been demonstrated that the barbiturate-induced enhancement of GABA-binding can be inhibited by picrotoxin but not by the benzodiazepine antagonist, RO 15-1788. This lends support to the view that the picrotoxin binding site is the one at which the barbiturates bind and is involved in the mediation of GABA activity by barbiturates.

A number of investigators have noted that barbiturates can effectively enhance the binding of benzodiazepine agonists to the benzodiazepine binding site on the $GABA_A$ receptor.[69,91,158,194-196] This enhancement of benzodiazepine binding was shown to be inhibited by picrotoxin.[67,68,77,179,196-198] Interestingly, the barbiturates also appear to allosterically inhibit the binding of benzodiazepine inverse agonists to the benzodiazepine binding site.[190]

Importantly, the allosteric action of barbiturates on the GABA and benzodiazepine binding sites show both chemical specificity and stereospecificity which correlates well with the sedative-hypnotic activity of the barbiturates as well as with their ability to enhance GABA agonist activity *in vivo* and *in vitro*.[184,185,189-191,193,199-205] In this regard it has been demonstrated that the ability of barbiturates to enhance the binding of GABA and benzodiazepine agonists to the specific GABA and benzodiazepine binding sites as well as the ability of the barbiturates to inhibit the effects of GABA and benzodiazepine antagonists, and benzodiazepines inverse agonists, generally follows in the order of: dimethyl barbituric acid (DMBB) $\rangle$ secobarbital $\rangle$ (−) pentobarbital $\rangle$ (+) pentobarbital $\rangle$ cyclohexylidene-ethyl barbituric acid (CHEB) = (−) hexobarbital $\rangle$ phenobarbital $\rangle$ metharbital.[67,68,75,77,186,194]

Finally, it has been noted that GABA and pentobarbital were both able to enhance the uptake of chloride ion into synaptic membrane preparations while bicuculline, picrotoxin, and convulsant barbiturates inhibited this chloride ion uptake.[78,160,206] Furthermore, it was demonstrated that pentobarbital enhanced the efflux of $^{36}Cl^-$ from preloaded synaptosomal preparations and that this efflux was inhibited by pretreatment with either picrotoxin or bicuculline.[161,162] The ability of pentobarbital to enhance chloride efflux was well correlated with the relative density of TBPS binding sites in various brain regions.[161,162] Thus, it appears that pentobarbital, acting through a site which binds picrotoxin or TBPS may modulate the movement of chloride ion across synaptosomal membrane preparations.

The *in vitro* studies provide a good deal of support for the ability of benzodiazepine, barbiturates, and convulsant compounds to allosterically modulate the binding of ligands to the specific binding sites on the $GABA_A$ receptor. This allosteric alteration of binding, in turn, alters the activity of the chloride ion channel. *In vivo* investigations have also been conducted to characterize the alteration of the $GABA_A$ receptor as a function of the establishment of tolerance and physical dependence on barbiturates. These *in vivo* studies have yielded a variety of results regarding the role of the $GABA_A$ receptor in tolerance and dependence on barbiturates.

Mohler and colleagues[207] found that 30 d of once daily treatment with either diazepam or phenobarbital had no effect on benzodiazepine receptors in any area of the brain, but that a decrease in the number of GABA receptors in the striatum was observed. Weizman and colleagues[208] found a similar decrease in GABA binding in hippocampus as a consequence of the chronic administration of phenobarbital. Several other investigators also observed a decrease in the number of GABA binding sites in various brain areas of rats and mice chronically treated with pentobarbital.[24,209-211] On the other hand, a number of investigators failed to demonstrate any consistent alteration in the number of GABA binding sites as a result of chronic exposure to barbiturates.[73,212,213] In a similar manner, the effects of chronic barbiturate administration on the affinity of the GABA binding site has been variously reported to be an increase in the K_d of the high-affinity site,[73] an increased K_d of the low-affinity site,[211,213] or no alteration in affinity of the GABA site.[207,208,210]

Chronic barbiturate administration has also been shown to alter the binding of a variety of benzodiazepines to the benzodiazepine binding site. Mohler and colleagues[207] reported no alteration in the binding of [^{3}H]-diazepam to benzodiazepine receptors following chronic administration of either phenobarbital or diazepam. On the other hand, Miller and co-workers[214] reported that acute barbiturate administration resulted in increased binding to the benzodiazepine receptor as reflected by an increase in the affinity of the benzodiazepine receptor. Liljequist and Tabakoff[215] indicated that the chronic administration of phenobarbital to mice resulted in a decrease in the number of benzodiazepine binding sites in the cortex and the cerebellum. The decrease in binding was attributed to a decrease in the number of high-affinity benzodiazepine binding sites in the cerebellum as well as to a decrease in the number of low-affinity benzodiazepine binding sites in the cortex. There was also noted to be an inhibition of the ability of pentobarbital to enhance [^{3}H]-flunitrazepam binding to the benzodiazepine site. Likewise, the chronic administration of pentobarbital inhibited the ability of either GABA or pentobarbital to significantly enhance the binding of ^{3}H-diazepam to the benzodiazepine receptor.[216]

Finally, in pentobarbital-tolerant rats there was observed an increase in the number of binding sites for [^{35}S]-TBPS in the cerebellum but there was no observable change in the number of binding sites in the frontal cortex, striatum, or in the substantia nigra.[217] In addition, there was no change in the affinity of the TBPS binding site in any of the brain regions that were assayed. Twenty-four hours following the removal of the pentobarbital pellet from the tolerant animal, there was observed to be an increased density of [^{35}S]-TBPS binding sites in the frontal cortex, substantia nigra, and cerebellum, but not in the striatum. Again, there was no apparent change in the affinity of the [^{35}S]-TBPS binding sites in any of the brain regions.

As a result of the various *in vitro* and *in vivo* studies, it is evident that there is a good deal of support for the hypothesis that the effects of barbiturates are mediated through an interaction with the $GABA_A$ receptor complex. The exact mechanisms by which this interaction comes about and the role of the $GABA_A$ receptor in mediation of the establishment of tolerance and physical dependence is still somewhat unclear and contradictory. Thus there is a need to continue to investigate and explore more fully the dynamic interactions that occur between barbiturates and the specific sites which make up the $GABA_A$ receptor complex.

D. ROLE OF ACETYLCHOLINE, INDOLEAMINES, AND MONOAMINES IN BARBITURATE DEPENDENCE

While the bulk of the evidence to date points to the $GABA_A$ receptor as being the primary site for the action of barbiturates, and thus a primary site for processes underlying the development of functional tolerance and physical dependence, there is also evidence that points to the possible involvement of other neurotransmitter systems. Cholinergic and monoaminergic neurotransmitter systems have been investigated relative to their role in mediating the establishment of physical dependence on barbiturates.

A number of studies have indicated that barbiturates may influence the activity of cholinergic systems in the CNS.[18,218-220] Early studies indicated that following the acute administration of barbiturates there was noted to be increased levels of acetylcholine in several brain regions.[221,222] reduced rates of acetylcholine turnover,[222,223] a dose-dependent inhibition of K^+-stimulated acetylcholine release from midbrain,[224,225] and up to 50% inhibition of high-affinity choline uptake in synaptosomal preparations from rat hippocampus.[226] In addition, it has recently been demonstrated that, in rats undergoing dietary supplementation with up to 10 times the normal choline concentration for 35 d, there was a profound attenuation of both the sedative-hypnotic and the hypothermic effects of pentobarbital.[227] In attempting to explain these results, however, it was suggested that the choline-supplemented diet may have induced an increase in cerebral glucose metabolism which, in turn, may have produced a hyperactive state of behavior which may, ultimately, have been responsible for the attenuation of the action of pentobarbital.

The situation following the chronic administration of barbiturates was noted to be somewhat different than that observed with acute barbiturate administration. With chronic barbital administration, it was noted that acetylcholine levels were reduced in the striatum of barbital-dependent rats by the third day of the barbital abstinence period.[228-230] There were noted to be a continued reduction in acetylcholine levels through at least the twelfth day of abstinence. These investigators also noted that in rats experiencing either chronic barbital treatment or barbital abstinence, there was increased acetylcholine synthesis in all areas except striatum.[229-231]

Receptor binding studies using [^{3}H]-QNB as a ligand for muscarinic cholinergic receptors indicated that chronic barbital administration resulted in an increase in the number of muscarinic receptors with no change in the affinity of the receptors for the ligand.[229,232,233] The increase in [^{3}H]-QNB binding was found to occur from the second to the twenty-first day of abstinence. Peak increases in binding occurred at 0 to 3 d and 12 to 21 d in the striatum and from 2 to 9 d in cortex and midbrain. [^{3}H]-QNB binding was found to be lower in rats that had experienced episodes of spontaneous convulsive activity during abstinence.[233] On the other hand, Mohler and coworkers[207] had previously demonstrated that there were no observable changes in muscarinic cholinergic binding in rat brain following chronic phenobarbital administration. The exception to this was the observation of a 30% decline in muscarinic receptor binding in cerebellum. The decreased binding was found to be due to a decrease in the maximum number of binding sites with no change in affinity of the receptor for the [^{3}H]-QNB. Interestingly, it has recently been demonstrated that there is an allosteric barbiturate binding site on the nicotinic acetylcholine receptor isolated from the *Torpedo* electroplaque.[234-236] Studies have demonstrated that while the barbiturate binding sites located on both the acetylcholine receptor and on the $GABA_A$ receptor are similar, they are not identical. However, this does provide support for the idea that the $GABA_A$ receptor and the acetylcholine receptor may be derived from a common receptor superfamily.

Numerous investigators have also attempted to define the role of CNS monoamines (norepinephrine dopamine, and serotonin) in the development of tolerance and dependence on barbiturates. It has been demonstrated that pentobarbital reduces both the activity of serotonergic neurons as well as the rate of serotonin turnover in whole brain.[237,238] More recently it has been reported that acute pentobarbital administration to mice was noted to produce an increase in brain serotonin levels, a decrease in the levels of 5-hydroxyindoleacetic acid, and a decrease in serotonin turnover.[239] Interestingly, in pentobarbital-tolerant mice there was noted to be a significant decrease in serotonin synthesis.[239] However, it was also noted that the establishment of physical dependence on pentobarbital did not significantly alter the brain concentration of either serotonin or 5-hydroxyindoleacetic acid. Tagashira and colleagues[240] reported that there was an increase in serotonin synthesis in barbiturate-dependent rats which were experiencing abstinence-induced convulsions and that the administration of nitrazepam, with subsequent amelioration of convulsive activity, served to decrease the rate of serotonin synthesis.

Several investigators have noted that administration of *p*-chlorophenylalanine, with a subsequent reduction in serotonin synthesis, results in a delay in the rate of acquisition of tolerance to barbiturates.[239,241,242] More recently it was reported that electrolytic lesions of the median raphe nuclei, but not the dorsal raphe nuclei, also resulted in the delay in the acquisition of both acute and chronic tolerance to barbiturates.[243,244] Thus, although the evidence for the role of serotonin in the action of barbiturates is rather incomplete, the results of these investigations do indicate that serotonergic pathways may be involved in both the acquisition of tolerance as well as the establishment of physical dependence to barbiturates.

The results obtained from a number of investigations have suggested that the monoamines (such as norepinephrine and dopamine) may play an important role in both the development of barbiturate dependence as well as in the expression of withdrawal signs during abstinence. Tabakoff and colleagues[245] reported that functional noradrenergic systems were important in the establishment of tolerance to the hypnotic and hypothermic effects of phenobarbital. In addition, it was noted that treatment of phenobarbital-dependent mice with 6-hydroxydopamine (6-OHDA), in doses sufficient to lower norepinephrine levels but not dopamine levels, resulted in an increased incidence of convulsions during the abstinence period. This was in agreement with previously obtained results using ethanol-treated mice.[245,246] An earlier study had demonstrated that when both norepinephrine and dopamine levels were significantly reduced by administration of 6-OHDA there was both an earlier onset of withdrawal seizure activity as well as a significant increase in the number of convulsive episodes.[247]

A number of other investigators have also reported on the alteration of monoaminergic activity following the chronic administrations of barbiturates. Nabeshima and co-workers[248] noted that chronic pentobarbital administration resulted in both decreased levels of norepinephrine in medulla/pons and decreased dopamine levels in the striatum of tolerant mice. Morgan and colleagues[249] reported decreased norepinephrine levels in cerebral cortex, thalamus, and hypothalamus after 1 to 2 d of abstinence while dopamine levels in striatum were found to be increased at 2 d of abstinence. The depletion of norepinephrine by administration of either α-methyl-*p*-tyrosine (α-MPT) or FLA-63 was found to occur to a greater degree in both dependent mice maintained on barbiturates as well as in dependent rats during barbiturate abstinence.[19,248,249] Interestingly, it has also been demonstrated that α-MPT treatment of dependent rats resulted in a dose-dependent decrease in abstinence-induced convulsive activity.[250]

In order to more fully assess the influence of monoamines in the development of dependence on barbiturates. Tagashira and colleagues[251] administered either *p*-chlorophenylalanine (PCPA), tranylcypromine (TCP), α-MTP, methamphetamine, or reserpine to rats during the chronic administration of barbital. Following discontinuation of barbital administration, it was noted that rats treated with either PCPA or methamphetamine exhibited very mild signs of withdrawal while rats treated with either reserpine or α-MPT exhibited very severe abstinence signs. Other investigators also demonstrated that the abstinence signs in barbital-dependent mice were significantly decreased when the mice were treated with either timolol, propranolol, phenoxybenzamine, methyldopa, or l-dopa during the barbital administration period.[252,253] It is interesting to note that there has been reported to be an increase in number of β-adrenergic receptors as a result of the chronic administration of barbiturates.[254]

These studies indicate that during the chronic administration of barbiturates there tends to be a decreased release of norepinephrine, coupled with an increase in both the turnover of norepinephrine and adrenergic receptor density. It suggests that the monoamines may be involved in the establishment of tolerance and physical dependence as well as in the expression of the withdrawal syndrome following discontinuation of barbiturate administration in dependent subjects.

IV. SUMMARY AND CONCLUSIONS

Although abuse of the barbiturates has declined over the past two decades, particularly as the benzodiazepines have supplanted them in most therapeutic situations, they continue to be of clinical and toxicological importance. In fact, both animals and humans exposed to either barbiturates or benzodiazepines prefer the short-to-intermediate-acting barbiturates. In addition, albeit the current level of abuse of barbiturates is relatively low, it has remained constant for the better part of a decade and is not likely to disappear entirely in the foreseeable future.

The reinforcing properties and the physical dependence liability of the barbiturates are both readily demonstrable in experimental animals and in human subjects. Those properties, and the abstinence syndrome associated with discontinuation of drug, are qualitatively similar for the barbiturates and for other CNS depressants such as ethanol, methaqualone, and the benzodiazepines. There also exists a substantial degree of cross-tolerance and cross-dependence among those agents. The experimental models that have been developed in animals appear to have good predictive values for abuse potential and dependence liability in man.

The $GABA_A$ receptor complex probably represents the primary site for the action of CNS depressants such as benzodiazepines, barbiturates, and ethanol. Alterations of the neurochemical processes subserved by this receptor complex may represent a primary site for the establishment of functional tolerance and physical dependence on barbiturates. There are also indications that other neurotransmitter systems may be important for the development of physical dependence.

Future investigations should be designed to elucidate more fully the complex interactions which occur between the various neurochemical entities within the CNS. Clearly, an increased understanding of the interactions which may occur between the various entities which comprise the $GABA_A$ receptor is essential in order to more fully establish the role of this receptor complex in mediating the development of physical dependence on barbiturates and other CNS depressants.

ACKNOWLEDGMENTS

The authors wish to thank Stacey Anderson and Sussie Robinson for their help in preparation of this manuscript. This activity was supported, in part, by NIDA grants DA06476 and DA04244.

REFERENCES

1. **Johnson, L. D., O'Malley, P. M., and Bachman, J. G.,** Drug use, drinking and smoking. National Survey. Results from High School, College, and Young Adult Populations 1975-1988, DHHS Publ. No. (ADM) 89-1638. NIDA, Washington, D.C., U.S. Government Printing Office, 1989.
2. **Johnson, L. D., O'Malley, P. M., and Bachman, J. G.,** NIDA 1989 National High School Senior Drug Abuse Survey: Monitoring the Future, NIDA U.S. Government Printing Office (1990).
3. NIDA Statistical Series: Annual Data 1989, Data from the Drug Abuse Warning Network, DHHS Publ. No. AMD 90-1717, Washington, D.C., U.S. Government Printing Office, (1990).
4. **Seevers, M. H. and Tatum, A. L.,** Chronic experimental barbital poisoning, *J. Pharmacol. Exp. Ther.*, 42, 217, 1931.
5. **Fraser, H. F. and Isbell, H.,** Abstinence syndrome in dogs after chronic barbiturate medication, *J. Pharmacol. Exp. Ther.*, 112, 261, 1954.
6. **Deneau, G. A. and Weiss, S.,** A substitution technique for determining barbiturate-like physiological dependence capacity in the dog, *Pharmako psychiatr. Neuro-Psychopharmakol.*, 1, 270, 1968.
7. **Deneau, G. A.,** Preclinical assessment in the physiological dependence capacity of depressant drugs, in *Predicting Dependence Liability of Stimulant and Depressant Drugs*, Thompson, T. and Unna, K. R., Eds., University Park Press, Baltimore, MD, 1977, 29.
8. **Essig, C. F. and Flanary, H. G.,** Convulsive aspects of barbital sodium withdrawal in the cat, *Exp. Neurol.*, 3, 149, 1961.
9. **Okamoto, M., Rosenberg, H. C., and Boisse, N. R.,** Withdrawal characteristics following chronic pentobarbital dosing in the cat, *Eur. J. Pharmacol.*, 40, 107, 1976.
10. **Boisse, N. R. and Okamoto, M.,** Physical dependence to barbital compared to pentobarbital. I. "Chronically equivalent" dosing method, *J. Pharmacol. Exp. Ther.*, 204, 497, 1978.
11. **Boisse N. R. and Okamoto, M.,** Physical dependence to barbital compared to pentobarbital. II. Tolerance characteristics, *J. Pharmacol. Exp. Ther.*, 204, 507, 1978.
12. **Boisse, N. R. and Okamoto, M.,** Physical dependence to barbital compared to pentobarbital. III. Withdrawal characteristics, *J. Pharmacol. Exp. Ther.*, 204, 514, 1978.
13. **Boisse, N. R. and Okamoto, M.,** Physical dependence to barbital compared to pentobarbital. IV. Influence of elimination kinetics, *J. Pharmacol. Exp. Ther.*, 204, 526, 1978.
14. **Yanagita, T. and Takahashi, S.,** Dependence liability of several sedative-hypnotic agents evaluated in monkeys, *J. Pharmacol. Exp. Ther.*, 185, 307, 1973.

15. **Crossland, J. and Leonard, B. E.,** Barbiturate withdrawal convulsion in the rat, *Biochem. Pharmacol.,* (Suppl.) 12, 103, 1963.
16. **Essig, C. F.,** Barbiturate withdrawal in white rats, *Int. J. Neuropharmacol.,* 5, 103, 1966.
17. **Gildea, M. L. and Bourne, W. M.,** Effect of barbiturate withdrawal on pentylenetetrazol seizure threshold, *Commun. Psychopharmacol.,* 1, 123, 1977.
18. **Walstrom, G.,** The effects of atrophine on the tolerance and the convulsions seen after withdrawal from forced barbital drinking in the rat, *Psychopharmacology,* 59, 123, 1978.
19. **Morgan, W. W., Huffman, R. W., Pfeil, K. A., and Gonzales, E. G.,** Effect of synthesis inhibition on the levels of brain catecholamines in barbital dependent rats, *Psychopharmacology,* 56, 41, 1978.
20. **Tagashira, E., Izumi, T., and Yanaura, S.,** Experimental barbiturate dependence. I. Barbiturate dependence development in rats by drug-admixed food (DAF) method, *Psychopharmacology,* 57, 137, 1978.
21. **Tagashira, E., Urano, T., Hiramori, T., and Yanaura, S.,** Cross-physical dependence liability of psychotropic drugs in rats dependent on barbiturates, *Jpn. J. Pharmacol.,* 33, 659, 1983.
22. **Siew, C. and Goldstein, D. B.,** Osmotic minipumps for administration of barbital to mice: demonstration of functional tolerance and physical dependence, *J. Pharmacol. Exp. Ther.,* 204, 541, 1978.
23. **Hu, J. and Ho, I. K.,** Susceptibility to electro-shock convulsion after barbiturate pellet implantation in the mouse, *Pharmacologist,* 17, 190, 1975.
24. **Ho, I. K.,** Effects of acute and chronic administration of pentobarbital on GABA system, *Brain Res. Bull.,* 5(Suppl. 2), 913, 1980.
25. **Lin, S. C. and Sutherland, V. C.,** Regional distribution of barbital in the brain of mice during the development of tolerance and physical dependence, *Res. Commun. Chem. Pathol. Pharmacol.,* 18, 215, 1977.
26. **Jones, B. E., Prada, J. A., and Martin, W. R.,** A method for bioassay of physical dependence on sedative drugs in the dog, *Psychopharmacology,* 48, 7, 1976.
27. **Jones, B. E.,** Predicting abuse liability of depressant drugs, in *Predicting Dependence Liability of Stimulant and Depressant Drugs,* Thompson, T. and Unna, K. R., Eds., University Park Press, Baltimore, MD, 1977, 35.
28. **Jaffe, J. H. and Sharpless, S. K.,** The rapid development of physical dependence on barbiturates, *J. Pharmacol. Exp. Ther.,* 150, 140, 1965.
29. **Okamoto, M.,** Physical dependency characteristics of a long-acting barbiturate, barbital, compared to the short-acting pentobarbital. Cat as an experimental physical dependency model system, in *Predicting Dependence Liability of Stimulant and Depressant Drugs,* Thompson, T. and Unna, K. R., University Park Press, Baltimore, MD, 1977, 57.
30. **Deneau, G. A., Yanagita, T., and Seevers, M. H.,** Self-administration of psychoactive substances by the monkey: a measure of psychological dependence, *Psychopharmacologia,* 16, 30, 1969.
31. **Yanagita, T. and Takahashi, S.,** Development of tolerance to and physical dependence on barbiturates in rhesus monkeys, *J. Pharmacol. Exp. Ther.,* 172, 163, 1970.
32. **Winger, G., Stitzer, M. L., and Woods, J. H.,** Barbiturate-reinforced responding in rhesus monkeys: comparison of drugs with different duration of action, *J. Pharmacol. Exp. Ther.,* 195, 505, 1975.
33. **Griffiths, R. R., Lukas, S. E., Bradford, L. D., Brady, J. V., and Snell, J. D.,** Self-injection of barbiturates and benzodiazepines in baboons, *Psychopharmacology,* 75, 101, 1981.
34. **Stanton, E. J.,** Addiction and tolerance to barbiturates? The effects of daily administration and abrupt withdrawal of phenobarbital and pentobarbital in the albino rat, *J. Pharmacol. Exp. Ther.,* 57, 245, 1936.
35. **Martin, W. R., McNichols, L. F., and Cherian, S.,** Diazepam and pentobarbital dependence in the rat, *Life Sci.,* 31, 721, 1982.

36. **Yutrzenka, G. J., Patrick, G. A., and Rosenberger, W.,** Chronic intraperitoneal infusion of pentobarbital: a model of barbiturate dependence in the rat, *J. Pharmacol. Exp. Ther.*, 232, 111, 1985.
37. **Yutrzenka, G. J., Patrick, G. A., and Rosenberger, W.,** Substitution of temazepam and midazolam in pentobarbital-dependent rats, *Physiol. Behav.*, 46, 55, 1989.
38. **Yutrzenka, G. J., Patrick, G. A., and Rosenberger, W.,** Substitution of psychoactive drugs in pentobarbital-dependent rats, *Drug Alc. Depend.*, 26, 9, 1990.
39. **Isbell, H., Altschul, S., Kornetsky, C. H., Eisenman, A. J., Flanary, H. G., and Fraser, H. F.,** Chronic barbiturate intoxication, *Arch. Neurol. Psychiatr.*, 64, 1, 1950.
40. **Fraser, H. F., Shaver, M. R., Maxwell, E. S., and Isbell, H.,** Death due to withdrawal of barbiturates, *Ann. Intern. Med.*, 38, 1319, 1953.
41. **Fraser, H. F., Isbell, H., Eisenman, A. J., Wikler, A., and Pescors, F. T.,** Chronic barbiturate intoxication, *Arch. Intern. Med.*, 94, 34, 1954.
42. **Fraser, H. F., Wikler, A., Essig, C. F., and Isbell, H.,** Degree of physical dependence induced by secobarbital or pentobarbital, *J. Amer. Med. Assoc.*, 166, 126, 1958.
43. **Martin, W. R., Thompson, W. O., and Fraser, H. F.,** Comparison of graded single intramuscular doses of morphine and pentobarbital in man, *Clin. Pharmacol. Ther.*, 15, 623, 1974.
44. **McClane, T. K. and Martin, W. R.,** Subjective and physiologic effects of morphine, pentobarbital and meprobamate, *Clin. Pharmacol. Ther.*, 20, 192, 1976.
45. **Fraser, H. F. and Jasinski, D. R.,** The assessment of abuse potentiality of sedative-hypnotics (depressants), in *Handbook of Experimental Pharmacology*, Vol. 45, Martin, W. R., Ed., Springer-Verlag, New York, 1977, 589.
46. **Griffiths, R. R., Bigelow, G., and Liebson, I.,** Human drug self-administration: double blind comparison of pentobarbital, diazepam, chlorpromazine and placebo, *J. Pharmacol. Exp. Ther.*, 210, 301, 1979.
47. **Griffiths, R. R., Bigelow, G., Liebson, I., and Kaliszak, J. E.,** Drug preference in humans: double-blind choice comparison of pentobarbital, diazepam and placebo, *J. Pharmacol. Exp. Ther.*, 215, 649, 1980.
48. **McCleod, D. C. and Griffiths, R. R.,** Human progressive-ratio performance: maintenance by pentobarbital, *Psychopharmacology*, 79, 4, 1983.
49. **Carroll, M. E., Stotz, D. C., Kliner, D. J., and Meisch, R. A.,** Self-administration of orally-delivered methohexital in rhesus monkeys with phencyclidine or pentobarbital histories: effects of food deprevation and satiation, *Pharmacol. Biochem. Behav.*, 20, 145, 1984.
50. **Ator, N. A. and Griffiths, R. R.,** Oral self-administration of methohexital in baboon, *Psychopharmacology*, 79, 120, 1982.
51. WHO Technical Report, WHO Expert Committee on Drug Dependence, Twenty-Third Rept., p. 7, 1987.
52. **Taberner, P. V.,** The GABA system in functional tolerance and dependence following barbiturates, benzodiazepines or ethanol-correlation or causality? *Comp. Biochem. Physiol.*, 93A, 241, 1989.
53. **Cox, B. M.,** Drug tolerance and physical dependence, in *Principles of Drug Action*, 3rd ed., Pratt, W. B. and Taylor, P., Eds., Churchill Livingstone, New York, 1990, 639.
54. **Goldstein, D. B. and Goldstein, A.,** Possible role of enzyme inhibition and repression in drug tolerance and addiction, *Biochem. Pharmacol.*, 8, 48, 1961.
55. **Shuster, L.,** Repression and de-repression of enzyme synthesis as a possible explanation of some aspects of drug action, *Nature (London)*, 189, 314, 1961.
56. **Seevers, M. H. and Deneau, G. A.,** Physiological aspects of tolerance and physical dependence, in *Physiological Pharmacology: A Comprehensive Treatise*, Vol. 1, Root, W. S. and Hoffman, F. G., Eds., Academic Press, San Diego, 1963, 565.
57. **Klee, W. A. and Streaty, R. A.,** Narcotic receptor sites in morphine dependent rats, *Nature (London)*, 248, 61, 1974.

58. **Pert, C. B. and Snyder, S. H.,** Opiate receptor: demonstration in nervous tissues, *Science,* 179, 1011, 1973.
59. **Lord, J. A. H., Waterfield, A. A., Hughes, J., and Kosterlitz, H. W.,** Endogenous opioid peptides: multiple agonists and receptors, *Nature (London),* 267, 495, 1977.
60. **Dum, J., Meyer, G., Hollt, V., and Herz, A.,** *In vivo* opiate binding unchanged in tolerant/dependent mice, *Eur. J. Pharmacol.,* 58, 453, 1979.
61. **Schulz, R., Wuster, M., Krenss, H., and Herz, A.,** Selective development of tolerance without dependence in multiple opiate receptors of mouse vas deferens, *Nature (London),* 285, 242, 1980.
62. **Porreca, F. and Burks, T. F.,** Affinity of normorphine for its pharmacologic receptor in the naive and morphine-tolerant guinea pig ileum, *J. Pharmacol. Exp. Ther.,* 225, 668, 1983.
63. **Martin, W. R.,** Pharmacology of opioids, *Pharmacol. Rev.,* 35, 283, 1983.
64. **Chavkin, C. and Goldstein, A.,** Opioid receptor reserve in normal and morphine tolerant guinea pig ileum myenteric plexus, *Proc. Natl. Acad. Sci. U.S.A.,* 81, 7253, 1984.
65. **Brunn-Meyer, S. E.,** The GABA/benzodiazepine receptor-chloride ionophore complex: nature and modulation, *Prog. Neuro-Psychopharmacol. Biol. Psychiatry,* 11, 365, 1987.
66. **Ticku, M. K. and Olsen, R. W.,** Picrotoxin binding sites in brain, in *Methods in Neurobiology: Brain Neurotransmitter and Neuromodulator Receptor Methodologies,* Marangos, P. J., Campbell, I., and Cohen, R. C., Eds., Academic Press, San Diego, 1983.
67. **Olsen, R. W. and Leeb-Lundberg, F.,** Convulsant and anticonvulsant drug binding sites related to the GABA receptor ionophore system, in *Neurotransmitters, Seizures, & Epilepsy,* Morselli, P. L., Lloyd, K. G., Loscher, W., Meldrum, B. S., and Reynolds, E. H., Eds., Raven Press, New York, 1981, 151.
68. **Olsen, R. W. and Leeb-Lundberg, F.,** Convulsant and anticonvulsant drug binding sites related to GABA related chloride ion channels, in *Advances in Biochemical Psychopharmacology,* Vol. 26, GABA and Benzodiazepine Receptors, Costa, E., Chiara, D., and Gessa, G. L., Eds., Raven Press, New York, 1981, 93.
69. **Olsen, R. W.,** GABA-benzodiazepine-barbiturate receptor interactions, *J. Neurochem.,* 37, 1, 1981.
70. **Miller, L. G., Greenblatt, D. J., Barnhill, J. G., and Shader, R. I.,** Chronic benzodiazepine administration. I. Tolerance is associated with benzodiazepine receptor downregulation and decreased gamma-aminobutyric $acid_A$ receptor function, *J. Pharmacol. Exp. Ther.,* 246, 170, 1988.
71. **Miller, L. G., Greenblatt, D. J., Barnhill, J. G., and Shader, R. I.,** Differential modulation of benzodiazepine receptor binding by ethanol in LS and SS mice, *Pharmacol. Biochem. Behav.,* 29, 471, 1988.
72. **Crawley, J. N., Marangos, P. J., Stivers, J., and Goodwin, F. K.,** Chronic clonazepam administration induces benzodiazepine receptor subsensitivity, *Neuropharmacology,* 21, 85, 1982.
73. **Gray, P. L. and Taberner, P. V.,** Evidence for GABA tolerance in barbiturate-dependent and withdrawn mice, *Neuropharmacology,* 24, 437, 1985.
74. **Ticku, M. K.,** Benzodiazepine-GABA receptor-ionophore complex, *Neuropharmacology,* 22, 1459, 1980.
75. **Olsen, R. W., Stauber, G. B., King, R. G., Yang, J., and Dilber, A.,** Structure and function of the barbiturate modulated benzodiazepine/GABA receptor protein complex, in *GABA-ergic Transmission and Anxiety,* Biggio, G. and Costa, E., Eds., Raven Press, New York, 1986, 21.
76. **Ramenjaneyula, R. and Ticku, M. K.,** Binding characteristics and interactions of depressant drugs with [^{35}S]-t-butylbicyclophosphorothionate, a ligand that binds to the picrotoxinin site, *J. Neurochem.,* 42, 221, 1984.
77. **Leeb-Lundberg, F., Snowman, A., and Olsen, R. W.,** Barbiturate receptor sites are coupled to benzodiazepine receptors, *Proc. Natl. Acad. Sci. U.S.A.,* 77, 7468, 1980.

78. **Allan, A. M. and Harris, R. A.,** Anesthetic and convulsant barbiturates alter γ-aminobutyric acid-stimulated chloride flux across brain membranes, *J. Pharmacol. Exp. Ther.*, 238, 763, 1986.
79. **Essig, C. F.,** Anticonvulsant effect of amino-oxyacetic acid during barbiturate withdrawal in the dog, *Int. J. Neuropharmacol.*, 2, 199, 1963.
80. **Essig, C. F.,** Possible relation of brain gamma-aminobutyric acid (GABA) to barbiturate abstinence convulsion, *Arch. Int. Pharmacodyn. Ther.*, 176, 97, 1968.
81. **Crossland, J. and Turnbull, M. J.,** Gamma-aminobutyric acid and the barbiturate abstinence syndrome in rats, *Neuropharmacology*, 11, 733, 1972.
82. **Sutton, I. and Simmonds, M. A.,** Effects of acute and chronic pentobarbitone on the gamma-aminobutyric acid system in rat brain, *Biochem. Pharmacol.*, 23, 1801, 1974.
83. **Sutton, I. and Simmonds, M. A.,** Effects of acute and chronic ethanol on the γ-aminobutyric acid system in rat brain, *Biochem. Pharmacol.*, 22, 1685, 1973.
84. **Blagoeva, P., Longo, V. G., Masi, I., and deCarolis, A. S.,** Alterations in behavior, EEG, and brain cortical GABA content following prolonged administration and subsequent withdrawal of barbital in the rat, *Behav. Biol.*, 7, 755, 1972.
85. **Lennox, R. H., Wray, H. L., Kant, G. J., Hawkins, T. D., and Meyerhoff, J. L.,** Changes in brain levels of cyclic nucleotides and γ-aminobutyric acid in barbiturate dependence and withdrawal, *Eur. J. Pharmacol.*, 55, 150, 1979.
86. **Tzeng, S. and Ho, I. K.,** Effects of acute and continuous pentobarbital administration on the gamma-aminobutyric acid system, *Biochem. Pharmacol.*, 26, 699, 1977.
87. **Ho, I. K., Flint, B. A., Gilliland, T.,** Comparative effect of diazepam and valproate on pentylenetetrazol (PTZ) induced convulsions in pentobarbital dependent mice, *Fed. Proc. Fed. Am. Soc. Exp. Biol.*, 38, 434, 1979.
88. **Nabeshima, T. and Ho, I. K.,** GABA release from mouse brain slices following acute and chronic pentobarbital administration and during withdrawal, *Res. Commun. Subst. Abuse*, 3, 483, 1982.
89. **Sandoval, M. R. L. and Palermo-Neto, J.,** GABA-ergic influences of barbital withdrawal induced convulsions, *Gen. Pharmacol.*, 17, 431, 1986.
90. **Yutrzenka, G. J., Patrick, G. A., and Rosenberger, W.,** Chronic pentobarbital (PB) administration alters GABA levels in rat brain, *Pharmacologist*, 25, 199, 1983.
91. **Olsen, R. W.,** The GABA postsynaptic membrane receptor-ionophore complex: site of action of convulsant and anticonvulsant drugs, *Mol. Cell Biochem.*, 39, 261, 1981.
92. **Olsen, R. W., Wong, E. H. F., Stauber, G. B., and King, R. G.,** Biochemical pharmacology of the GABA/benzodiazepine receptor/ionophore protein, *Fed. Proc. Fed. Am. Soc. Exp. Biol.*, 43, 2773, 1984.
93. **Elderfrawi, A. T. and Elderfrawi, M. F.,** Receptors for γ-aminobutyric acid and voltage-dependent chloride channels as targets for drugs and toxicants, *FASEB J.*, 1, 262, 1987.
94. **Enna, S. J. and Gallagher, J. P.,** Biochemical and electrophysical characteristics of mammalian GABA receptors, *Int. Rev. Neurobiol.*, 24, 181, 1983.
95. **Roberts, E.,** γ-aminobutyric acid and nervous system function — a perspective, *Biochem. Pharmacol.*, 23, 2637, 1974.
96. **Simmonds, M. A.,** Classification of inhibitory amino acid receptors in the mammalian nervous system, *Med. Biol.*, 64, 301, 1986.
97. **McCormick, D. A.,** GABA as an inhibitory neurotransmitter in human cerebral cortex, *J. Neurophysiol.*, 62, 1018, 1989.
98. **Bowery, N. G. and Hudson, A. L.,** γ-Aminobutyric acid reduces the evoked release of [^{3}H]-noradrenaline from sympathetic nerve terminals, *Br. J. Pharmacol.*, 66, 108P, 1979.
99. **Bowery, N. G., Hill, D. R., Hudson, A. L., Doble, A., Middlemiss, D. N., Shaw, J., and Turnbull, M.,** (−)Baclofen decreases neurotransmitter release in the mammalian CNS by an action at a novel GABA receptor, *Nature (London)*, 283, 92, 1980.

100. **Bowery, N. G., Doble, A., Hill, D. R., Hudson, A. L., Shaw, J. S., Turnbull, M. J., and Warrington, R.,** Bicuculline-insensitive GABA receptors on peripheral autonomic nerve terminals, *Eur. J. Pharmacol.*, 71, 53, 1981.
101. **Newberry, N. R. and Nicoll, R. A.,** Direct hyperpolarizing action of baclofen on hippocampal pyramidal cells, *Nature (London),* 308, 450, 1984.
102. **Newberry, N. R. and Nicoll, R. A.,** Comparison of the action of baclofen with γ-aminobutyric acid on rat hippocampal pyramidal cells *in vitro, J. Physiol.*, 360, 161, 1985.
103. **Dunlap, K.,** Two types of γ-aminobutyric acid receptor on embryonic sensory neurones, *Br. J. Pharmacol.*, 74, 579, 1981.
104. **Desarmenien, M., Feltz, P., Occhipinti, G., Santangelo, F., and Schlichter, R.,** Coexistence of GABA-A and GABA-B receptors of Aδ and C primary afferents, *Br. J. Pharmacol.*, 81, 327, 1984.
105. **Curtis, D. R., Game, C. J. A., Johnston, G. A. R., and McCulloch, R. M.,** Central effects of β-(*p*-chlorophenyl)-γ-aminobutyric acid, *Brain Res.*, 70, 493, 1974.
106. **Davies, J. and Watkins, J. C.,** The action of β-phenyl-GABA derivatives on neurones of the cat cerebral cortex, *Brain Res.*, 70, 501, 1974.
107. **Enna, S. J. and Kabon, E. W.,** Receptor regulation: evidence for a relationship between phospholipid metabolism and neurotransmitter receptor mediated cAMP formation in brain, *Trends Pharmacol. Sci.*, 8, 21, 1987.
108. **Kuriyama, K. and Taguchi, J. I.,** Purification of γ-aminobutyric acid receptor, benzodiazepine receptor and Cl^- channel from bovine cerebral cortex by benzodiazepine affinity gel column chromatography, *Neurochem. Int.*, 10, 253, 1987.
109. **Sigel, E., Stephenson, F. A., Mamalaki, C., and Barnard, E. A.,** A γ-aminobutyric acid/benzodiazepine receptor complex of bovine cerebral cortex: purification and partial characterization, *J. Biol. Chem.*, 258, 6965, 1983.
110. **Sigel, E., Stephenson, F. A., Mamalaki, C., and Barnard, E. A.,** The purified GABA/benzodiazepine/barbiturate receptor complex: four types of ligand-binding sites, and the interactions between them, are preserved in a single isolated protein complex, *J. Recept. Res.*, 4, 175, 1984.
111. **Sigel, E., Baur, R., Malherke, P., and Mohler, H.,** The rat β_1-subunit of the GABA-A receptor forms a picrotoxin-sensitive anion channel open in the absence of GABA, *FEBS Lett.*, 257, 377, 1989.
112. **Schoch, P., Richards, J. G., Haring, P., Takacs, B., and Mohler, H.,** Structural analysis and localization of GABA/benzodiazepine/TBPS-receptor complex using monoclonal antibodies, in *GABA-ergic Transmission and Anxiety,* Biggio, G. and Costa, E., Eds., Raven Press, New York, 1986, 11.
113. **Casalotti, S. O., Stephenson, F. A., and Barnard, E. A.,** Separate subunits for agonist and benzodiazepine binding in the gamma-aminobutyric acid-A receptor oligomer, *J. Biol. Chem.*, 261, 15013, 1986.
114. **Mamalaki, C., Stephenson, F. A., and Barnard, E. A.,** The $GABA_A$/benzodiazepine receptor is a heterotetramer of homologous α and β subunits, *EMBO J.*, 6, 561, 1987.
115. **Schofield, P. R., Darlison, M. G., Fujita, N., Burt, D. R., Stephenson, F. A., Rodriguez, H., Rhee, L. M., Ramachandran, J., Reale, V., Glencorse, T. A., Seeburg, P. H., and Barnard, E. A.,** Sequence and functional expression of the $GABA_A$ receptor shows a ligand-gated super-family, *Nature (London),* 328, 221, 1987.
116. **Kirkness, E. F. and Turner, A. J.,** The gamma-aminobutyrate/benzodiazepine receptor from pig brain. Enhancement of gamma-aminobutyrate receptor binding by the anesthetic propanidid, *Biochem. J.*, 233, 259, 1986.
117. **Kirkness, E. F. and Turner, A. J.,** The gamma-aminobutyrate/benzodiazepine receptor from pig brain. Purification and characterization of the receptor complex from cerebral cortex and cerebellum, *Biochem. J.*, 233, 265, 1986.
118. **Stephenson, F. A., Mamalaki, C., Casalotti, S. O., and Barnard, E. A.,** The GABAA receptor and its antibodies, *Biochem. Soc. Symp.*, 52, 33, 1986.

119. **Schofield, P. R.,** The $GABA_A$ receptor: molecular biology reveals a complex picture, *Tr. Pharmacol. Sci.,* 10, 476, 1989.
120. **Maksay, G. and Ticku, M. K.,** CNS depressants accelerate the dissociation of ^{35}S-TBPS binding and GABA enhances their displacing potencies, *Life Sci.,* 43, 1331, 1988.
121. **Maksay, G. and Ticku, M. K.,** Dissociation of [^{35}S]-*t*-butylbicyclophosphorthionate binding differentiates convulsant and depressant drugs that modulate GABA-ergic transmission, *J. Neurochem.,* 44, 480, 1988.
122. **Trifletti, R. R., Snowman, A. M., and Snyder, S. H.,** Barbiturate recognition site on the GABA/benzodiazepine receptor complex is distinct from the picrotoxin/TBPS recognition site, *Eur. J. Pharmacol.,* 106, 441, 1985.
123. **Guidotti, A., Toffano, G., and Costa, E.,** An endogenous protein modulates the affinity of GABA and benzodiazepine receptors in rat brain, *Nature (London),* 275, 553, 1978.
124. **Supavilai, P., Mannonen, A., and Karobath, M.,** Modulation of GABA binding sites by CNS depressants and CNS convulsants, *Neurochem. Int.,* 4, 259, 1982.
125. **Popot, J. L. and Chaneux, J. P.,** Nicotinic receptor of acetylcholine: structure of an oligomeric integral membrane protein, *Physiol. Rev.,* 64, 1162, 1984.
126. **Kubo, T., Noda, M., Takai, T., Tanabe, T., Kayano, T., Shimizu, S., Tanaka, K., Takahashi, H., Hirose, T., Inayama, S., Kikuno, R., Miyata, T., and Numa, S.,** Primary structure of delta subunit precursor of calf muscle acetylcholine receptor deduced from cDNA sequence, *Eur. J. Biochem.,* 149, 5, 1985.
127. **Grenningloh, G., Rienitz, A., Schmitt, B., Methfessel, C., Zensen, M., Beyreuther, K., Gundelfinger, E. D., and Betz, H.,** The strychnine-binding subunit of the glycine receptor shows homology with nicotinic acetylcholine receptors, *Nature (London),* 328, 215, 1987.
128. **Dingledine, R., Myers, S. J., and Nicholas, R. A.,** Molecular biology of mammalian amino acid receptors, *FASEB J.,* 4, 2636, 1990.
129. **Pritchett, D. B., Sontheimer, H., Shivers, B. D., Ymer, S., Kettenmann, H., Schofield, P. R., and Seeburg, P. H.,** Importance of a novel GABAA receptor subunit for benzodiazepine pharmacology, *Nature (London),* 338, 582, 1989.
130. **Shivers, B. D., Killisch, I., Sprengel, R., Sontheimer, H., Kohler, M., Schofield, P. R., and Seeburg, P. H.,** Two novel $GABA_A$ receptor subunits exist in distinct neuronal subpopulations, *Neuron,* 3, 327, 1989.
131. **Ymer, S., Draguhn, A., Kohler, M., Schofield, P. R., and Seeburg, P. H.,** Sequence and expression of a novel GABAA receptor alpha subunit, *FEBS Lett.,* 258, 119, 1989.
132. **Ymer, S., Schofield, P. R., Draguhn, A., Werner, P., Kohler, M., and Seeburg, P. H.,** GABAA receptor beta subunit heterogeneity: functional expression of cloned cDNAs, *EMBO J.,* 8, 1665, 1989.
133. **Ymer, S., Schofield, P. R., Shivers, B. D., Pritchett, D. B., Luddens, H. A., Kohler, M., Werner, P., Sontheimer, H., Kettenman, H., and Seeburg, P. H.,** Molecular studies of the GABAA receptor, *J. Protein Chem.,* 8, 352, 1989.
134. **Verdoorn, T. A., Draguhn, A., Ymer, S., Seeburg, P. H., and Sakman, B.,** Functional properties of recombinant rat GABAA receptors depend upon subunit composition, *Neuron,* 4, 919, 1990.
135. **Lolait, S. J., O'Carroll, A. M., Kusano, K., Muller, J. M., Brownstein, M. J., and Mahan, L. C.,** Cloning and expression of a novel rat GABAA receptor, *FEBS Lett.,* 246, 145, 1989.
136. **Sato, T. N. and Neale, J. H.,** Immunological identification of multiple alpha-like subunits of the gamma-aminobutyric acidA receptor complex purified from neonatal rat cortex, *J. Neurochem.,* 53, 1089, 1989.
137. **Chan-Palay, V. and Palay, S. L.,** Ultrastructural localization of γ-aminobutyric acid receptors in the mammalian central nervous system by means of [^{3}H]-muscimol binding, *Proc. Natl. Acad. Sci. U.S.A.,* 75, 2977, 1978.

138. **Gee, K. W., Walmsley, J. K., and Yamamura, H. I.,** Light microscopic autoradiographic identifications of picrotoxin/barbiturate binding sites in rat brain with [^{35}S] t-butyl-bicyclophosphothionate, *Eur. J. Pharmacol.,* 89, 323, 1983.
139. **Lawrence, L. J., Palmer, C. J., Gee, K. W., Wang, X., Yamamura, H. I., and Casida, J. E.,** t-[^{3}H]Butylbicycloorthobenzoate: new radioligand probe for the γ-aminobutyric acid-regulated chloride ionophore, *J. Neurochem.,* 45, 798, 1985.
140. **O'Connor, L. H. and McEwen, B. S.,** Autoradiographic localization of GABA-regulated chloride ionophore binding sites using [^{3}H]t-butylbicycloorthobenzoate, *Eur. J. Pharmacol.,* 120, 141, 1986.
141. **McCabe, R. T. and Wamsley, J. K.,** Benzodiazepine receptor function. II. Autoradiographic localization of subcomponents of the macromolecular GABA receptor complex, *Life Sci.,* 39, 1937, 1986.
142. **deBlas, A. L., Vitorica, J., and Friedlich, P.,** Localization of the $GABA_A$ receptor in the rat brain with a monoclonal antibody to the 57,000 M_r peptide of the $GABA_A$ receptor/benzodiazepine receptor/Cl^- channel complex, *J. Neurosci.,* 8, 602, 1988.
143. **Vitorica, J., Park, D., Chin, G., and deBlas, A. L.,** Monoclonal antibodies and conventional antisera to the $GABA_A$ receptor/benzodiazepine receptor/Cl^- channel complex, *J. Neuroscience,* 8, 615, 1988.
144. **Vitorica, J., Park, D., Chin, G., and deBlas, A. L.,** Characterization with antibodies of the γ-aminobutyric $acid_A$/benzodiazepine receptor complex during development of the rat brain, *J. Neurochem.,* 54, 187, 1990.
145. **Enna, S. J. and Snyder, S. H.,** Properties of γ-aminobutyric acid (GABA) receptor binding in rat brain synaptic membrane fractions, *Brain Res.,* 100, 81, 1975.
146. **Zukin, S. R., Young, A. B., and Snyder, S. H.,** Gamma-aminobutyric acid binding to receptor sites in the rat central nervous system, *Proc. Natl. Acad. Sci. U.S.A.,* 71, 4802, 1974.
147. **Beaumont, K., Chilton, W., Yamamura, H. I., and Enna, S. J.,** Muscimol binding in rat brain: association with synaptic GABA receptors, *Brain Res.,* 148, 153, 1978.
148. **Mohler, H. and Okada, T.,** Properties of γ-aminobutyric acid receptor binding with (+)-[^{3}H]-bicuculline methiodide in rat cerebellum, *Mol. Pharmacol.,* 14, 256, 1978.
149. **Mohler, H. and Okada, T.,** Properties of ^{3}H-diazepam binding to benzodiazepine receptors in rat cerebral cortex, *Life Sci.,* 26, 2101, 1977.
150. **Mohler, H. and Okada, T.,** Benzodiazepine receptor: demonstration in the central nervous system, *Science,* 198, 849, 1977.
151. **Squires, R. F. and Braestrup, C.,** Benzodiazepine receptors in rat brain, *Nature (London),* 266, 732, 1977.
152. **Braestrup, C., Nielsen, M., Krogsgaard-Larsen, P., and Falch, E.,** Partial agonists for brain GABA/benzodiazepine receptor complex, *Nature (London),* 280, 331, 1979.
153. **Mohler, H., Burkhard, W. P., Keller, H. H., Richards, J. G., and Haefely, W.,** Benzodiazepine antagonist R015-1788: binding characteristics and interaction with drug-induced changes in dopamine turnover and cerebellar cGMP levels, *J. Neurochem.,* 37, 714, 1981.
154. **Ticku, M. K., Ban, M., and Olsen, R. W.,** Binding of [^{3}H]-α-dihydropicrotoxinin, a gamma-aminobutyric acid synaptic antagonist, to rat brain membranes, *Mol. Pharmacol.,* 14, 391, 1978.
155. **Ticku, M. K., VanNess, P., Haycock, J. W., Levy, W. B., and Olsen, R. W.,** Dihydropicrotoxinin binding sites in rat brain: comparison to GABA receptors, *Brain Res.,* 150, 642, 1978.
156. **Ticku, M. K. and Olsen, R. W.,** Picrotoxinin binding sites at the GABA synapse: a target for drug action, *Brain Res. Bull.,* 5(Suppl.2), 213, 1980.
157. **Supavilai, P. and Karobath, M.,** [^{35}S]-*t*-butylbicyclophosphorothionate binding sites are constituents of the γ-aminobutyric acid benzodiazepine receptor complex, *J. Neurosci.,* 4, 1193, 1984.

158. **Richter, J. A. and Yamamura, H. I.,** Effects of pentobarbital on t-[^{35}S]butylbicyclophosphorothionate, and [^{3}H]flunitrazepam binding to membrane bound and solubilized preparation from rat forebrain, *J. Pharmacol. Exp. Ther.*, 233, 125, 1985.
159. **Harris, R. A. and Allan, A. M.,** Functional coupling of γ-aminobutyric acid receptors to chloride channels in brain membrane, *Science,* 228, 1108, 1985.
160. **Allan, A. M. and Harris, R. A.,** γ-aminobutyric acid agonists and antagonists alter chloride flux across brain membranes, *Mol. Pharmacol.*, 29, 497, 1986.
161. **Schwartz, R. D., Skolnick, P., Hollingsworth, E. B., and Paul, S. M.,** Barbiturate and picrotoxin-sensitive chloride efflux in rat cerebral cortical synaptoneurosomes, *FEBS Lett.*, 175, 193, 1984.
162. **Schwartz, R. D., Jackson, J. A., Weigert, D., Skolnick, P., and Paul, S. M.,** Characterization of barbiturate-stimulated chloride efflux from rat brain synaptoneurosomes, *J. Neurosci.*, 5, 2953, 1985.
163. **Skerrit, J. H., Willow, M., and Johnston, G. A. R.,** Diazepam enhancement of low affinity GABA binding to rat brain membranes, *Neurosci. Lett.*, 29, 63, 1982.
164. **Costa, E. and Guidotti, A.,** Molecular mechanisms in the receptor action of benzodiazepines, *Annu. Rev. Pharmacol. Toxicol.*, 19, 531, 1979.
165. **Matsumoto, K. and Fukuda, H.,** Stimulation and protective effects of benzodiazepines on GABA receptors labeled with [^{3}H]muscimol, *Life Sci.*, 30, 935, 1982.
166. **Wastek, G. J., Speth, R. C., Reisene, T. D., and Yamamura, H. I.,** The effect of gamma-aminobutyric acid on [^{3}H]flunitrazepam binding in the rat brain, *Eur. J. Pharmacol.*, 50, 445, 1978.
167. **Tallman, J. F., Thomas, J. W., and Gallager, D. W.,** GABA-ergic modulation of benzodiazepine binding site sensitivity, *Nature (London),* 274, 383, 1978.
168. **Karobath, M. and Sperk, G.,** Stimulation of benzodiazepine receptor binding by gamma-aminobutyric acid, *Proc. Natl. Acad. Sci. U.S.A.*, 76, 1004, 1979.
169. **Tallman, J. F., Paul, S. M., Skolnick, P., and Gallager, D. W.,** Receptors for the age of anxiety: pharmacology of the benzodiazepines, *Science,* 207, 274, 1980.
170. **Marangos, P. J. and Martino, A. M.,** Studies on the relationship of γ-aminobutyric acid-stimulated diazepam binding and the γ-aminobutyric acid receptor, *Mol. Pharmacol.*, 20, 16, 1981.
171. **Davies, W. C. and Ticku, M. K.,** Pentobarbital enhances [^{3}H]diazepam binding to soluble receptors at the benzodiazepine-GABA receptor ionophore complex, *Neurosci. Lett.*, 23, 209, 1981.
172. **Ito, Y. and Kuriyama, K.,** Some properties of solubilized GABA receptors, *Brain Res.*, 236, 351, 1982.
173. **Gallagher, D. W., Thomas, J. W., and Tallman, J. F.,** Effect of GABAergic drugs on benzodiazepine binding site sensitivity in rat cerebral cortex, *Biochem. Pharmacol.*, 27, 2745, 1978.
174. **Martin, I. L. and Candy, J. M.,** Facilitation of benzodiazepine binding by sodium chloride and GABA, *Neuropharmacology,* 17, 993, 1978.
175. **Azanza, M. J.,** Benzodiazepines, barbiturates and baclofen interactions within the GABA receptor, *Gen. Pharmacol.*, 12, 123, 1981.
176. **Fujimoto, M., Hiran, K., and Okabayashi, T.,** Comparison of the effects of GABA and chloride ion on the affinities of ligands for the benzodiazepine receptor, *Life Sci.*, 30, 51, 1982.
177. **Marangos, P. J. and Crawley, J. N.,** Chronic benzodiazepine treatment increases (^{3}H)-muscimol binding in mouse brain, *Neuropharmacology,* 2, 81, 1982.
178. **Chiu, T. H. and Rosenberg, H. C.,** GABA receptor-mediated modulation of ^{3}H-diazepam binding in rat cortex, *Eur. J. Pharmacol.*, 56, 337, 1979.
179. **Leeb-Lundberg, F., Snowman, A., and Olsen, R. W.,** Perturbation of benzodiazepine receptor binding by pyrazolopyridines involve picrotoxinin/barbiturate receptor sites, *J. Neurosci.*, 1, 471, 1981.

180. **Leeb-Lundberg, F., Napias, C., and Olsen, R. W.,** Dihydropicrotoxinin binding sites in mammalian brain: interactions with convulsant and depressant benzodiazepines, *Brain Res.,* 216, 399, 1981.
181. **Massotti, M. and Guidotti, A.,** Endogenous regulators of benzodiazepine recognition sites, *Life Sci.,* 27, 847, 1980.
182. **Greenlee, D. V., Van Ness, P. C., and Olsen, R. W.,** Gamma-aminobutyric acid binding in mammalian brain: receptor-like specificity of sodium independent sites, *J. Neurochem.,* 31, 933, 1978.
183. **Krogsgaard-Larsen, P., Hjeds, H., Curtis, D. R., Lodge, D., and Johnston, G. A. R.,** Dihydromuscimol, thiomuscimol and related heterocyclic compounds as GABA analogues, *J. Neurochem.,* 32, 1717, 1979.
184. **Olsen, R. W. and Snowman, A. M.,** Chloride-dependent enhancement by barbiturates of GABA receptor binding, *J. Neurosci.,* 2, 1812, 1982.
185. **Whittle, S. R. and Turner, A. J.,** Differential effects of sedative and anticonvulsant barbiturates on specific [^{3}H]GABA binding to membrane preparations from rat brain cortex, *Biochem. Pharmacol.,* 31, 2891, 1982.
186. **Olsen, R. W.,** Drug interactions at the GABA receptor ionophore complex, *Annu. Rev. Pharmacol. Toxicol.,* 22, 245, 1982.
187. **Squires, R. F., Casida, J. E., Richardson, M., and Saederup, E.,** [^{35}S]-t-Butylbicyclophosphorothionate binds with high affinity to brain specific sites coupled to γ-aminobutyric acid-A and ion recognition sites, *Mol. Pharmacol.,* 23, 326, 1983.
188. **Ticku, M. K. and Olsen, R. W.,** Cage convulsants inhibit picrotoxinin binding, *Neuropharmacology,* 18, 315, 1979.
189. **Schulz, D. W. and MacDonald, R. L.,** Barbiturate enhancement of GABA-mediated inhibition and activation of chloride ion conductance: correlation with anticonvulsant and anesthetic actions, *Brain Res.,* 209, 177, 1981.
190. **Wong, E. H. F., Snowman, A. M., Leeb-Lundberg, L. M. F., and Olsen, R. W.,** Barbiturates allosterically inhibit GABA antagonist and benzodiazepine inverse agonist binding, *Eur. J. Pharmacol.,* 102, 205, 1984.
191. **Wong, E. H. F., Leeb-Lundberg, L. M. F., Teichberg, V. I., and Olsen, R. W.,** γ-Aminobutyric acid activation of $^{36}Cl^-$ flux in rat hippocampal slices and its potentiation by barbiturates, *Brain Res.,* 303, 267, 1984.
192. **Thyagarajan, R., Ramanyaneyulu, R., and Ticku, M. K.,** Enhancement of diazepam and γ-aminobutyric acid binding by (+)etomidate and pentobarbital, *J. Neurochem.,* 41, 578, 1983.
193. **Simmonds, M. A.,** Distinction between the effects of barbiturates, benzodiazepines and phenytoin on responses to γ-aminobutyric acid receptor activation and antagonism by bicuculline and picrotoxin, *Br. J. Pharmacol.,* 73, 739, 1981.
194. **Leeb-Lundberg, F. and Olsen, R. W.,** Interactions of barbiturates of various pharmacological categories with benzodiazepine receptors, *Mol. Pharmacol.,* 21, 320, 1982.
195. **Miller, L. J., Deutsch, S. I., Greenblatt, D. J., Paul, S. M., and Shader, R. I.,** Acute barbiturate administration increases benzodiazepine receptor binding *in vivo, Psychopharmacology,* 96, 385, 1990.
196. **Skolnick, P., Rice, K. C., Barker, J. L., and Paul, S. M.,** Interaction of barbiturates with benzodiazepine receptors in the central nervous system, *Brain Res.,* 233, 143, 1982.
197. **Supavilai, P. and Karobath, M.,** Action of pyrazolopyridines as modulators of [^{3}H]-flunitrazepam binding to the GABA/benzodiazepine receptor complex of the cerebellum, *Eur. J. Pharmacol.,* 70, 183, 1981.
198. **Asano, T. and Ogasawara, N.,** Chloride-dependent stimulation of GABA and benzodiazepine receptor binding by pentobarbital, *Brain Res.,* 225, 212, 1981.
199. **Bowery, N. G. and Dray, A.,** Reversal of the action of amino acid antagonists by barbiturates and other hypnotic drugs, *Br. J. Pharmacol.,* 63, 179, 1978.
200. **Evans, R. H.,** Potentiation of the effects of GABA by pentobarbitone, *Brain Res.,* 171, 113, 1979.

201. **Barnes, D. M., White, W. F., and Dichter, M. A.,** Etazolate (SQ 20009): electrophysiology and effects on [^{3}H]flunitrazepam binding in cultured cortical neurons, *J. Neurosci.,* 3, 762, 1983.
202. **Haefely, W. and Polc, P.,** Electrophysiological studies on the interaction of anxiolytic drugs with GABAergic mechanisms, in *Anxiolytics: Neurochemical, Behavioral, and Clinical Perspectives,* Malick, J. B., Enna, S. J., and Yamamura, H. I., Eds., Raven Press, New York, 1983, 113.
203. **Willow, M. and Johnston, G. A. R.,** Pharmacology of barbiturates: electrophysiological and neurochemical studies, *Int. Rev. Neurobiol.,* 24, 15, 1983.
204. **Study, R. E. and Barker, J. L.,** Diazepam and (−)-pentobarbital: fluctuation analysis reveals different mechanisms for potentiation of γ-aminobutyric acid responses in cultured central neurons, *Proc. Natl. Acad. Sci. U.S.A.,* 78, 180, 1981.
205. **Iadarola, M. J., Fanelli, R. J., McNamara, J. O., and Wilson, W. A.,** Comparison of the effects of diphenylbarbituric acid, phenobarbital, pentobarbital and secobarbital on GABA-mediated inhibition and benzodiazepine binding, *J. Pharmacol. Exp. Ther.,* 232, 127, 1985.
206. **Huidobro-Toro, J. P., Bleck, V., Allan, A. M., and Harris, R. A.,** Neurochemical actions of anesthetic drugs on the γ-aminobutyric acid receptor-chloride channel complex, *J. Pharmacol. Exp. Ther.,* 242, 963, 1987.
207. **Mohler, H., Okada, T., and Enna, S. J.,** Benzodiazepine and neurotransmitter receptor binding in rat brain after chronic administration of diazepam or phenobarbital, *Brain Res.,* 156, 391, 1978.
208. **Weizman, A., Fares, F., Pick, C. G., Yanai, J., and Gavish, M.,** Chronic phenobarbital administration effects GABA and benzodiazepine receptors in the brain and periphery, *Eur. J. Pharmacol.,* 169, 235, 1989.
209. **Sonawane, B. R., Yaffe, S. J., and Shapiro, B. H.,** Changes in mouse brain diazepam receptor binding after phenobarbital administration, *Life Sci.,* 27, 1335, 1980.
210. **Sivam, S. P., Nabeshima, T., and Ho, I. K.,** Acute and chronic effects of pentobarbital in relation to postsynaptic GABA receptors: a study with muscimol, *J. Neurol. Res.,* 7, 37, 1982.
211. **Saunders, P. A., Ito, Y., Baker, M. L., Hume, A. S., and Ho, I. K.,** Pentobarbital tolerance and withdrawal: correlation with effects on the GABA-A receptor, *Pharmacol. Biochem. Behav.,* 37, 343, 1990.
212. **Tatsuoka, Y., Kato, Y., Yoskida, K., and Imura, H.,** *In vivo* effects of phenytoin and phenobarbital on GABA receptors in rat cerebral cortex and cerebellum, *Neurosci. Lett.,* 46, 255, 1984.
213. **Nordberg, A., Wahlstrom, G., and Eriksson, B.,** [^{3}H]Quinuclidinyl benzilate binding in rat cortex in the abstinence after long-term barbital treatment, *Acta Physiol. Scand.,* 126, 153, 1986.
214. **Miller, L. G., Greenblatt, D. J., Roy, R. B., Summer, W. R., and Shader, R. I.,** Chronic benzodiazepine administration. II. Discontinuation syndrome is associated with upregulation of gamma-aminobutyric acidA receptor complex binding and function, *J. Pharmacol. Exp. Ther.,* 246, 177, 1988.
215. **Liljequist, S. and Tabakoff, B.,** Binding characteristics of ^{3}H-flunitrazepam and CL-218-872 in cerebellum and cortex of C57B1 mice made tolerant to and dependent on phenobarbital or ethanol, *Alcohol,* 2, 215, 1985.
216. **Roth-Schechter, B. F., Ebel, C., and Mallorga, P.,** Chronic pentobarbital diminishes GABA- and pentobarbital-enhanced ^{3}H-diazepam binding to benzodiazepine receptors, *Eur. J. Pharmacol.,* 87, 169, 1983.
217. **Ito, Y., Saunders, P. A., Lim, D. K., and Ho, I. K.,** Binding characteristics of *t*-[^{35}S]butylbicyclophosphorothionate in discrete brain regions of rats made tolerant to and dependent on pentobarbital, *J. Neurochem.,* 52, 1093, 1989.

218. **Wahlström, G. and Nordberg, A.,** Interaction between atropine and hexobarbital in the abstinence after chronic barbital treatment in the rat, in *Advances in Behavioral Biology: Cholinergic Mechanisms,* Vol. 25, Pepper, G. and Landinsky, H., Eds., Plenum Press, New York, 1981, 819.
219. **Wahlström, G.,** Addictive drugs and some neurotransmitters (acetylcholine and gamma-amino butyric acid), in *The Basis of Addiction,* Fishman, J., Ed., Dahlem Konferenzen, Berlin, 1978, 353.
220. **Wahlström, G.,** Drinking and cholinergic mechanism in the abstinence after chronic barbital treatment, *Scand. J. Psychol.,* (Suppl. I), 79, 1982.
221. **Modak, A. T., Weintraub, S. T., McCoy, T. H., and Stavinoha, W. B.,** Use of 300-msec microwave irradiation for enzyme inactivation: a study of effects of sodium pentobarbital on acetylcholine concentration in mouse brain regions, *J. Pharmacol. Exp. Ther.,* 197, 245, 1976.
222. **Nordberg, A.,** Apparent regional turnover of acetylcholine in mouse brain, *Acta Physiol. Scand.,* (Suppl.), 445, 1, 1977.
223. **Cheney, D. L. and Costa, E.,** Pharmacological implications of brain acetylcholine turnover measurement in rat brain nuclei, *Annu. Rev. Pharmacol. Toxicol.,* 17, 369, 1977.
224. **Richter, J. A. and Werling, L. A.,** K^+-stimulated acetylcholine release inhibition by several barbiturates and chloral hydrate but not by ethanol, chlordiazepoxide or 11-OH-Δ^9-tetrahydrocannabinol, *J. Neurochem.,* 32, 935, 1979.
225. **Richter, J. A. and Jackson, S. K.,** [^{3}H]-acetylcholine and [^{3}H]5-hydroxytryptamine release from rat midbrain slices and the effects of calcium and phenobarbital, *Neurochem. Res.,* 5, 719, 1980.
226. **Richter, J. A., Gormley, J. M., Hoffman, J. R., and Sim, J. R.,** High-affinity choline uptake in the hippocampus: its relationship to the physiological state produced by administration of barbiturate and other treatments, *J. Neurochem.,* 39, 1440, 1982.
227. **Wecker, L., Rothermel, S., and Cawley, G.,** Chronic choline supplementation attenuates the behavioral effects of pentobarbital, *Pharmacol. Biochem. Behav.,* 28, 469, 1987.
228. **Nordberg, A. and Wahlström, G.,** Effect of long-term forced oral barbital administration on endogenous acetylcholine in different regions of rat brain, *Eur. J. Pharmacol.,* 43, 237, 1977.
229. **Nordberg, A. and Wahlström, G.,** Changes in cholinergic function in rat brain late in the abstinence after chronic barbital treatment, *Drug Alcohol Depend.,* 7, 51, 1981.
230. **Nordberg, A. and Wahlström, G.,** Chronic barbital treatment and cholinergic mechanisms in brain, in *Cholinergic Mechanisms: Phylogenetic Aspects, Central and Peripheral Synapses and Clinical Significance,* Pepeu, G. and Ladinsky, H., Eds., Plenum Press, New York, 1981, 809.
231. **Nordberg, A. and Wahlström, G.,** Regional biosynthesis of acetylcholine in brain following forced oral chronic barbitone treatment to rat, *J. Neurochem.,* 32, 371, 1979.
232. **Nordberg, A., Wahlström, G., and Larsson, C.,** Increased number of muscarinic binding sites in brain following chronic barbiturate treatment to rat, *Life Sci.,* 26, 231, 1980.
233. **Nordberg, A. and Wahlström, G.,** Changes in populations of cholinergic binding sites in brain after chronic exposure to barbital in rats, *Brain Res.,* 246, 105, 1982.
234. **Roth, S. H., Forman, S. A., Braswell, L. M., and Miller, K. W.,** Actions of pentobarbital enantiomers on nicotinic cholinergic receptors, *Mol. Pharmacol.,* 36, 874, 1989.
235. **Dodson, B. A., Braswell, L. M., and Miller, K. W.,** Barbiturates bind to an allosteric regulatory site on nicotinic acetylcholine receptor-rich membranes, *Mol. Pharmacol.,* 32, 119, 1987.
236. **Dodson, B. A., Urh, R. R., and Miller, K. W.,** Relative potencies for barbiturate binding to the *Torpedo* acetylcholine receptor, *Br. J. Pharmacol.,* 101, 710, 1990.

237. **Corrodi, H., Fuxe, K., and Hokfelt, T.,** The effect of some psychoactive drugs on central monoamine neurons, *Eur. J. Pharmacol.,* 1, 363, 1967.
238. **Lidbrink, P., Corrodi, H., and Fuxe, K.,** Benzodiazepines and barbiturates: turnover changes in central 5-hydroxy-tryptamine pathways, *Eur. J. Pharmacol.,* 26, 35, 1974.
239. **Nabeshima, T. and Ho, I. K.,** Acute and chronic effects of pentobarbital on 5-hydroxytryptamine in mouse brain, *Prog. Neuro-Psychopharmacol.,* 5, 373, 1981.
240. **Tagashira, E., Hiramori, T., Urano, T., Nakao, K., and Yanaura, S.,** Participant of serotonin turnover rate in the brain on barbital withdrawal convulsions, *Japan. J. Pharmacol.,* 32, 159, 1982.
241. **Frankel, D., Khanna, J. M., LeBlanc, A. E., and Kalant, H.,** Effect of p-chlorophenyl-alanine on the acquisition of tolerance to ethanol and pentobarbital, *Psychopharmacology,* 44, 247, 1975.
242. **Lê, A. D., Khanna, J. M., Kalant, H., and LeBlanc, A. E.,** Effect of p-chlorophenylalanine on the acquisition of tolerance to barbital, *Can. J. Physiol. Pharmacol.,* 58, 12, 1980.
243. **Campanelli, C., Lê, A. D., Khanna, J. M., and Kalant, H.,** Effect of raphe lesions on the development of acute tolerance to ethanol and pentobarbital, *Psychopharmacology,* 96, 454, 1988.
244. **Khanna, J. M., Campanelli, C., Lê, A. D., and Kalant, H.,** Effect of raphe lesions on the development of chronic tolerance to pentobarbital and cross-tolerance to ethanol, *Psychopharmacology,* 91, 473, 1987.
245. **Tabakoff, B., Yanai, J., and Ritzmann, R. F.,** Brain noradrenergic systems as a prerequisite for developing tolerance to barbiturates, *Science,* 200, 449, 1978.
246. **Tabakoff, B. and Ritzmann, R. F.,** The effects of 6-hydroxydopamine on tolerance to and dependence on ethanol, *J. Pharmacol. Exp. Ther.,* 203, 319, 1977.
247. **Morgan, W. W.,** Effect of 6-hydroxydopamine pretreatment on spontaneous convulsions induced by barbital withdrawal, *Experientia,* 32, 489, 1976.
248. **Nabeshima, T., Fujimori, K., and Ho, I. K.,** Effect of acute or chronic pentobarbital administration on the steady state levels and the turnover rate of catecholamines in discrete brain areas of mice, *Prog. Neurol. Psychopharmacol.,* 5, 121, 1981.
249. **Morgan, W. W., Pfiel, K. A., and Gonzales, E. G.,** Catecholamine concentration in discrete brain areas following the withdrawal of barbital dependent rats, *Life Sci.,* 20, 493, 1977.
250. **Morgan, W. W., Pfiel, K. A., and Gonzales, E. G.,** Effect of α-methyl-*p*-tyrosine on the incidence of spontaneous convulsions observed following barbital withdrawal, *Neuropharmacology,* 17, 115, 1978.
251. **Tagashira, E., Hiramori, T., Urano, T., and Yanaura, S.,** Formation of physical dependence on barbiturates and cerebral monoamines, *Japan. J. Pharmacol.,* 33, 415, 1983.
252. **Germany, A., Merino, A., and Contreras, E.,** Adrenergic mechanisms on barbital dependence in mice, *Cell. Mol. Biol.,* 31, 369, 1985.
253. **Tagashira, E., Hiramori, T., Urano, T., Nakao, K., and Yanaura, S.,** Effects of propranolol on barbital dependence formation and withdrawal signs, *Japan. J. Pharmacol.,* 33, 133, 1983.
254. **Waddington, W., Riffee, W. R., Belknap, J. K., and Shepard, J. R.,** Barbiturate dependence in mice: evidence for β-adrenergic receptor proliferation in brain, *Res. Commun. Chem. Pathol. Pharmacol.,* 20, 207, 1978.

Chapter 14

A NEW APPROACH TO DEVELOPMENTAL NEUROTOXICITY: DESMETHYLIMIPRAMINE AS A PROTOTYPE

Cheryl F. Dreyfus

TABLE OF CONTENTS

I. INTRODUCTION

Numerous studies have now indicated that the fetus is susceptible to a wide variety of maternally administered drugs.[1-3] In fact, as many as 10% of developmental abnormalities appear to be due to maternal exposure to environmental toxins and other agents.[4] Yet, in very few instances do we understand the full extent of drug or toxin action or the molecular bases of these effects that result in birth defects. This is particularly true with respect to agents affecting the brain and spinal cord. Although developmental deficiences have been reported after exposure to a variety of psychoactive agents, the cellular and molecular bases of their actions remain obscure.

In treatment of depressive disease, one of these psychoactive agents, the tricyclic antidepressant drug, desmethylimipramine (DMI) is commonly used in adults, as well as in children.[5-7] In the adult, DMI is best known for its actions on catecholaminergic neurons of the brain. Catecholaminergic neurons project widely in the central nervous system (CNS), and have been implicated in attention disorders.[8-14] DMI crosses the placental barrier, is excreted in breast milk, and is concentrated in the fetal and neonatal brain.[15,16] Surprisingly, however, we know little of its actions during early development, or how these influences in the fetus and neonate influence later brain function. This is undoubtedly due to the complexity and inaccessibility of the CNS, particularly in the developing animal.

The goal of this chapter is to address the problem of assessing the actions of DMI during brain development. First, this chapter summarizes the currently known effects of DMI in the adult brain and relates these to what is known of the actions of DMI in development. These data suggest that it has been difficult to precisely define the cellular and molecular bases of drug action in the intact and interacting CNS.

This problem is further examined in the second part of the chapter that goes on to introduce a new experimental approach. This approach, involving the use of tissue culture, overcomes many of the problems associated with the study of the intact brain. Selected brain regions are grown for extended periods in an easily manipulated and well-characterized environment. Isolation of neuronal cell groups in culture permits the direct actions of psychoactive agents to be defined. The culture paradigm allows better control of drug dosage and times of exposure than would be possible in the intact brain. Therefore, critical developmental periods and stage-specific toxicity can be described. This approach appears promising in identification of the molecular loci of action of psychoactive agents, such as DMI, on the developing CNS. This information can then be used to explain how drug manipulations of developing brain systems can result in behavioral deficits in the mature nervous system.

II. ACTIONS OF DESMETHYLIMIPRAMINE ON THE MATURE BRAIN

In the adult, as noted above, DMI is best known for its influence on noradrenergic neurons. In particular, actions have been described on the brain nucleus locus coeruleus (l.c.), the largest brain noradrenergic nucleus. Noradrenergic neurons synthesize the neurotransmitter, norepinephrine (NE) and are distinguished by a well-characterized, specific, high-affinity NE uptake process. When administered acutely, DMI blocks this uptake process,[17-19] resulting in the increase of postsynaptic concentrations of norepinephrine and the potentiation of transmitter function.[19,20] It has been proposed that increased levels of NE interact with inhibitory α_2 receptors on l.c. neurons.[21-24] This triggers a feedback mechanism which ultimately results in decreased firing rate,[14,22,25-27] decreased activity of the NE biosynthetic enzyme, tyrosine hydroxylase (TH),[28] and a decrease in TH mRNA,[29] as well as a reduction in NE levels[29] and turnover.[30-33] That these actions of DMI appear to be mediated through α_2-receptors is indicated by the observations that effects of DMI are mimicked by clonidine, a powerful α_2-receptor agonist,[22,34] and are reversed by α_2-receptor blockers.[22,27]

Desmethylimipramine has been widely used as an antidepressant agent. However, the relationship of the above well-studied actions of DMI to its effects on depression are only partly understood and remain a subject of intense investigation.[26,28,30,35-39] The controversy in the literature relates to a number of observations that have been difficult to reconcile. For example, it has been proposed that the action of DMI as an antidepressant is related to inhibition of NE uptake. However, a variety of agents that block NE uptake lack antidepressant activity.[30,40] On the other hand, drugs with antidepressant activities have been shown to lack the effects of DMI on uptake.[41-43]

Perhaps more important to this controversy is the problem of the temporal disparity evident between pharmacological and clinical effects of DMI. DMI affects NE uptake acutely.[17-19] However, the therapeutic, antidepressant effects of the drug are not apparent initially, but develop after chronic treatment.[44] It is not understood why this discrepancy exists.[28,36,37] One popular theory states that antidepressant actions of DMI may be related not to uptake, but to effects on brain β-adrenergic receptors. Chronic treatment with DMI results in a decrease of postsynaptic β-receptors[30,39,45-50] and a subsensitivity of NE-stimulated adenylate cyclase activity.[38,51-53] These actions of DMI exhibit a time course similar to that of its clinical effects.[30-54] However, again, inconsistencies emerge. Many clinically effective antidepressant drugs do not downregulate β-adrenergic receptors[46,55-58] or influence noradrenergic sensitivity.[59] Moreover, effects of DMI are not limited to β-receptors. DMI has now been reported to interact with a number of receptor sites[33,38] including α_1- (References 43,47,60) and α_2-adrenergic receptors[19,61-64] as well as receptors for substance *P*,[65] dopamine,[66] benzodiazepine,[67] and serotonin.[43,47,68,69] Alterations in serotonergic neuron function[70,71] have been noted. DMI also interferes with corticotropin-releasing factor function in the l.c.[26]

Chronic actions of this tricyclic antidepressant drug may thus extend beyond the catecholaminergic brain systems to involve the complex interactions that exist among the brain's neurotransmitter and neuromodulatory systems. How these interactions are affected by DMI to result in antidepressant actions is still under study. However, these multiple actions of chronic DMI administration point to the problems of evaluating the molecular bases of drug action within the complexity of the intact CNS. This problem is magnified when assessing actions in the developing system.

III. EFFECTS OF TRICYCLIC ANTIDEPRESSANT DRUGS ON THE DEVELOPING BRAIN

The effects of exposing the maturing brain to tricyclic antidepressant drugs have not been examined extensively. Yet, it is clear that these agents may have access to the developing animal and have the potential to influence brain function. For example, tricyclic antidepressants are now known to cross the placental barrier and accumulate in the fetal CNS. In pharmacokinetic studies, concentrations of imipramine (IMI) and its metabolites, 2-hydroxyimipramine, desmethylimipramine, and 2-hydroxydesipramine, have been assessed in embryonic day 18 (E18) fetal tissues after injection of IMI into a pregnant rat. In particular, both IMI and DMI were concentrated in the fetal brain. Moreover, DMI appeared to be the major agent to which the brain was exposed. Levels of DMI were greater than those of IMI, and DMI remained in fetal tissues for a longer period of time.[15]

Exposure to the tricyclic antidepressants during development is not limited to the prenatal period. A number of studies suggest that neonates and children are potential targets of these agents. Thus, IMI and DMI are used to treat maternal depression that occurs in the postpartum period, and both agents have the potential to be transferred to the neonate from a nursing mother. These antidepressant drugs are excreted in human breast milk in concentrations similar to those found in plasma.[16] Furthermore, IMI is administered to children, and is commonly used to treat nocturnal enuresis, phobia associated with school, and depression.[5,6]

Surprisingly, in spite of the many possible avenues of exposure to these drugs, little is known of how they influence ontogeny. Yet, these drugs may have critical effects. They have been reported to affect function of central monoaminergic neurons of the l.c., a neural system implicated in such diverse functions as vigilance and attention,[8-11] memory and learning,[12,13] and anxiety.[14] In particularly, a variety of studies indicate that the l.c. may control the transfer of afferent information within the cerebrum and cerebellum by enhancing the signal-to-noise ratio. The l.c. has been proposed to be a filter that separates relevant and irrelevant data and, therefore, focuses attention.[8,14,72-76] Drug actions that modulate these brain functions, indeed, may influence brain behavior.

Further, there is reason to suspect that the young l.c. is susceptible to drug action at these times. Immature l.c. neurons that exhibit the NE biosynthetic enzymes, tyrosine hydroxylase and dopamine-β-hydroxylase, also display a DMI-sensitive, high-affinity NE uptake system.[8,77-82] β-Adrenergic binding sites are present in the brain during early fetal development.[83] The data suggest that antidepressants may elicit effects noted in adults in this young noradrenergic system.

Moreover, action of the drug may not be limited to the maturing l.c., but may extend to other brain regions as well. The developing l.c. projects widely throughout the CNS to reach target structures.[84,85] It has been noted that the l.c. matures days before its targets do.[86] This has prompted the hypothesis that targets may be dependent on locus input for normal maturation. Consequently, alterations in locus function may result in alterations in target function. Although still open to debate,[87-89] in fact, it has been reported that innervation of the primary visual cortex by noradrenergic neurites is essential for shifts in occular dominance during development of the postnatal animal.[90,91] In addition, other work indicates that lesion of the l.c. results in increased density of synapses in the visual cortex,[92] or the development of abnormal pyramidal cells.[93,94] In a related study, elimination of the l.c. appears to alter the innervation of target regions by other afferent systems.[12] An action of DMI on developing l.c. neurons, then, may affect not only the l.c. but also its target regions.

In summary, although it is known that DMI blocks uptake in developing l.c. acutely, it is unclear what implication this blockade has on subsequent maturation of a functioning brain or on its behavior. Yet, work on the role of the l.c. suggests that such blockade may critically influence brain maturation. A variety of investigations have, therefore, evaluated the actions of antidepressants on developing brain.

In these investigations, IMI has been administered to rats for periods of time throughout ontogeny, during the third trimester,[95] and second and third trimesters,[96,97] or even before mating and during gestation,[98] or during nursing.[99] Initially, effects on the mother and her litter size were evaluated. Although IMI-treated mothers were found to gain significantly less weight than control animals during pregnancy,[96] the percentage giving birth and the length of gestation were unaffected.[96] Litter size was affected in some,[98] but not other studies.[96] These actions of IMI appear to be specifically related to the drug. Control experiments, in which mothers were fed a restricted diet, suggest that these changes on the developing litter are not related to possible undernourishment due to the drug.[99]

When fetuses were evaluated, distinct alterations were apparent after IMI treatment. Birth weight and growth rate of the offspring were found to be reduced and these effects extended beyond the time of drug administration.

Thus, when IMI was present only during the third trimester, a 20% reduction in birth weight and significantly reduced body weight were evident through 40 days of postnatal development. In addition, a 10% infant mortality rate was observed.[95]

The above studies not only evaluated gross indices of weight and mortality, but behavior as well. These experiments indicate that exposure to the antidepressant drugs during maturation may well influence behavior. For example, drug-treated offspring were found to be behaviorally unresponsive and interacted for significantly less time than controls with other animals and their environment.[100] This impairment in behavior was most obvious when animals were raised in an enriched environment.[101] In other work, pups exposed to IMI during gestation exhibited delayed surface-righting reflexes and altered development of negative geotaxis,[96] again suggesting a developmental deficit in behavior.

In related experiments, antidepressant drug actions were evaluated on the noradrenergic system, specifically. In most cases, no changes in norepinephrine or normetanephrine, an NE metabolite, were recorded in response to IMI.[102] This is in contrast to observations in the adult (see above). The data have been confusing because no correlation is readily apparent between actions on the noradrenergic system and the noted changes in behavior.

More interesting observations were apparent, however, when the density of β-adrenoceptors were evaluated. Prenatal exposure,[96] or postnatal exposure[103] to IMI resulted in a decrease in β-receptors, measured by [^{3}H]-dihydroalprenolol binding. Moreover, this effect was prolonged beyond the time of drug application. After prenatal exposure, effects noted at postnatal day 14 (P14) were still observed 16 days later. Affinity for binding was increased at this time as well.[96] These effects on β-receptors were similar to those observed in adults. Interestingly, however, unlike the case with the neonates, the decrease in adult binding sites does not persist for more than 1 week after treatment has ceased.[54] Prenatal exposure to antidepressants appears to produce effects that are similar to, but more marked than, those observed in adult animals.[104]

How are these prolonged effects mediated in the fetus? One possibility is that continuous treatment with IMI or DMI results in a change in the density of [^{3}H]-IMI binding sites. This has been reported to occur in adult animals.[105,106] It is also noted in response to the serotonin uptake blockers, chlorimipramine and fluoxetine, and to the selective monoamine oxidase inhibitors, clorgyline and deprenyl, when these agents are applied prenatally and effects are assessed at P25.[107] However, it is less clear when evaluating DMI. Prenatal DMI does not appear to influence the density of IMI binding sites in P25 animals.[104,107] On the other hand, postnatal treatment with DMI has been reported to decrease IMI sites in P9 animals.[103] Further studies are necessary to explore this apparent inconsistency and evaluate the molecular bases of prolonged drug actions.

In total, these studies on the effects of tricyclic antidepressants on development are incomplete. These agents have access to the brain during maturation and potentially influence widely projecting, early developing neural systems. Some evidence suggests that early exposure to IMI and DMI may have actions on the brain that extend into adult life and affect behavior, but this requires further elucidation. Virtually no work has been directed to understanding molecular mechanisms underlying the direct and long-term actions of IMI on specific brain systems. Part of the problem lies with the complexity of the brain and the difficulties associated with such study.

IV. THE USE OF CULTURES TO DEFINE DESMETHYLIMIPRAMINE ACTION

One approach to this problem is the use of tissue culture to evaluate specific actions on selected populations. For example, brain cultures can be grown as explants, cultures in which a carefully dissected brain region is maintained intact, but isolated from the rest of the CNS. Alternatively, brain regions can be grown as dissociated neuronal/glial, pure neuronal, or pure glial cultures.[108] In this way, direct effects on particular cellular components of a brain region can be detected. For example, such an approach has been taken in studies evaluating NE uptake in glial cells.[109] The astrocytic component of glia grown from neonatal rat brain exhibits a high-affinity uptake system for NE. Moreover, NE uptake is inhibited by DMI. This observation suggests a new mechanism by which DMI may influence brain neurons. By inhibiting NE uptake into glial elements, DMI may prolong NE action in the brain, or make it more accessible to metabolism by extracellular enzymes.

Use of tissue cultures to study the neural crest precursor cells of the peripheral nervous system similarly has suggested new mechanisms by which DMI may modulate neural function during development. DMI blocks the expression of NE and its enzymes, tyrosine hydroxylase and dopamine-β-hydroxylase, in young neural crest cells.[110] These data suggest actions of DMI on the early development of the peripheral nervous system that may have profound effects on subsequent function of this system. Similar actions of this antidepressant agent must now be evaluated in neurons of the CNS.

V. USE OF BRAIN CULTURES OF THE LOCUS COERULEUS TO EVALUATE ACTIONS OF DESMETHYLIMIPRAMINE

To begin evaluation of the actions of DMI on brain development we have been examining its effects on the l.c., grown as explants and dissociated cultures. Our initial experiments have defined the developmental potential of l.c. of various maturational stages in an explant culture paradigm.[81] DMI may function at a variety of developmental periods in brain ontogeny. For example,

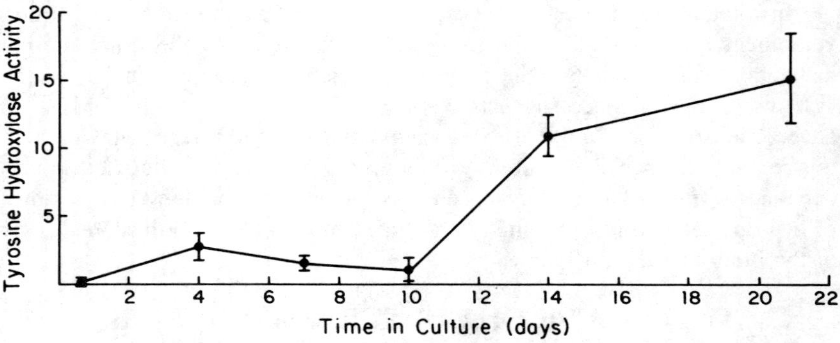

FIGURE 1. Development of tyrosine hydroxylase (TH) catalytic activity in E12 l.c. cultures. At least five pairs of explant cultures were evaluated at varying culture times. TH activity is expressed as mean picomoles DOPA produced per l.c. pair ±SEM. (From Dreyfus, C. F. et al., *Dev. Biol.*, 97, 48, 1983. With permission.)

it may affect neurons during phases of mitosis, migration, or synaptogenesis. By growing locus neurons from a range of stages, we are able to determine how central neurons respond to a variety of extrinsic signals throughout ontogeny and are able to define critical ontogenetic periods. This approach will permit us to examine stage-specific molecular mechanisms underlying the actions of the tricyclic antidepressants. However, to assess the impact of DMI on our model l.c. culture system, it is first important to understand the normal developmental pattern of locus cells at particular maturities.

A. STAGE E12: CELL MITOSIS AND MIGRATION WITHOUT AFFERENT INNERVATION

Embryonic day-12 locus cells, dissected from a mouse, provide the opportunity to study effects of DMI on immature brain cells. At E12, presumptive locus cells in the mouse are completing their last divisions and moving away from the ventricle toward their definitive brain locations.[86,111,112] These cells have not yet received detectable presynaptic innervation,[113] and do not appear to respond to presynaptic agonists or antagonists.[113a] In addition, they have not yet projected to target structures.[77,85,114-118] However, they do exhibit the specific, high-affinity NE uptake system. This system is saturable, temperature-dependent, and inhibited by DMI.[82] Explanation of E12 locus results in the development of the NE uptake system in culture. This permits evaluation of molecular effects of exogenous agents in an accessible, well-defined environment.

Unlike NE uptake, TH and DBH are not detectable at E12 *in vivo*.[81] However, upon explanation of the locus, they appear *de novo* and subsequently exhibit developmental increases (Figure 1). Examination of these NE characters, therefore, provides us with the opportunity to evaluate toxic effects on initial appearance as well as maturation of phenotypic traits.

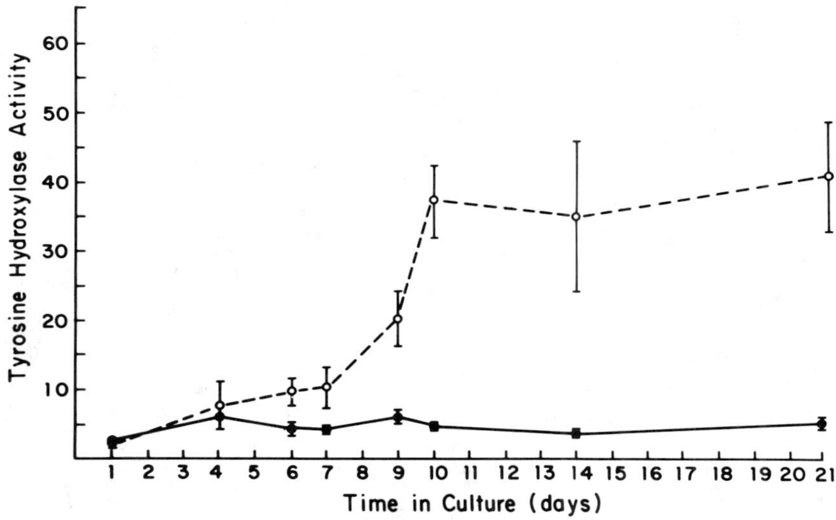

FIGURE 2. Development of tyrosine hydroxylase (TH) activity in E18 cultures. For each time point, one side of the brainstem was cultured for the noted time in culture and the contralateral brainstem was frozen at explantation. TH catalytic activity was assayed simultaneously in the time 0 controls (solid line) and the cultured explants (dashed line). At least five pairs of cultures or controls were examined at each time point. Results are expressed as in Figure 1. (From Dreyfus, C. F. et al., *Dev. Biol.*, 97, 48, 1983. With permission.)

The E12 cultures, then, constitute a system with which to examine effects of exogenous agents on immature cells devoid of detectable afferent or target interactions. In this system, moreover, drug effects on noradrenergic traits exhibiting developmental increases can be compared with effects on initial expression of such characters.

B. STAGE E18: MATURE DENSELY PACKED LOCUS, INNERVATED AND EXTENDING NEURITES TO TARGETS

E18 locus provides the opportunity to test effects of agents, such as DMI, on relatively mature brain cells. E18 locus cells exhibit all NE traits *in vivo*.[77,80,81,114,118] These cells, however, differ from younger cells in appearing more mature morphologically.[81] Cells are larger and more tightly clustered. They have clearly received synaptic contacts[113] and project to target structures in the brain.[77,85,114-118]

Explants of E18 locus exhibit developmental increases in noradrenergic traits with time in culture. Moreover, E18 cells in culture appear more mature than their younger counterparts. TH levels per locus cell are higher than those observed in E12 cultures (Figure 2). In addition, immunocytochemical analysis has localized TH to clustered, morphologically mature cells which are distinctly different than cells observed in E12 explants[81] (Figures 3 and 4).

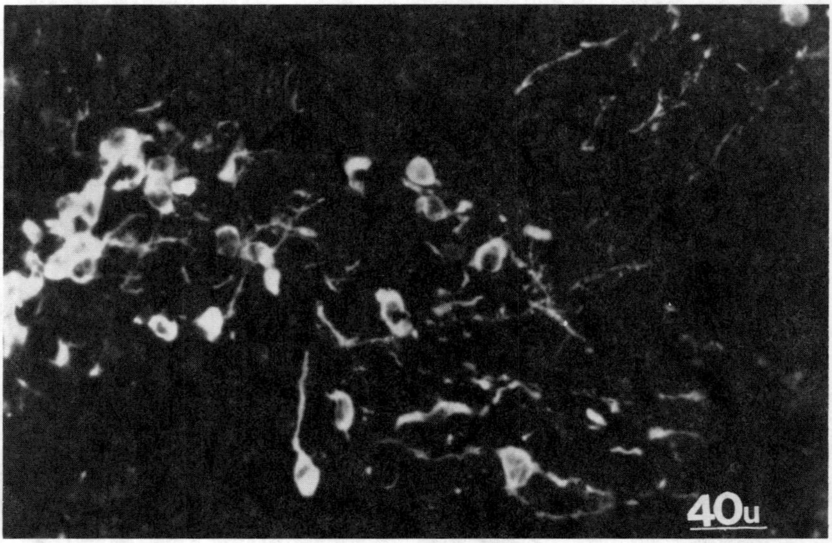

FIGURE 3.. Immunocytochemical detection of tyrosine hydroxylase (TH) cells in E12 l.c. cultures. Cells are loosely associated within the explant. (From Dreyfus, C. F., et al., *Dev. Biol.*, 97, 48, 1983. With permission.)

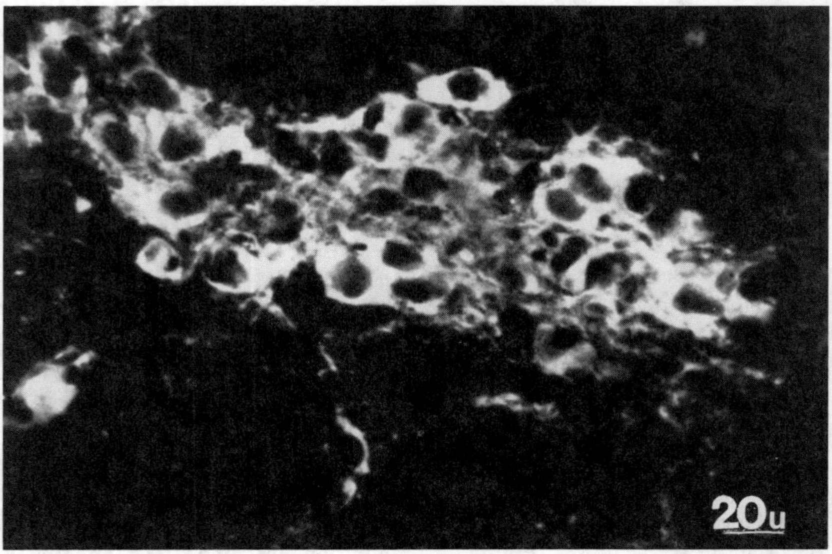

FIGURE 4. Immunocytochemical detection of tyrosine hydroxylase (TH) cells in E18 cultures. Cells are densely packed within the explant. (From Dreyfus, C. F. et al., *Dev. Biol.*, 97, 48, 1983. With permission.)

By explanting the locus at different embryonic ages, then, critical developmental stages and apparent capabilities have been identified. We have begun to use cultures prepared from these stages to study the effects of DMI and define critical ontogenetic periods.

C. EFFECTS OF DESMETHYLIMIPRAMINE ON LOCUS CULTURES OF DIFFERENT AGES

In our work we have been growing l.c. explants from E12 and E18 mice. Previous study indicated that these locus explants can be maintained for up to 5 weeks in culture. In the present experiments, cultures were grown for 1 week and then exposed to DMI (1 or 10 μM) for an additional week. Work *in vivo* (see above) suggested that treatment with DMI decreases TH catalytic activity in adult brain.[28,29] To define similar actions in culture, explants were evaluated for TH catalytic activity.

Preliminary study suggests that treated cultures appear normal and no gross changes were apparent after drug treatment. Distinct differences were noted, however, when E12 and E18 cultures were assessed for TH activity. This activity in the more immature, E12, explants appeared to be unaffected by DMI treatment. However, TH in the E18 group was significantly decreased 65% by the antidepressant agent.[82] This effect appeared to be selective. Treated E18 cultures could not be distinguished from controls. Cultures appeared healthy and exhibited extensive neuritic outgrowth.

These data suggest that critical DMI-insensitive and -sensitive stages may exist during locus ontogeny.

What may be the molecular basis for the lack of a DMI effect on the young E12 l.c. in culture? One possibility is that DMI recognition sites are not yet present. However, when E12 l.c. was examined *in vivo* or after growth in explant culture, high-affinity uptake of NE was inhibited by treatment with DMI.[82] These data suggest that DMI sites are present on E12 l.c. neurons. The lack of effect on TH activity must be related to some other immaturity in young l.c. cells.

In summary, distinct stage differences may exist between E12 and E18 l.c. E18 l.c. neurons that appear mature, as evaluated by a variety of criteria, respond to DMI treatment as do adult neurons. Drug treatment depresses TH activity. In contrast, TH activity in the less mature E12 locus neurons is apparently unresponsive to this drug.

Mechanisms underlying these differences in responsivity remain to be defined. However, these studies indicate that the well-defined, easily manipulated culture conditions may be favorable for such study. Using explants, critical periods may be identified. Subsequent work that makes use of dissociated cultures grown as neuronal or glial cell dissociates, can then be employed to define the cellular basis of drug action.

VI. PROSPECTS

In summary, actions of DMI have been extensively defined on adult l.c. neurons. DMI inhibits l.c. firing rates,[14,22,25-27] TH activity,[28] and NE turnover.[30-33] Neurons of the l.c. project throughout the brain and have widespread actions in the CNS.[8-14] These neurons have been implicated in such functions as learning, memory, and attention and have been associated with such disorders as Alzheimer's disease.[119-121] Actions on this cell group may have profound effects on the entire neuraxis.

In fact, chronic treatment with DMI does have profound effects on brain function resulting in antidepressant actions.[26,28,30,35-39] However, in spite of extensive research, the molecular and cellular bases of these actions still remain unclear. No doubt, this is due to the difficulty in defining direct effects of pharmacological agents on interacting systems.

The problem becomes magnified when evaluating effects of this drug on the developing animal. DMI has access to the developing brain from the mother via the placenta and breast milk.[15,16] Moreover, in humans the drug is given directly to children.[5-7] No gross defects have been reported in the human.[122,123] Yet it is possible that subtle influences on the widespread l.c. may elicit actions that affect the maturing brain for years to come. Indeed, developmental studies in the rat suggest that DMI may influence behavior and that effects of drug treatment may extend for prolonged periods after cessation of drug administration. It has been difficult to examine the cellular and molecular bases of these actions because of the inaccessibility of the developing system.

We propose the use of a tissue culture paradigm as a new approach to the problem of assessing DMI effects during ontogeny. Our cultures have defined susceptible periods to this antidepressant agent. Moreover, they offer the possibility of examining direct drug actions on neural systems, dissected from the complexity of the intact animal. Information gained in culture may then be applied to the *in vivo* system.

It should be noted that the importance of this approach extends beyond the evaluation of tricyclic antidepressant agents. Although it is clear that a number of psychoactive agents adversely affect brain development, underlying mechanisms remain obscure. Moreover, although some reports have indicated actions of psychoactive drugs in various culture systems (i.e., References 124 to 126), the molecular bases of these actions at critical ontogenetic periods remain to be described. Culture work offers the possibility of delineating the cellular bases of drug action during development in a well-defined and accessible environment. These studies allow us to define the molecular pathogenesis that may be associated with toxic agents during specific periods in embryogenesis.

ACKNOWLEDGMENTS

The work described was supported in part by NIH grant-DA05132. I thank Ms. Renee Crawford, Ms. Mary Ginsburg, Dr. Stella Elkabes, Dr. Aniek Schoups, Dr. Howard Mount, Dr. Wilma Friedman, and Dr. Ira B. Black for valuable discussion and critical review of this paper.

REFERENCES

1. **Boethius, G.,** Recording of drug prescriptions in the county of Jamtland, Sweden. II Drug exposure of pregnant women in relation to course and outcome of pregnancy, *Eur. J. Clin. Pharmacol.,* 12, 37, 1977.
2. **Leonard, B. E.,** Effect of psychotropic drugs administered to pregnant rats on the behavior of the offspring, *Neuropharmacology,* 20, 1234, 1981.
3. **Kalter, H. and Warkany, J.,** Congenital malformations, *N. Engl. J. Med.,* 308, 491, 1983.
4. **Wilson, J. G.,** in *Handbook of Teratology,* Plenum Press, New York, 1977, 309.
5. **Preskorn, S. H., Weller, E. B., Weller, R. A., and Glotzback, E.,** Plasma levels of imipramine and adverse effects in children, *Am. J. Psychiatry,* 140, 1332, 1983.
6. **Puig-Antich, J., Perel, J. M., Lupatkin, W., Chambers, W. J., Shea, C., Tabrizi, M. A., and Stiller, R. L.,** Plasma levels of imipramine (IMI) and desmethylimipramine (DMI) and clinical response in prepubertal major depressive disorder, *J. Am. Acad. Child. Psychiatry,* 18, 616, 1979.
7. **Yaffe, S. J.,** *Pediatric Pharmacology: Therapeutic Principles in Practice,* Grune and Stratton, New York, 1980.
8. **Foote, S. L., Bloom, F. E., and Aston-Jones, G.,** Nucleus locus coeruleus: new evidence of anatomical and physiological specificity, *Physiol. Rev.,* 63, 844, 1983.
9. **Mason, S. T.,** Noradrenaline in the brain: progress in theories of behavioral function, *Prog. Neurobiol.,* 16, 263, 1981.
10. **Jones, B. E. and Moore, R. Y.,** Ascending projection of the locus coeruleus in the rat. II. Autoradiographic study, *Brain Res.,* 127, 23, 1977.
11. **Jouvet, M.,** The role of monoamines and acetylcholine containing neurons in the regulation of the sleep-waking cycle, *Ergeb. Physiol. Biol. Chem. Exp. Pharmak.,* 64, 166, 1972.
12. **Amaral, D. G. and Foss, J. A.,** Locus coeruleus lesions and learning, *Science,* 188, 377, 1975.
13. **Ogren, S. O., Archer, T., and Ross, S. B.,** in *Catechcholamines. Neuropharmacology and Central Nervous Systems: Theoretical Aspects,* Usdin, E., Carlsson, A., Dahlstrom, A., and Engel, J., Eds., Alan R. Liss, New York, 1984, 285.
14. **Olpe, H. R., Jones, R. S. G., and Steinmann, M. W.,** The locus coeruleus: actions of psychoactive drugs, *Experientia (Basel),* 39, 242, 1983.
15. **DeVane, C. L. and Simpkins, J. W.,** Pharmacokinetics of imipramine and its major metabolites in pregnant rats and their fetuses following a single dose, *Drug Metab. Dispos.,* 13, 438, 1985.
16. **Sovner, R. and Orsulak, P. J.,** Excretion of imipramine and desipramine in human breast milk, *Am. J. Psychiatry,* 136, 451, 1979.
17. **Carlsson, A., Fuxe, K., Hamberger, B., and Lindquist, M.,** Biochemical and histochemical studies on the effects of imipramine, *Acta Physiol. Scand.,* 67, 481, 1966.

18. **Glowinski, J. and Axelrod, J.,** Inhibition of uptake of tritiated noradrenaline in intact rat brain by imipramine and related compounds, *Nature (London),* 204, 1318, 1964.
19. **Iversen, L. L.,** *The Uptake and Storage of Noradrenaline in Sympathetic Nerves,* Cambridge University Press, New York, 1967.
20. **Bunney, W. E. and Davis, J. M.,** Norepinephrine in depressive reactions: a review, *Arch. Gen. Psychiatr.,* 13, 483, 1965.
21. **Taube, H. D., Starke, K., and Borowski, E.,** Presynaptic receptor systems on the noradrenergic neurones of rat brain, *Naunyn-Schmiedeberg's Arch. Pharmacol.,* 299, 123, 1977.
22. **Svensson, T. H. and Usdin, T.,** Feedback inhibition of brain noradrenaline neurons by tricyclic antidepressants: α-receptor mediation, *Science,* 202, 1089, 1978.
23. **Cedarbaum, J. M. and Aghajanian, G. K.,** Catecholamine receptors on locus coeruleus neurons: pharmacological characterization, *Eur. J. Pharmacol.,* 44, 375, 1977.
24. **Huang, Y. H., Maas, J. W., and Hy, G. H.,** The time course of noradrenergic pre- and postsynaptic activity during chronic desipramine treatment, *Eur. J. Pharmacol.,* 68, 41, 1981.
25. **Nyback, M. V., Walters, J. R., Aghajanian, G. K. and Roth, R. H.,** Tricyclic antidepressants: effects on the firing rate of brain noradrenergic neurons, *Eur. J. Pharmacol.,* 32, 302, 1975.
26. **Valentino, R. J., Curtis, A. L., Parris, D. G., and Wehby, R. G.,** Antidepressant actions on brain noradrenergic neurons, *J. Pharmacol. Exp. Ther.,* 253, 833, 1990.
27. **Haskins, T., Moyer, J. A., Muth, E. A., and Sigg, E. B.,** DMI, WY 45,030, WY-45,881, and Ciramadol inhibit locus coeruleus neuronal activity, *Eur. J. Pharmacol.,* 115, 139, 1985.
28. **Segal, D. S., Kuczenski, R., and Mandell, A. J.,** Theoretical implications of drug-induced adaptive regulation of a biogenic amine hypothesis of affective disorders, *Biol. Psychiatry,* 9, 147, 1974.
29. **Nestler, E. J., McMahon, A., Sabaan, E. L., Tallman, J. F., and Duman, R. S.,** Chronic antidepressant administration decreases the expression of tyrosine hydroxylase in the rat locus coeruleus, *Proc. Natl. Acad. Sci. U.S.A.,* 87, 7522, 1990.
30. **Sulser, F., Vetulani, J., and Mobley, P. L.,** Mode of action of antidepressant drugs, *Biochem. Pharmacol.,* 27, 257, 1978.
31. **Nielsen, M. and Baestrup, C.,** Desipramine and some other antidepressant drugs decrease the major norepinephrine metabolite, 3,4-dihydroxy-phenylglycol-sulphate in the rat brain, *Naunyn-Schmiedeberg's Arch. Pharmacol.,* 300, 93, 1977.
32. **Rosloff, B. N. and Davis, J. M.,** Decrease in brain NE turnover after chronic DMI treatment: no effect of iprindole, *Psychopharmacology,* 56, 335, 1978.
33. **Heninger, G. R. and Chaney, D. S.,** in *Psychopharmacology: The Third Generation of Progress,* Meltzer, H. Y., Ed., Raven Press, New York, 1987, 535.
34. **Marwaha, J., Kehne, J. H., Commissaris, R. L., Lakoski, J., Shaw, W., and Davis, M.,** Spinal clonidine inhibits neural firing in locus coeruleus, *Brain Res.,* 276, 379, 1983.
35. **Siever, L. J.,** in *Psychopharmacology: The Third Generation of Progress,* Meltzer, H. Y., Ed., Raven Press, New York, 1987, 493.
36. **Stone, E. A.,** Subsensitivity to norepinephrine as a link between adaptation to stress and antidepressant therapy: an hypothesis, *Res. Commun. Psychol. Psychiatry Behav.,* 4, 241, 1979.
37. **Stone, E. A.,** Problems with the current catecholamine hypothesis of antidepressive agents: speculations leading to a new hypothesis, *Behav. Brain Sci.,* 6, 535, 1983.
38. **Sugrue, M. F.,** Do antidepressants possess a common mechanism of action? *Biochem. Pharmacol.,* 32, 1811, 1983.

39. **Ventulani, J., Stawarz, R. J., Dingell, J. V., and Sulser, F.,** A possible common mechanism of action of tricylic antidepressant treatments. Reduction in the sensitivity of noradrenergic cyclic AMP generating systems in the rat limbic forebrain, *Naunyn-Schmiedeberg's Arch. Pharmacol.*, 239, 109, 1976.
40. **Post, R. M., Kopin, J., and Goodwin, F. K.,** The effects of cocaine to depressed patients, *Am. J. Psychiatry*, 131, 511, 1974.
41. **Lahti, R. A. and Maickel, R. P.,** The tricyclic antidepressants-inhibition of norepinephrine uptake as related to potentiation of norepinephrine and clinical efficacy, *Biochem. Pharmacol.*, 20, 482, 1971.
42. **Zis, A. D. and Goodwin, F. K.,** Novel antidepressants and the biogenic amine hypothesis of depression. The case for iprindole and mianserine, *Arch. Gen. Psychiatry*, 36, 1097, 1979.
43. **Hall, H. and Ogren, S.-O.,** Effects of antidepressant drugs on different receptors in the brain, *Eur. J. Pharmacol.*, 70, 393, 1981.
44. **Klein, F. and Davis, J. M.,** *Diagnosis and Drug Treatment of Psychiatric Disorders*, Williams & Wilkins, Baltimore, MD, 1969.
45. **Banerjee, S. P., Kung, L. S., Riggi, S. J., and Chande, S. K.,** Development of beta-adrenergic receptor subsensitivity by antidepressants, *Nature (London)*, 268, 455, 1977.
46. **Sellinger-Barnette, M. M., Mendels, J., and Frazer, A.,** The effect of psychoactive drugs on beta-adrenergic receptor binding sites in rat brain, *Neuropharmacology*, 19, 447, 1980.
47. **Maggi, A., U'Prichard, D. C., and Enna, S. J.,** Differential effects of antidepressant treatment on brain monoaminergic receptors, *Eur. J. Pharmacol.*, 61, 91, 1980.
48. **Sarai, K., Frazer, A., Brunswick, D., and Mendels, J.,** Desmethylimipramine-induced decrease in β-adrenergic receptor binding in rat cerebral cortex, *Biochem. Pharmacol.*, 27, 2179, 1978.
49. **Minneman, K. P., Wolfe, B. B., and Molinoff, P. B.,** Selective changes in the density of β_1-adrenergic receptors in rat striatum following chronic drug treatment and adrenalectomy, *Brain Res.*, 252, 309, 1982.
50. **Wolfe, B. B., Harden, T. K., Sporn, J. K., and Molinoff, P. B.,** Presynaptic modulation of beta-adrenergic receptors in rat cerebral cortex after treatment with antidepressants, *J. Pharmacol. Exp. Ther.*, 207, 446, 1978.
51. **Charney, D., Menkes, D., and Heninger, G.,** Receptor sensitivity and the mechanism of action of antidepressant treatment, *Arch. Gen. Psychiatry*, 38, 1160, 1981.
52. **Sulser, F., Janowsky, A. J., Okada, F., Manier, D. H., and Mobley, P. L.,** Regulation of recognition and action function of the norepinephrine (NE) receptor-coupled adenylate cyclase system, *Neuropharmacology*, 22, 425, 1983.
53. **Vetulani, J. and Sulser, F.,** Action of various antidepressant treatments reduces reactivity of noradrenergic cyclic AMP-generating system in limbic forebrain, *Nature (London)*, 257, 495, 1975.
54. **Pandey, G. N. and Davis, J. M.,** Treatment with antidepressants and down-regulation of beta-adrenergic receptors, *Drug Dev. Res.*, 3, 393, 1983.
55. **Clements-Jewery, S.,** The development of cortical β-adrenoceptor subsensitivity in the rat by chronic treatment with trazodone, doxepin and mianserine, *Neuropharmacology*, 17, 779, 1978.
56. **Mishra, R., Janowsky, A., and Sulser, F.,** Action of mianserin and zemelidine on the norepinephrine receptor coupled adenylate cyclase system in brain: subsensitivity without reduction in beta adrenergic receptor binding, *Neuropharmacology*, 19, 883, 1980.
57. **Ferris, R. M. and Beaman, O. J.,** Bupropion: a new antidepressant drug, the mechanism of action of which is not associated with down-regulation of postsynaptic β-adrenergic, serotonergic (5-HT_2), α_2-adrenergic, imipramine and dopaminergic receptors in brain, *Neuropharmacology*, 22, 1257, 1983.

58. **Suranyi-Cardotte, B. E., Dam, T.-V., Bodnoff, S., and Quirion, R.,** Variability of chronic antidepressant treatments on beta-adrenergic receptor sites, *Prog. Neuro.-Psychopharmacol. Biol. Psychiatry,* 9, 765, 1985.
59. **Hauser, K., Olpe, H. R., and Jones, R. S.,** Trimipramine, a tricyclic antidepressant exerting atypical actions on the central noradrenergic system, *Eur. J. Pharmacol.,* 111, 23, 1985.
60. **Menkes, D. C., Kehne, J. H., Gallager, D. W., Aghajanian, G. K., and Davis, M.,** Functional supersensitivity of CNS α_1-adrenoceptors following chronic antidepressant treatment, *Life Sci.,* 33, 181, 1981.
61. **Asakura, M., Tsukamoto, T., and Hasegawa, K.,** Modulation of rat brain α_2-and β-adrenergic receptor sensitivity following long term treatment with antidepressants, *Brain Res.,* 235, 192, 1982.
62. **Crews, F. T. and Smith, S. H.,** Presynaptic α-receptor subsensitivity after long-term antidepressant treatment, *Science,* 202, 322, 1978.
63. **Johnson, R. W., Reisine, T., Spornitz, S., and Yamamura, H. I.,** Effects of desipramine and yohimbine on α_2- and β-adrenoreceptor sensitivity, *Eur. J. Pharmacol.,* 67, 123, 1980.
64. **Spyraki, C. and Fibiger, H. C.,** Functional evidence for subsensitivity of noradrenergic alpha$_2$ receptors after chronic desipramine treatment, *Life Sci.,* 27, 1863, 1980.
65. **Jones, R. S. G. and Olpe, H. R.,** Cortical neuron responsiveness to substance P is enhanced following chronic but not acute administration of desipramine, *Eur. J. Pharmacol.,* 67, 123, 1983.
66. **Spyraki, C. and Fibiger, H. C.,** Behavioral evidence for supersensitivity of postsynaptic dopamine receptors in the mesolimbic system after chronic administration of desipramine, *Eur. J. Pharmacol.,* 74, 195, 1981.
67. **Suranyi-Cadotte, B. E., Dam, T.-V., and Quirion, R.,** Antidepressant-anxiolytic interaction: decreased density of benzodiazepine receptors in rat brain following chronic administration of antidepressants, *Eur. J. Pharmacol.,* 106, 673, 1984.
68. **Peroutka, S. J. and Snyder, S. H.,** Chronic antidepressant treatment decreases spriop-eridol-labeled serotonin receptor binding, *Science,* 210, 88, 1980.
69. **Bergstrom, D. A. and Kellar, K. J.,** Adrenergic and serotonergic receptor binding in rat brain after chronic desmethylimipramine treatment, *J. Pharmacol. Exp. Ther.,* 209, 256, 1979.
70. **Janowsky, A., Okada, R., Manier, D. H., Applegate, C. D., Sulser, F., and Steranka, L. R.,** Role of serotonergic input in the regulation of the β-adrenergic receptor-coupled adenylate cyclase system, *Science,* 218, 900, 1982.
71. **De Montigny, C. and Aghajanian, G. K.,** Tricyclic antidepressants: long-term treatment increases responsivity of rat brain fore-brain neurons to serotonin, *Science,* 202, 1303, 1978.
72. **Segal, M. and Bloom, F. E.,** The action of norepinephrine in the rat hippocampus. III Hippocampal cellular responses to locus coeruleus stimulation in the awake rat, *Brain Res.,* 107, 499, 1976.
73. **Moises, H. C., Waterhouse, B. D., and Woodward, B. J.,** Locus coeruleus stimulation potentiates Purkinje cell responses to afferent input: the climbing fiber system, *Brain Res.,* 222, 43, 1981.
74. **Segal, M. and Bloom, F. E.,** The action of norepinephrine in the rat hippocampus. I. Iontophoretic studies, *Brain Res.,* 72, 79, 1974.
75. **Waterhouse, B. D., Moises, H. C., and Woodward, D. J.,** Noradrenergic modulation of somatosensory cortical neuronal responses to iontophoretically applied putative neurotransmitters, *Exp. Neurol.,* 69, 30, 1980.
76. **Waterhouse, B. D., Moises, H. C., Yeh, H. H., and Woodward, D. J.,** Norepinephrine enhancement of inhibitory synaptic mechanisms in the cerebellum and cerebral cortex: mediation by β-adrenergic receptors, *J. Pharmacol. Exp. Ther.,* 221, 495, 1982.

77. **Coyle, J. T.,** Biochemical aspects of neurotransmission in the developing brain, *Int. Rev. Neurobiol.,* 20, 65, 1977.
78. **Coyle, J. T. and Axelrod, J.,** Development of the uptake and storage of L-[^{3}H] norepinephrine in the rat brain, *J. Neurochem.,* 18, 2061, 1971.
79. **Coyle, J. T. and Axelrod, J.,** Dopamine-β-hydroxylase in the rat brain: developmental characteristics, *J. Neurochem.,* 19, 449, 1972.
80. **Coyle, J. T. and Axelrod, J.,** Tyrosine hydroxylase in rat brain: developmental characteristics, *J. Neurochem.,* 19, 1117, 1972.
81. **Dreyfus, C. F., Markey, K. A., Goldstein, M., and Black, I. B.,** Development of catecholaminergic phenotypic characters in the mouse locus coeruleus in vivo and in culture, *Dev. Biol.,* 97, 48, 1983.
82. **Dreyfus, C. F.,** Differential effects of desmethylimipramine (DMI) on the brain nucleus locus coeruleus (l.c.) cultured at two developmental periods, *Soc. Neurosci. (Abstr.),* 17, 1435, 1991.
83. **Bruinink, A. and Lichtensteiger, W.,** β-adrenergic binding sites in fetal rat brain, *J. Neurochem.,* 43, 578, 1984.
84. **Schlumpf, M., Shoemaker, W. J., and Bloom, F. E.,** Innervation of embryonic rat cerebral cortex by catecholamine-containing fibers, *J. Comp. Neurol.,* 192, 361, 1980.
85. **Loy, R. and Moore, R. Y.,** Ontogeny of the noradrenergic innervation of the rat hippocampal formation, *Anat. Embryol.,* 157, 243, 1979.
86. **Lauder, J. M. and Bloom, F. E.,** Ontogeny of monoamine neurons in the locus coeruleus, raphe nucleus and substantia nigra of the rat. I. Cell differentiation, *J. Comp. Neurol.,* 155, 469, 1974.
87. **Ebersole, P., Parnavelas, J. G., and Blue, M. E.,** Development of the visual cortex of rats treated with 6-hydroxydopamine in early life, *Anat. Embryol.,* 162, 489, 1981.
88. **Wendlandt, S., Crow, T. J., and Stirling, R. V.,** The involvement of the noradrenergic system arising from the locus coeruleus in the postnatal development of the cortex in the rat brain, *Brain Res.,* 125, 1, 1977.
89. **Lidov, H. G. W. and Molliver, M. E.,** The structure of cerebral cortex in the rat following prenatal administration of 6-hydroxydopamine, *Develop. Brain Res.,* 3, 81, 1982.
90. **Kasamatsu, T. and Pettigrew, J. D.,** Depletion of brain catecholamines: failure of ocular dominance shift after monocular occlusion in kittens, *Science,* 194, 206, 1976.
91. **Kasamatsu, T., Pettigrew, J. D., and Ary, M.,** Restoration of visual cortical plasticity by local microperfusions of norepinephrine, *J. Comp. Neurol.,* 185, 163, 1979.
92. **Parnevelos, J. G. and Blue, M. E.,** The role of the noradrenergic system on the formation of synapses in the visual cortex of the rat, *Develop. Brain Res.,* 3, 140, 1982.
93. **Felten, D. L., Hallman, H., and Jonsson, G.,** Evidence for a neurotrophic role of noradrenaline neurons in the postnatal development of rat cerebral cortex, *J. Neurocytol.,* 11, 119, 1982.
94. **Maeda, T., Tohyama, M., and Shimizu, N.,** Modification of postnatal development of neocortex in rat brain with experimental deprevation of locus coeruleus, *Brain Res.,* 70, 515, 1974.
95. **Simpkins, J. W., Field, F. P., Torosian, G., and Soltis, E. E.,** Effects of prenatal exposure to tricyclic antidepressants on adrenergic responses in progeny, *Develop. Pharmacol. Ther.,* 8, 17, 1985.
96. **Jason, K. M., Cooper, T. B., and Friedman, E.,** Prenatal exposure to imipramine alters early behavioral development and beta-adrenergic receptors in rats, *J. Pharmacol. Exp. Ther.,* 217, 461, 1981.
97. **Ali, S. F., Buelke-Sam, J., Newport, G. D., and Slikker, W., Jr.,** Early neurobehavioral and neurochemical alterations in rats prenatally exposed to imipramine, *Neurotoxicology,* 7, 365, 1986.
98. **Singer, G. and Coyle, I. R.,** The effect of imipramine administered before and during pregnancy on litter size in the rat, *Psychopharmacologia,* 32, 337, 1973.

99. **Broitman, S. T. and Donoso, A. O.,** Effects of chronic imipramine and chlorimipramine oral administration on maternal behavior and litter development, *Psychopharmacology,* 56, 93, 1978.
100. **Coyle, I. R. and Singer, G.,** The interactive effects of pre-natal imipramine exposure and post-natal rearing conditions on behavior and histology, *Psychopharmacologia,* 44, 253, 1975.
101. **Coyle, I. R. and Singer, G.,** The interaction of post-weaning housing conditions and prenatal drug effects on behavior, *Psychopharmacologia,* 41, 237, 1975.
102. **Tonge, S. R.,** Catecholamine concentrations in discrete areas of the rat brain after the pre- and neonatal administration of phenylcyclidine and imipramine, *J. Pharm. Pharmacol.,* 25, 164, 1973.
103. **Racagni, G., Macchetti, I., Calderini, G., Battistella, A., and Brunello, N.,** Temproal sequence of changes in central noradrenergic system of rat after prolonged antidepressant treatment: receptor desensitization and neurotransmitter interactions, *Neuropharmacology,* 22, 415, 1983.
104. **Del Rio, J., Montero, D., and de Ceballos, M. L.,** Long-lasting changes after perinatal exposure to antidepressants, *Prog. Brain Res.,* 73, 173, 1988.
105. **Raisman, R., Briley, M. S., and Langer, S. Z.,** Specific tricyclic antidepressant binding sites in rat brain characterized by high-affinity ^{3}H-imipramine binding, *Eur. J. Pharmacol.,* 61, 373, 1980.
106. **Kinnier, W. J., Chuang, D. M., and Costa, E.,** Down regulation of dihydroalprenolol and imipramine binding sites in brain of rats repeatedly treated with imipramine, *Eur. J. Pharmacol.,* 67, 289, 1980.
107. **Montero, D., de Ceballos, M. L., and Del Rio, J.,** Down regulation of ^{3}H-imipramine binding sites in rat cerebral cortex after prenatal exposure to antidepressants, *Life Sci.,* 46, 1619, 1990.
108. **Dreyfus, C. F. and Black, I. B.,** in *Methods in Neurosciences,* Vol. 2, Corn, P. M., Ed., Academic Press, San Diego, 1990, 3.
109. **Kimelberg, H. K. and Pelton, E. W.,** High-affinity uptake of [^{3}H]norepinephrine by, primary astrocyte cultures and its inhibition by tricyclic antidepressants, *J. Neurochem.,* 40, 1265, 1983.
110. **Sieber-Blum, M.,** Inhibition of the adrenergic phenotype in cultured neural crest cells by norepinephrine uptake inhibitors, *Develop. Biol.,* 136, 372, 1989.
111. **Taber-Pierce, E.,** Time of origin of neurons in the brain stem of the mouse, *Prog. Brain Res.,* 40, 53, 1973.
112. **Rothman, T. P., Specht, L. A., Joh, T. H., Teitelman, G., Gershon, M. D., and Reis, D. J.,** Catecholamine biosynthetic enzymes are expressed in replicating cells of the peripheral but not central nervous system, *Proc. Natl. Acad. Sci. U.S.A.,* 77, 621, 1980.
113. **Lauder, J. M. and Bloom, F. E.,** Ontogeny of monoamine neurons in the locus coeruleus, raphe nuclei and substantia nigra of the rat. II. Synaptogenesis, *J. Comp. Neurol.,* 163, 469, 1974.

113a. **Dreyfus, C.,** personal observation.

114. **Golden, G. S.,** Prenatal development of the biogenic amine systems of the mouse brain, *Develop. Biol.,* 33, 300, 1973.
115. **Seiger, A. and Olson, L.,** Late prenatal ontogeny of central monoamine neurons in the rat: fluorescence histochemical observations, *Z. Anat. Entwicklungsgesch.,* 140, 281, 1973.
116. **Levitt, P. and Moore, R. Y.,** Development of the noradrenergic innervation of neocortex, *Brain Res.,* 162, 243, 1979.
117. **Sievers, J., Lolova, I., Jenner, S., Klemm, H. P., and Sievers, H.,** Morphological and biochemical studies on the ontogenesis of the nucleus locus coeruleus, *Bibl. Anat.,* 19, 53, 1981.

118. **Specht, L. A., Pickel, V. M., Joh, T. H., and Reis, D. J.,** Light microscopic immunocytochemical localization of tyrosine hydroxylase in prenatal rat brain. I. Early ontogeny, *J. Comp. Neurol.,* 199, 233, 1981.
119. **Bondareff, W., Mountjoy, C. Q., and Roth, M.,** Selective loss of neurons of origin of adrenergic projection to cerebral cortex (nucleus locus coeruleus) in senile dementia, *Lancet,* 1, 783, 1981.
120. **Perry, E. K., Tomlinson, B. E., Blessed, G., Perry, R. H., Cross, A. J., and Crow, T. T.,** Neuropathological and biochemical observations on the noradrenergic system in Alzheimer's disease, *J. Neurol. Sci.,* 51, 279, 1981.
121. **Yates, C. M., Ritchie, M., Simpson, J., Maloney, A. F. J., and Gordon, A.,** Norepinephrine in Alzheimer type dementia and Down's syndrome, *Lancet,* 2, 39, 1981.
122. **Rachelefsky, G. S., Flynt, J. W., Ebbin, A. J., and Wilson, M.,** Possible teratogenicity of tricyclic antidepressants, *Lancet,* 1, 838, 1972.
123. **Scanlon, F. J.,** Use of antidepressant drugs during the first trimester, *Med. J. Aust.,* 2, 1077, 1969.
124. **Mathers, D. A. and Barker, J. L.,** Spontaneous voltage and current fluctuations in tissue cultured mouse and dorsal root ganglion cells, *Brain Res.,* 293, 35, 1984.
125. **Charness, M. E., Gordon, A. S., and Diamond, I.,** Ethanol modulation of opiate receptors in cultured neural cells, *Science,* 222, 1246, 1983.
126. **Dow, K. E. and Riopelle, R. J.,** Ethanol neurotoxicity: effects on neurite formation and neurotrophic factor production *in vitro, Science,* 228, 591, 1985.

Chapter 15

ENKEPHALIN-DEGRADING ENZYMES AS MODELS TO PREPARE ANTINOCICEPTIVE OR OTHER PHARMACOLOGICAL AGENTS

Neville Marks and Martin J. Berg

TABLE OF CONTENTS

I. INTRODUCTION

This chapter provides an overview on the metabolism of smaller opioid peptides with emphasis on the use of degrading enzymes as model to design useful antinociceptive or other pharmacological agents. The use of enzymes as models has led to the successful introduction of orally active drugs for treatment of hypertension (captopril and enalapril for suppression of ACE, and lovastin for suppression of hydroxymethylglutaryl-CoA reductase), MAO inhibitors for treatment of affective disorders, and accounts for the use of AZT (reverse transcriptase) and penicillins (transpeptidases) as antiviral or antibacterial agents.[1] Based on these successes, the following criteria appear to apply to enkephalins and antinociception:

1. Identification of a specific biological pathway linked to opioid peptide metabolism, and clarification of the role of enkephalins in peptidergic pathways related to the response to noxious stimuli (pain).
2. Selection of a unique degrading enzyme. Although this has been hampered by the lack of specificity of enkephalin-degrading enzymes and their multiplicity (Figure 1), some progress has been made on their localization and kinetics (K_m, K_{cat}).
3. Finding inhibitors that are stable in tissues or in blood, are long acting, and capable of penetrating to target sites in the CNS, and active orally; have acceptable toxicological profiles and dose-response curves; and not be subject to abuse (tolerance, physical dependence).

Although no inhibitor satisfies all criteria, several show promise as naloxone-reversible analgesic agents, especially when coadministered parentally or centrally with enkephalins or other opioids.[2-12] It can be noted that enkephalins themselves are proteolytic products formed by cleavage of three precursors at sites adjacent to the active domains[13-17] (Figure 2A,B). The pathways postulated for processing are similar to those proposed for other prohormones, such as insulin, parathyroid hormone, and VIP (vasoactive intestinal peptide). These involve the coordinate action of endo- and exopeptidases, accompanied by posttranslational modifications that are required for targeting and sorting (glycosylation, phosphorylation, sulfation, *N*-acetylation, C-terminal amide formation, etc.)[13,15,16,18-20] Processing enzymes are themselves potential targets for drug design, since they influence levels of tissue peptides or amounts secreted in response to functional changes. As a result of such processing, tissues contain a spectrum of opioid peptide intermediates, including forms with *N*-acetyl- or des-Tyr-, that have varying opioid or nonopioid properties.[21]

Most studies have emphasized the use of membrane-bound enzymes as models to design inhibitors based on the premise that these enzymes are localized in the vicinity of opioid receptors. These include ‘‘ectoenzymes’’

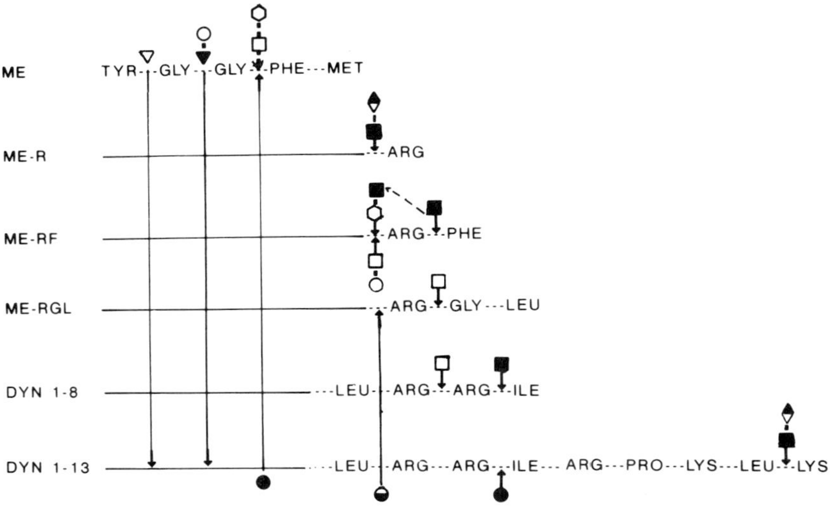

FIGURE 1. Multiplicity of brain enzymes acting on shorter enkephalins. **Site-1** — ▽ aminopeptidases (Table 1); **Site-2** — ○ diaminopeptidase aminopeptidase (DAP-I or DAP-III), ▼ cathepsin L (Table 2); **Site-3** — □ angiotensin converting enzyme, ● metalloendopeptidase 24.11, ⬡ cathepsin B (Table 4); **Site-5** — ■ carboxypeptidase A (acidic, neutral), ⧫ carboxypeptidase E, ◒ metalloendopeptidase 24.15 (Table 5). Note absence of enzymes acting on Site-4. Some of the enzymes act also at other bonds of C-terminally extended enkephalins. Metalloendopeptidase 24.15 acts only on $Dyn_{1\text{-}8}$ or $Dyn_{1\text{-}13}$ for the series shown. The order of enzymes listed is identical to those described in more detail in Tables 1, 2, 4, and 5.

having membrane-spanning domains plus extracellular ones for catalysis, or others that are loosely anchored or associated with membranes. The oft-cited "soluble" and "membrane-bound" enzymes are operational terms dependent on methods chosen for extraction. This can be clarified through sequencing and examination of hydrophobicity to identify the putative membrane-spanning or other domains.

For purposes of description, candidate enzymes are grouped by sites of cleavage using smaller enkephalins, as shown in Figure 1. Isolation of enzymes is essential for studies on their kinetics, specificity, and mechanisms of catalysis, or for purposes of eliciting specific antisera for immunolocalization. Since most are metalloenzymes, inhibitors can be designed using general principles formulated earlier for angiotensin-converting enzyme (ACE), carboxypeptidase A, and prokaryotic thermolysin.[2,22] Lysosomal enzymes also should be considered since these may play supplementary roles in processing (exocytosis) or following internalization of peptide or receptor-peptide complexes (endocytosis).[19,20,23]

Some comments apply to the nomenclature used for naming enzymes metabolizing the shorter enkephalins, since this has led to considerable confusion. Although the terms "enkephalinase A or B" have appeared, these

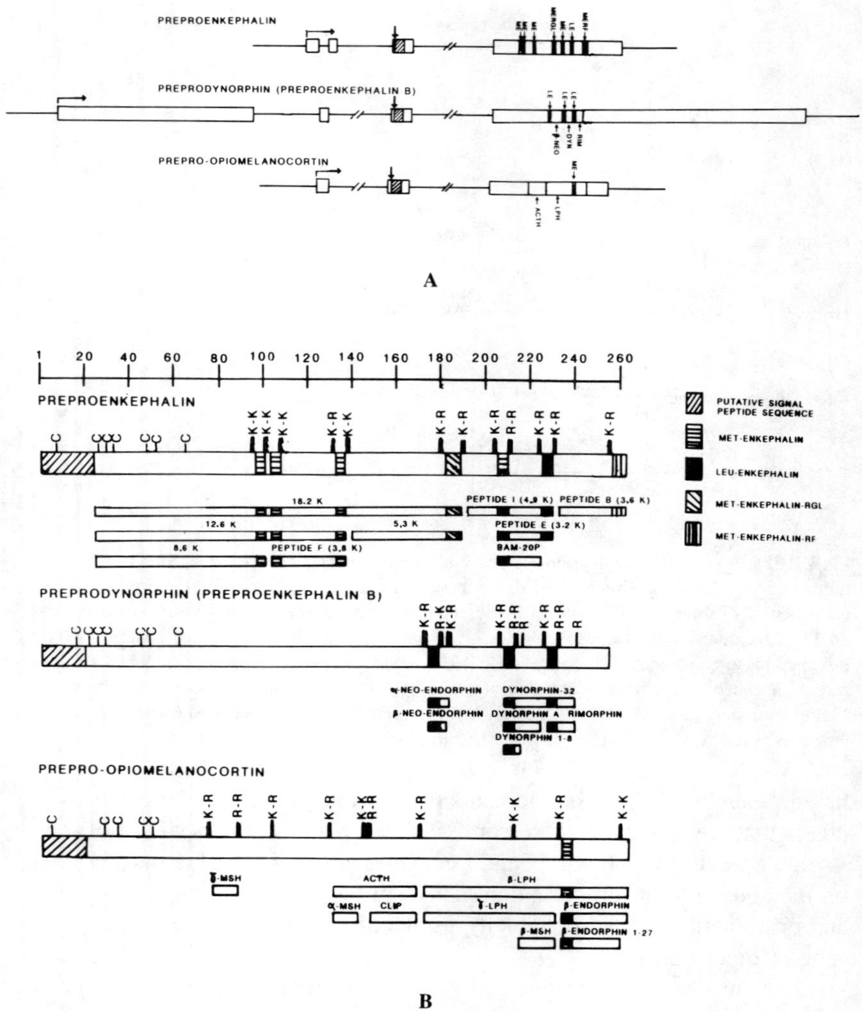

FIGURE 2. (A) Structural comparison of genes for human pro-opiomelanocortin (POMC), proenkephalin, and prodynorphin. Lines, introns; blocks, exons. Horizontal arrows represent the 5′-terminus for transcription; vertical arrows indicate the start point for translation of the signal peptide. For other details see (B). (B) Schematic representation of translated opioid precursors showing products formed after processing and found in tissues. The heavy vertical lines above the precursors indicate sites of processing at either single or pairs of basic residues.[13] The boxes represent the signal sequence or copies of selected opioid peptides cited in the text as models to study breakdown (see key, upper left-hand side). The "C" on the N-termini represent position of cysteine residues. Adapted from 14.

enzymes are nonspecific, or are identical to known exo- or endopeptidases. Since degradation is not accompanied in all cases by inactivation, the terminology must take into account the biological properties of the substrate and product. This is illustrated by formation of a delta agonist (Leu-enkephalin) by inactivation of the kappa agonists Dyn_{1-8} or Dyn_{1-13}.[17,24] Dependent on the receptor of interest, this can be viewed as either inactivation or conversion. Receptor subtypes were proposed first by Martin et al.[25] in studies on pharmacological profiles of morphine (μ), ketocyclazocine (κ) and *N*-allylnormetazocine (SQ 10,047) (σ) in chronic spinal dogs. Following the discovery of opioid peptides, a new receptor (δ) was identified using their effects on vas deferens contraction compared to guinea pig ileum.[26] Although the functional significance of receptor subtypes is unclear, they appear to be linked to diverse effects associated with pain, hyperalgesia, eating, locomotion, the cardiovascular system, behavior, the immune system, or pathways mediated by dopamine or catecholamines, among others.[27-30] The discovery of opioid peptides stimulated research on neuropeptidases,[31-34] and led to some notable findings of general application to polypeptide and protein turnover, including:

1. A given precursor can yield more than one species of active peptide dependent on the anatomical sites of processing. Pro-opiomelanocortin (POMC) can yield ACTH, MSH, lipotropin, and endorphin, with the ratio of products varying in pituitary regions.[16] Similarly, prodynorphin yields a large number of intermediates with kappa or delta receptor properties[17] (Figure 2A,B).
2. A given gene contains multiple copies of opioids such as Met- or Leu-enkephalin (Figure 2A,B).
3. Opioid peptides were instrumental for discovery of delta receptors.
4. Early recognition that peptides are rapidly degraded led to synthesis of biostable analogs, some of which, unexpectedly, had altered agonist properties; i.e., DAMGO (D-Ala^2, MePhe^4, Gly-ol^5 enkephalin) is a stable mu agonist.[35,36]
5. Bioassays are required to differentiate opioid from nonopioid effects using naloxone reversibility for evaluation.[26] Thus, the response to pain is multifaceted and requires a battery of tests, such as the use of tail flick or hot-plate latencies, and vocalization.[37,38]
6. Inhibitors of enkephalin degradation are useful as probes to examine enzyme catalysis, and to map peptidergic pathways in spinal cord or brain.

In the absence of relevant bioassays, opioid peptides were overlooked in studies on tissue homogenates or merely were regarded as breakdown products of little consequence.[39,40] Opioid peptides constitute some of the most powerful pharmacological agents described, with properties akin to neurotransmitters or neuromodulators. The discovery of "opiate receptors" binding alkaloids was made prior to that of endogenous opioid peptides now believed to be the native ligands. Although such receptors have not been sequenced, they appear to be members of a family of G-linked proteins.[41-43] The following description applies to enzymes acting on shorter enkephalins which, for the most part, are preferential delta or kappa agonists.

II. SITE-1 CLEAVAGE (AMINOPEPTIDASES)

Several aminopeptidases of brain inactive opioid peptides at site-1 with release of Tyr and Gly-Gly-Phe-Leu (Met) as products.[33] Tyr is essential for expression of typical naloxone-reversible properties of opioid peptides in bioassay.[36] Cleavage may result in formation of des-Tyr opioids, such as des-Tyr-γ-endorphin and -dynorphin, having nonopioid properties of interest in motor dysfunction or as drugs for use in altering behavior.[21,44,45] In the case of shorter enkephalins, there is no rational classification for site-1 enzymes. For the most part, these are nonspecific but can be assigned to one or more categories acting on synthetic substrates containing acid, neutral, or basic amino acids on the N-termini (Table 1). Aminopeptidases acting on peptides with N-terminal acidic amino residues generally are not regarded as relevant. Others considered more relevant are enzymes referred to in the literature as aminopeptidase or arylamidase B, aminopeptidase M or N (AP-M, -N), leucine aminopeptidase (LAP), puromycin-sensitive aminopeptidase (PSA) I or II, or those referred to as MI and MII[33,46-50] (Table 1). Although such enzymes are nonspecific, PSA (''aminoenkephalinase'') or MII show a higher affinity for pentapeptide enkephalins or Tyr-2-naphthylamide as substrates.[47,48,51,52] Such enzymes show a higher sensitivity to puromycin as an inhibitor (Table 1). While aminopeptidases are diffusely localized in brain,[52,53] recent studies on PSAs indicate different localization for the I and II forms in glia and neurons, and that they are probably derived from different mRNAs.[54] Membrane-associated and soluble aminopeptidase (MII), however, appear to be structurally identical based on N-termini (P-E-K-R-P-F-E-R-L-P-T-E-V-P-I-N-Y-), peptide maps, and immunoreactivity.[47] In addition, brain, like other tissues or bacteria, contains LAP (leucine aminopeptidase, E.C. 3.4.11.1), a polymeric enzyme M_r 350 kDa.[33] Although 80% of brain aminopeptidases appear to be cytosolic, SPM-associated enzymes account for a more rapid degradation of enkephalin than other membrane-bound enzymes (see Table 3; Figure 3).[55] Since rate of degradation decreases with peptide length, this may account for higher potencies of larger peptides such as Dyn_{1-13} or Dyn_{1-17} (Table 3).[49] The higher potency of enkephalins containing D-amino acids at positions 2 (-Ala, -Met) and/or 5 (-Pro, -Met) probably is the result of their biostability.[36,56-60]

Aminopeptidase M (E.C. 3.4.11.2, Table 1) differs from other site-1 enzymes since it occurs largely on cerebral microvessels.[61-63] It has been cloned and shown to be homologous in different tissues and to human myeloid

TABLE 1
Site-1 Cleavage (Tyr[1]—Gly[2]-Gly-Phe-Leu)

E.C.	3.4.11.6	3.4.11.	3.4.11.2
M_r (kDa)	60	100	140
Specificity	X-2NA, X-NMec	X = Arg or Tyr	X = Arg, Leu, Ala, etc.
	Arylamides (X-NA), coumarins (X-NMec), X = Arg, Lys, Leu, Met, etc.	Enkephalin	
Substrates	Peptides R^1-R^2—-R^n. R^1 is unprotected and not an acidic amino acid; R^2 is not Pro		
Inhibitors	Chelating reagents (EDTA, 1, 10-*o*-phenanthroline); derivatives of amino acids or peptides with –SH (S^-), phosphorous (O^-), chlormethyl ketones, or aminoaldehydes (O^-); leuthiol, phethiol, or its thioester carbaphethiol, kelatorphan; microbiol agents (bestatin, amastatin, puromycin, bacitracin, actinonin)		
Puromycin sensitivity	K_i = 100—1000 μM	K_i = 1 μM	K_i = 100—1000 μM
Comments (brain)	Cytosolic	Cytosolic or membrane	Vasculature
Names	MI, arylamidase, arginyl or AP-B	(Aminoenkephalinase), cytosolic (PSAI or II), or membrane-associated (MII)	AP-M or -N, CD13 (gp150)

Note: NA, naphthylamides; -NMec, *N*-methylcoumarin; PSA, puromycin-sensitive enzymes; AP, aminopeptidase; CD13, human myeloid plasma membrane glycoprotein homologous to AP-M. See text for a discussion of terminology. For structure of inhibitors see text or Refs. 7, 22, 33, 47-49, 54, 68—74, 88, 90—92. Some of the inhibitors shown to act also on sites-2 and -3 (see Figure 1, and Tables 2 and 4). Bestatin or phethiol interaction with the putative metallocenter of AP are depicted in Figure 4). The table omits leucine aminopeptidase (LAP, E.C. 3.4.11.1) since no detailed studies are available for this brain AP.[33]

plasma membrane glycoprotein CD13 (gp150). It consists of single polypeptide with 967 residues, including a 24-amino acid hydrophobic sequence at the N-terminus that may serve as a signal peptide and a stable membrane-spanning segment.[64-67]

A. INHIBITORS AND PHARMACOLOGY

Microbially derived agents inhibiting site-1 cleaving enzymes include puromycin, bestatin, amastatin, bacitracin, and actinonin, among others (Table 1). Several of these, at higher concentrations, also act on site-2 or -3 enzymes, which may amplify their pharmacological properties. For example, actinonin acts on site-1 (IC_{50}, 0.39 μM), site-2 (1.1), and site-3 (5.6), is analgesic in rats ip, and potentiates the analgesia induced by ic, administered

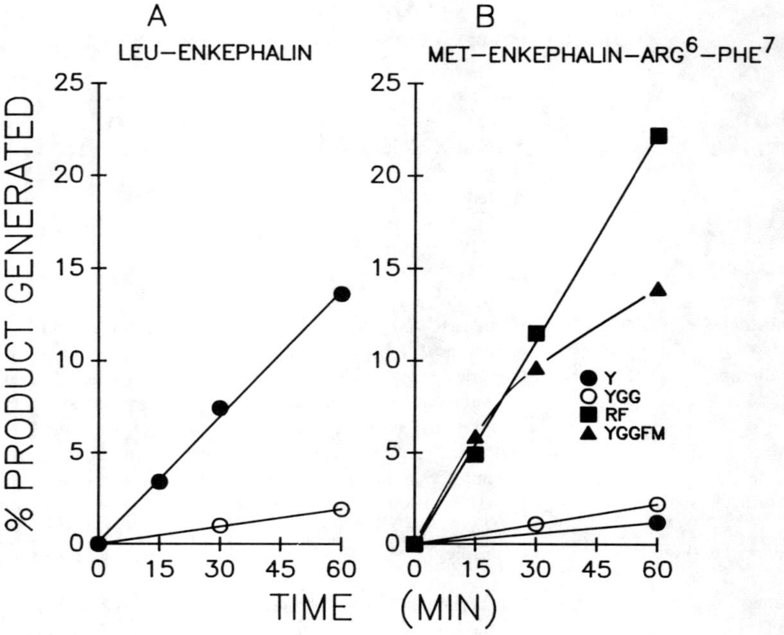

FIGURE 3. Comparison of Leu-enkephalin (A) and Met-enkephalin-Arg-Phe (B) breakdown by rat striatal membranes with time of incubation. Note rapid release of Tyr compared to Tyr-Gly-Gly for the pentapeptide indicating cleavage by SPM-associated aminopeptidase(s). For the heptapeptide, breakdown was accompanied by release of Met-enkephalin and C-terminal dipeptide blocked by captopril indicating action by ACE.[55] Note lower release of Tyr for the heptapeptide indicating lower activity of aminopeptidase toward C-terminally extended enkephalins.[49]

Met-enkephalin.[68] In rat plasma or serum, actinonin-sensitive enzymes, along with others, appear to account for enkephalin degradation.[59,60,69]

Puromycin and bacitracin, among the earliest agents shown to be ''analgesic,'' are also characterized by their effects on more than one degrading enzyme, or, in the case of puromycin, on protein synthesis. There have been efforts to modify site-1 inhibitors to make them more selective agents: replacement of methoxy-Phe by Tyr, Leu, β-alanine, methoxy-Tyr-Gly, or the use of aminonucleoside itself reduced potency toward purified brain arylamidase.[33] For bestatin, the Lys-substituted analogs were shown to inhibit soluble brain aminopeptidases more potently than membrane-associated forms.[70-74] Bestatin contains an unusual plant-derived α-hydroxy-β-amino acid, and it is thought that the hydroxy group permits formation of a five-member chelate with the enzyme zinc in conjunction with other amino or carbonyl groups.[22]

There are a large number of studies on antinociceptive properties of bestatin, puromycin, and bacitracin alone or in combination with opioid peptides.[2,70-75] Systemically, inhibitor and opioid peptides are given in doses of

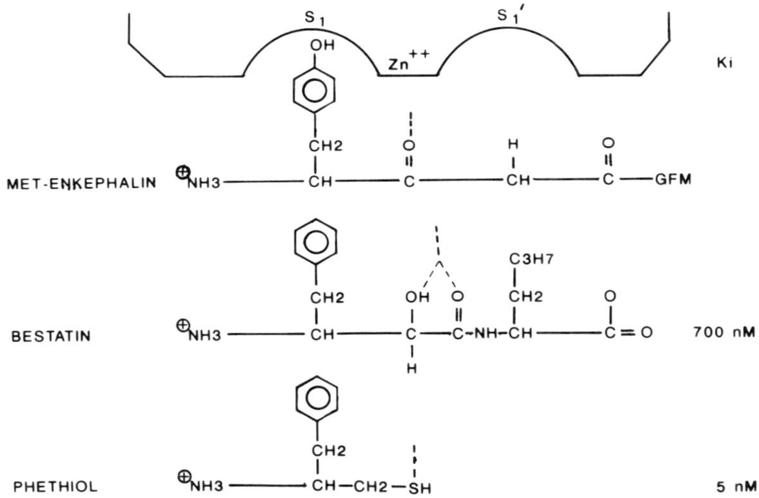

FIGURE 4. Binding of Met-enkephalin to putative sites of a metalloaminopeptidase containing Zn^{2+} at the active. Included for comparison are the binding of bestatin and the sulfhydryl derivative of phenylalanine (Phethiol) with K_i's, shown on the left-hand side. Adapted from References 7 and 22.

5 to 10 mg/kg as contrasted to 10^3-fold lower concentrations when given ic, icv, or directly into the spinal cord. In the case of bestatin, inhibitor administered ic enhances the antinociception induced by dynorphin B, or by β-endorphin;[11,76] however, it does not enhance nonopioid effects of dynorphin or des-Tyr-dynorphin on motor dysfunction.[11] Injected intraspinally, it increases recovery of iR-Met-enkephalin from perfusates.[76] Bestatin also affects metabolism of other biologically active peptides such as angiotensin II or III, and their action on cortical neurons, ventricular arrhythmias, and their pressor and dipsogenic responses.[77-82] Bestatin nonspecifically affects protein turnover by suppressing the final common pathway leading to the release of amino acids to cellular pools.[83,84]

Recognizing the need for selectivity, a very large number of site-1 inhibitors have been synthesized including amino acids or peptides substituted with –SH such as leucinethiol, phethiol, or kelatorphan (Figures 4, 5) or derivatives formed with boronic acid, chlormethyl ketones, aminoaldehyde, or phosphoryl groups.[22,85-92] The S^- or O^- promotes interaction with the enzyme metallo- subsite containing Zn^{2+}. Many of these agents are antinociceptive when administered ic or ip.[73,76] To improve penetrability and stability at sites of action, the use of a prodrug has been advocated. The thioester form phethiol (carbaphethiol) is metabolized by tissue esterases. When administered to rat at 10 mg/kg iv, it induced a naloxone-reversible analgesia, or alternatively, enhanced the analgesia induced by acetorphan, a site-3 inhibitor (see below and Figure 5).[7]

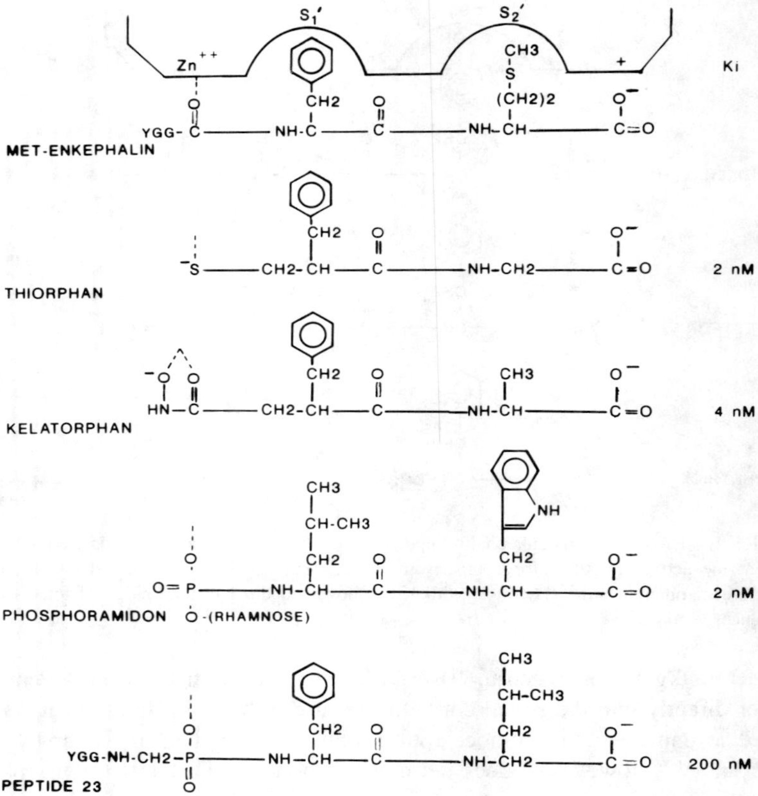

FIGURE 5. Binding of Met-enkephalin to putative enzyme sites of 24.11. Included in the figure for comparison are binding of site-specific inhibitors with K_i's, shown on the left-hand sides (see Table 4 or text for details).

III. SITE-2 CLEAVAGE

A. DIAMINOPEPTIDASES

Cleavage of Gly-Gly of opioids occurs more slowly at site-2 than at other sites, and as a result, has received less attention (Table 3). Enzymes acting on enkephalins with free α-amino groups are DAP-I (cathepsin C), a Cl^-- and –SH-dependent lysosomal enzyme active at pH 6.0, and DAP-III, a cytosolic enzyme active at pH 8.0.[33] Cathepsin C was shown to release Tyr-Gly from enkephalin, but not its *N*-acylated form.[93] Enkephalinase B (identical to DAP-III) acts on enkephalin and other peptides (angiotensin III), and is found in brain-soluble or membrane fractions.[94] Purified rat brain DAP-III is inhibited by chelating agents EDTA or 1,10-*o*-phenanthroline with resoration of activity by divalent metals such as Zn^{2+}.[94] However, it is insensitive to

TABLE 2
Relative Rates for Generation of Free Tyrosine (Aminopeptidase Cleavage) and Tyr-Gly-Gly (Endopeptidase 24.11) from Dynorphins by Washed SPM

Peptide	Tyr	Tyr-Gly-Gly	Y/YGG
YGGFL	1.0	0.09	11.1
YGGFLR	0.49	0.06	8.2
YGGFLRRI	0.23	0.01	23.0
YGGFLRRIRPKLK	0.04	0.01	4.0
YGGFLRRIRPKLKWDRQ	0.02	0.01	2.0

Note: Ratio of products formed by incubation of washed rat brain SPM with Leu-enkephalin (Dyn_{1-5}) and longer dynorphins with up to 17 residues.[49,55] Tyr release 8.2 μmol/mg of membrane protein was blocked by 100 μ*M* bestatin. Tyr-Gly-Gly release was blocked by 1 μ*M* thiorphan but not by captopril (see also Figure 3 for comparison with Met-enkephalin-Arg-Phe).

other agents acting on metalloenzymes, such as, bestatin, amastatin, captopril, or phosphoramidon.[33,94] Naturally occurring agents inhibiting DAP-III include propioxatins and actinonin.[68,95-97] Actinonin given ip or ic potentiates analgesia of Met-enkephalin given ic in mice, based on tail flick latency.[68] Propioxatins are *N*-acyl-Pro-Val peptides containing hydroxamate, which probably contributes to action at the putative metalloenzyme center,[97] and may also account for inhibition of kelatorphan toward DAP-III (see Section IV).

B. CATHEPSIN L (E.C. 3.4.22.15)

Lysosomal enzymes are of interest in terms of their roles in protein catabolism and in processes linked to exo- and endocytosis.[20,23] Cathepsin L is active at pH 5 to 6 in the degradation of enkephalins, with the release of Tyr-Gly (Table 2).[93] Based on its higher affinity for Gly-Gly of enkephalins (K_m 82.5 μ*M*), it is considered to be more relevant for inactivation than cathepsin B cleavage at site 3 (450 μ*M* (Table 4).[93,98]

Lysosomal enzymes are potently inhibited by endogenous proteins with M_r 12 to 60 kDa; they are known as cystatins and are present in the brain at high concentrations.[99] The 12-kDa cystatin is of pathological interest since mutated forms occur in cerebrovascular deposits of Alzheimer's patients with familial amyloid angiopathy.[100] It has been postulated that cystatins retard protein turnover and contribute to abnormal protein depositions. The 60-kDa forms have been identified as kininogens, yielding kinins on treatment with kallikrein, or, in the case of rat tissue, to yield T-kinin (ISRPPGFSPFR) on treatment with trypsin.[99] Kinins and kallikreins are involved in a large number of processes including the cascades linked to blood clotting and complement formation, or for peripheral responses to pain or inflammation (acute phase

TABLE 3
Site-2 Cleavage (Tyr-Gly^2—Gly^3-Phe-Leu)

Name	Diaminopeptidase-III (DAP-III)	Cathepsin L
E.C.	3.4.14	3.4.22.15
M_r (kDa)	82-86	28
pH optima	8.5-9.0	5.0-5.5
Specificity	Arg-Arg-NMec, Met-enkephalin-Arg	Z-Phe-Arg-NMec, enkephalins,[a] Dns-Y-G-G-F-$(NO)_2$P[b]
Inhibitors	Chelating agents, EDTA, propriox-atins, actinonin, kelatorphan	Leupeptin, dipeptidyl diazome-thanes, E-64, cystatins, YGGFLQVVAG[c], kininogens
Comments	Cytosolic "enkephalinase B"	Lysosomal, –SH dependent

Note: Actinonin and propioxatins are microbial peptides containing a hydroxamate group for interacting with a putative metallocenter. For structure of inhibitors and other comments see text or References 67, 68, 93—97, 223, 225. Cathepsin C (DAP-I) acting at site 2 is not included since there are no studies on enzyme purified from brain.[93]

[a] *N*-Acylated or enkephalins with 1 to 13 residues act as substrates.[93]
[b] A dansylated enkephalin-like peptide internally quenched with nitrophenyl group (hydrolysis at Gly-Gly accompanied by release of fluorescence).[223]
[c] Leu-enkephalin linked to a cystatin consensus sequence.[103]

response).[101] Structural analysis of cystatins shows that they contain a glycyl-rich "substrate binding site" and one or more consensus sequences consisting of QVVAG or QLVSG.[102] The Gly-Gly site of enkephalin vulnerable to cathepsin L, appears to be homologous to the substrate binding site since linking it directly to QVVAG converts it from a substrate to an inhibitor.[103] This approach can provide a rational basis to develop novel agents utilizing opioid peptides and their ability to bind specific receptors at target sites.

In addition to endogenous proteins, cysteine proteinases also are inhibited by leupeptin or E-64 derived from plant sources, or synthetic peptidyl diazomethanes, but these have not been examined for their analgesic potential.[93] Leupeptin acts on a variety of enzymes, including serine proteinases, and is not solely a specific inhibitor of cysteine proteinases.

IV. SITE-3 CLEAVAGE

A. METALLOENDOPEPTIDASE (E.C. 3.4.24.11)

The Gly-Phe site-3 is vulnerable to a brain metalloendopeptidase that is similar, in most respects, to that identified earlier in kidney brush border cells (Table 4).[104] The enzyme is nonspecific, degrading insulin, glucagon, bradykinin, neurotensin, substance P, ANF, endothelins, chemotactic factors, CCKs, somatostatin, and VIP, among others.[2,105-113] It preferentially cleaves bonds adjacent to hydrophobic residues, permitting the design of specific substrates and inhibitors, some of which are listed in Table 4.

TABLE 4
Site-3 Cleavage (Tyr-Gly-Gly^3-Phe^4-Leu)

Name	Peptidyl dipeptidase	Metalloendopeptidase	Cathepsin B
E.C.	3.4.15.1	3.4.24.11	3.4.22.1
pH optima	8.0	7.5	6.0—6.5
M_r (kDa)	180	96—100	28
Specificity	$-R^1-R^2---R^3COOH$ ($-R^2-$ is not Pro)	$-R^1-R^2-$(R^1, R^2 hydrophobic)	Acts on shorter opioids as a peptidyl dipeptidase
	Angiotensin I_5 substance P, dynorphin A_{1-8},dynorphin A_{1-10}	Angiotensin II, substance P, dynorphin$_{1-8}$, dynorphin$_{1-13}$, enkephalins.[a] ANF, CCK-7, CCK-8, endothelins, chemotactic peptides, kinins, neurotensin, VIP, insulin, glucagon	Met-Enk-Arg-GlyLeu, Met-Enk-Arg-Phe, dynorphin$_{1-8}$, dynorphin$_{1-13}$, Met-Enk-Arg-Phe NH_2, myelin basic pr. insulin B chain, glucagon
	Hipp-His-Leu	Gln-Ala-Ala-Phe-2NA, Succ-Ala$_2$-Phe-NMec	Z-Arg-Arg-NMec, Z-Lys-Arg-Arg-NMec
Inhibitors	EDTA, 1,10-*o*-phenanthroline; snake venom, nonapeptide, captopril, MK-421, MK-521	EDTA, 1,10-*o*-phenanthroline	Leupeptin, E-64, cystatins, kininogens, Leu-enk-QVVAG[c]
		Thiorphan, acetorphan, phelorphan, synorphan *(S)*,[b] rimorphan *(R)*,[b] azidothiorphan, phosphoramidon, phosphonamides (#23), kelatorphan, actinonin, D-Phe-AHPA, cFP-AAF-pAB,[d] cFP-AAY-pAB, SCH-32,615, SCH-34,826,[e] SQ 28,603, UK-69,578, UK-70,300	
Comments	Ectoenzyme, Cl^- dependent	Ectoenzyme	Lysosomal, -SH Dependent
	Contains Zn^{2+}	Contains Zn^{2+} Val, Arg, His at active center	His, Cys at active center
K_m Leu-enk (Site-3)	550 μM	100 μM	470 μM
Comments	Inhibitors are antihypertensive agents	Inhibitors are antinociceptive, or enhance coadministered opioid peptides	

TABLE 4 (continued)
Site-3 Cleavage (Tyr-Gly-Gly^3-Phe^4-Leu)

Name	Peptidyl dipeptidase	Metalloendopeptidase	Cathepsin B
Other names	Kininase II, angiotensin-converting enzyme (ACE), dipeptidyl carboxypeptidase	Neutral endopeptidase (NEP), enkephalinase, enkephalinase A, CALLA	

Note: The interaction of thiorphan, kelatorphan, phosphoramidon, and the phosphonamide (compound 23) with the putative metallocenter of 24.11 are depicted in Figure 5. For structure of other inhibitors see text or References 2, 69, 118, 137-147, 150, 151, 155, 156, 161, 194, 225—227. ACE and cathepsin B preferentially release C-terminal dipeptides except for substance P (endopeptidase cleavage). Several of the inhibitors at higher concentrations also suppress site-1 or -2 metalloenzymes (Tables 1 and 2) although SCH 32,615 and some newer kelatorphan derivatives appear to be more specific.[226,227] The abbreviation CALLA denotes the ''common acute lymphocytic leukemic antigen'' (CD10) that is homologous to metalloendopeptidase 24.11.[116]

[a] Cleaves all enkephalins or those with C-terminal extensions at the Gly^3-Phe^4 bonds.
[b] *S* and *R* isomers of acetorphan (*N*-acetylated and esterified thiorphan).
[c] Leu-enkephalin linked to the consensus sequence of cystatins.
[d] *N*-carboxylalkyl inhibitors: *N*-[19*(R-S)*-carboxyl-3-phenylpropyl]-Ala-Ala-Phe-*p*-aminobenzate (Y = tyrosine for the second example). See Table 5 for comparison with inhibitor suppressing the phosphoramidon-insensitive 24.15.
[e] SCH 34,826 is an esterified prodrug of SCH 34,615 a 2,4-dibenzylglutaric acid derivative {*N*-[L-(-1-carboxy-2-phenyl)ethyl]-L-phenylalanyl-β-alanine}. Orally (1 mg/kg po) potentiates effects of injection D-Ala^2-Met^5-enkephalinamide.[225,226]

Metalloendopeptidase 24.11 is an integral membrane glycoprotein requiring detergents or other treatments for extraction, and can be purified by lentil lectin, or other affinity procedures.[107,110] The M_r varies slightly according to its glycosylation, and has structural homologies to other metalloenzymes such as carboxypeptidases A, B, and E. Sequence analysis indicates the presence of a single transmembrane-spanning domain near the N-terminal, but the absence of a typical signal sequence.[114-116] Mapping of the active center shows a role for Val^{573} for binding at P1′ of substrates, and the presence of Arg^{102} and His at the active center.[117-121]

1. Enzyme Distribution

Metalloendopeptidase 24.11 is present on plasma membrane of synapses, bronchial epithelium, endothelial cells, male genital tract, prostate and epididymis, fibroblasts, adrenal and salivary glands, PMN, and leukocytes.[2,122,123] In pre-B leukemic lymphocytes, it was also identified as the common acute lymphocytic leukemic antigen (CALLA,CD10).[116,124-127] Immunohistochemical staining shows its presence in gray matter of the corpus striatum, caudate-putamen, globus pallidus, olfactory tubercle, nucleus interpeduncularis and substantia nigra of brain, the substantia gelatinosa of spinal cord, Schwann cell membranes of peripheral nerve, choroid plexus, pia mater, and ependymal membranes.[53,128,129] In the rat striatum, the immunostaining patterns resemble those found for striosomes containing Met-enkephalin or other neuropeptides and acetylcholinesterase.[53,129,130] Mapping studies using labeled 24.11 inhibitor or mRNA probes have reached similar conclusions on localization with some evidence for colocalization of 24.11 in regions containing μ and δ receptors.[129,131]

2. Inhibitors and Structure

Principles for developing specific inhibitors follow those applied to other metalloenzymes discussed earlier. X-ray crystallography of enzyme-substrate or -inhibitor complexes show similar three-dimensional geometries for carboxypeptidases A and B and thermolysin involving a Glu residue at the active site, a Zn^{2+} atom and its interaction with neighboring residues, and a molecule of water.[22,132-134] Phosphoramidon, a microbially derived inhibitor of thermolysin, was the first agent available for differentiating 24.11 from other metalloenzymes[22] (Table 4). Prior to learning that 24.11 was "thermolysin-like," inhibitors were designed on the assumption that the site-3 enzyme resembled ACE or carboxypeptidase A.[134–136] In the case of ACE, the C-prolyl domain of snake venom-derived inhibitors *(Botropus jaracacus)* provided a model to prepare succinyl proline and then, the considerably more potent mercapto-analog captopril.[134] In the case of 24.11, the mercapto-analogs of the C-terminal enkephalin dipeptide Phe-Met (Leu) led to synthesis of thiorphan and related inhibitors active in the nanomolar or subnanomolar range.[2,22,135-136] Subsequently, hundreds of compounds have been synthesized with the aim of selectively inhibiting 24.11, including dipeptides containing an O^- found in –COOH, phosphoryl, phosphonamide, hydroxamate, aldehyde or ketones, *N*-carboxylalkyl-, haloacetyl *N*-hydroxyamino acids, and 2,4-dibenzylglutaric acid groups, or others referred to as UK-69,578, UK-70,300, SCH-32,615 or its esterified prodrug form SCH-34,826, and SQ-28,603[69,137-147] (Table 4). Inhibitors containing SH^- include thiorphan, acetorphan (synorphan and retorphan are the *S* and *R* isomers), and phelorphan. The interaction of phosphoramidon, kelatorphan, thiorphan, and compound 23 (phosphonamide) with the metallocenter of 24.11 are illustrated in Figure 5. Kelatorphan, containing a hydroxamate group, is more selective for 24.11 compared to ACE, but at higher concentrations can affect activity of site-1 and -2 enzymes.[2,85,140,148] *N*-carboxylalkyl peptide inhibitors were used recently to clarify the role of 24.11 and 24.15 for inactivation of LH-RH (pEHWSYGLRPG-NH_2) in pituitary: inhibitors showed that 24.15 acted preferentially by cleavage of -Tyr^5 -Gly- rather than 24.11, acting at the -Gly^6 -Leu-.[149] There have been attempts to improve penetrability of inhibitors by preparation of thioester and/or inhibitors such as acetorphan and SCH-34,826, and orally active agents UK-79,300 or SCH-34,826, or ones more stable in tissues such as a retro-inverso kelatorphan[2,150-153] (Table 4).

3. Inhibitors and Pharmacology

Studies with site-3 inhibitors have received the most attention in terms of their use as drugs. The strategies adopted follow closely those described earlier for site-1 and -2 inhibitors. Recent studies have emphasized the use of site-3 inhibitors for influencing or potentiating the action of substance P, tachykinins, and ANF. No coherent picture has emerged on therapeutic applications because of the varying methodologies, routes of administration, and methods for evaluating biological effects.

Kelatorphan increases recovery of Met-enkephalin in the substantia gelatinosa of cat spinal cord following microelectrophoresis, or potentiates analgesic effects of ic administered Met-enkephalin in normal or arthritic rats.[9] Prior conditioning (stress) of animals enhances effects in rats based on tail-flick latencies.[154] Acetorphan surppresses naloxone-induced withdrawal symptoms in morphine-dependent mice.[8] Thiorphan, phelorphan (mercaptoacetyl-Phe-Phe), or kelatorphan, or an analog (RB 38A) given icv increases analgesia in mice or rats, elevates recovery of iR-Met-enkephalin in synaptic fractions, and attenuates the withdrawal symptoms in chronic morphine-dependent rats.[155-157] Thiorphan icv or acetorphan ip increases analgesia induced by transcranial electrostimulation in rats.[158] Acetorphan given ip or SCH-34,826 given po potentiates analgesia induced by low-frequency electrical stimulation in mice.[145] Thiorphan with bestatin introplanar or SCH-34,826 given ip prolongs the naltrexone-reversible analgesia in rats induced by stress and Freund's complete adjuvant injected into hind paws.[159] There is evidence that acetorphan is less subject to abuse based on the absence of reinforcing effects in rhesus monkeys trained to discriminate morphine from saline.[160] Kelatorphan coapplied with morphine or DAGO, a μ agonist, or DTLET, a δ agonist, to the nucleus accumbens or lateral ventricle modifies self-stimulatory behavior in rats.[161] SCH-32,615 microinjected into the periaqueductal gray (PAG), a known morphine-sensitive site,[38] the ventral medulla, or the amygdala, enhances naloxone-reversible analgesia as determined by hot-plate and tail-flick latencies.[143] SCH-34,826, the prodrug form of SCH-32,615, potentiates injected D-Ala^2-Met^5-enkephalinamide up to 4 h, and at 100 times the dose used to improve antinociception, is without respiratory or gastrointestinal side effects.[226]

Angiotensin III, a substrate for metalloenzymes, induces a naloxone-reversible analgesia icv based on hot-plate latencies probably by competitive inhibition of site-1 to -3 enzymes.[162] Other biological effects of inhibitors, some of which are mediated by nonopioid pathways, relate to food intake,[163] esophageal motility,[164] gastric and gall bladder secretion,[165-167] contraction of the urinary bladder,[168] locomotion,[169-171] increase in dopamine turnover in striatum and olfactory tubercle or serotonin in the cortex, decrease in noradrenaline in the cortex, or antidepressant effects in mice.[2,172,173] UK-79,300 enhances ANF in plasma but suppresses plasma renin and aldosterone. SCH-34,826 or UK-69,578 administered orally (*N*-carboxylmethyl dipeptide with a cycloalkane P1′ residue, and replacement of aza link with a methylene group between P1 and P10 residues) potentiates the biological responses of ANF (antihypertensive, diuretic, natriuretic) in rodents, pointing to application for treatment of blood pressure and heart disease.[141,147] Acetorphan given orally is active in humans, resulting in elevation of plasma iR-ANF and concomitant increases in urinary volume and sodium excretion.[147,174,175]

B. ANGIOTENSIN-CONVERTING ENZYME (E.C.3.4.15.1)

ACE has been the subject of extensive studies, largely because of its pivotal role in the renin-angiotensin system and its vasoactive properties.[176] In brain it is found in highest concentrations in the circumventricular organs, substantia nigra, striatum, and meninges.[177,178] ACE is membrane-bound, requring detergents for extraction, and can be purified by ion-exchange and affinity procedures. Nucleotide sequences of ACE mRNA from human kidney or endothelial cells show the presence of 1306 residues beginning with a signal peptide of 29 amino acids, and a hydrophobic sequence near the carboxyl-terminal believed to act as an anchor.[179] The sequence reveals a high degree of homology between two large domains, implying that the molecule is derived via gene duplication. One of these domains contains Zn^{2+} with homology to the active centers of thermolysin, metalloendopeptidase 24.11, and collagenase. Although purified enzyme cleaves site 3 of pentapeptide enkephalins, its low affinity compared to 24.11 excludes it as a relevant inactivating enzyme (Table 4). In contrast, it acts with higher affinity toward C-terminally extended enkephalins, and is a candidate for their conversion to generate smaller active forms (Figure 1).[33,180] In rat striatal SPM, the C-terminal dipeptide of the heptapeptide is the major point of cleavage, resulting in the conversion to pentapeptide. This cleavage is sensitive to captopril[55] (see Figure 3).

Since tachykinins and substance P are implicated in pain sensory mechanisms, there has been interest in their metabolism and effects of inhibitors. ACE is identical to kininase II, an enzyme known to inactivate kinins with release of C-terminal dipeptide, and substance P by cleavage at an internal site.[33] However, in rat substantia nigra slices, substance P is inactivated by metalloendopeptidase 24.11 rather than ACE.[181-183] This is supported by the increased recovery of iR-substance P in the presence of thiorphan, but not captopril, from dorsal horn slices of guinea pig treated with capsaicin.[184] Similarly, thiorphan or phosphoramidon potentiated electrically induced contraction of bronchi or trachea, mediated by capsaicin-sensitive C-fibers.[185-188]

1. Inhibitors and Pharmacology

Captopril is a –SH disubstituted peptide (*N*-1-*S*-carboxy-3-phenylpropyl-2-alanyl-L-proline) orally active as a hypotensive agent.[134] It acts primarily by suppression of circulatory ACE, although inhibition of other metalloenzymes cannot be ruled out.[189,176] Captopril, given at mg ip, induces analgesia in hot-plate latency tests, and attenuates withdrawal symptoms such as writhing, resulting from acetic acid.[189] It appears more effective than amastatin or phosphoramidon for enhancing dynorphin$_{1-8}$ assays on guinea pig gut but not rabbit or mouse vas deferens.[190]

C. CATHEPSIN B (E.C.3.4.22.1)

This site-3 enzyme is a membrane-bound glycoprotein acting as an exo-endopeptidase degrading opioid peptides and substance P, among others.[33,98] A role in processing is implicated by its presence in corticotrophs and melanotrophs of rat pituitary, in secretory granules of pancreas, and in tooth pulp.[191-193] It preferentially releases dipeptides from C-terminally extended peptides at rates greatly exceeding site-3 cleavage of pentapeptides. This is shown by K_m for breakdown of Leu-enkephalin versus C-terminal dipeptide cleavage of Met-enkephalin-Arg-Phe (450 and 58 μM, respectively).[98]

Cathepsin B, like other lysosomal cysteine proteinases, is potently inhibited by members of the cystatin superfamily present in high concentration in the CNS and in plasma and discussed in the section of cathepsin L.[99,194] Cystatin or synthetic peptides containing enkephalin and a consensus cystatin sequence are more active toward cathepsin L or papain.[103]

V. SITE-5 or -6 CLEAVAGE

A. CARBOXYPEPTIDASE A

Tissues contain carboxypeptidase A-like enzymes active at both low (cathepsin A) and physiological pH and preferentially removing C-terminal acidic amino acids. Brain cathepsin A is a lysosomally derived carboxypeptidase distinct from those active at physiological pH and inhibited by benzylsuccinic acid[33] (Table 5). Purified forms were shown to trim Met-enkephalin-Arg-Phe C-terminally with release of Phe and Arg.[195-196] The enzyme can be extracted from brain particulates with detergents and assayed with Z-Glu-Tyr or Z-Phe-Tyr. Since it is inactive toward pentapeptide enkephalins, it may have some role in conversion of precursor forms similar in concept to the role proposed for cathepsin B and ACE. In addition to C-terminally extended enkephalins, it degrades also other biologically active peptides, such as angiotensin I (Phe release).

Carboxypeptidase A active at physiological pH is also a metalloenzyme inhibited by agents containing SH– (2-benyl-3-mercaptopropanoic acid or SQ 14,603, K_i $10^{-8}M$),[197] or phosphoryl O^-(3-phosphonopropionic acids), but pharmacological properties of these site-specific agents have not been reported.[197,198]

1. Inhibitors and Pharmacology

There has been considerable interest in the use of carboxypeptidase inhibitors such as D-Phe (DPA, D-Leu, β-phenylpropionic acid, or hydrocinnamic acid as analgesic agents. High doses (100 to 250 mg/kg ip DPA) in mice were shown to be analgesic by the hot-plate method, with the effects

TABLE 5
Site-5 Cleavage (Enkephalin—Arg[6])

Name	Carboxypeptidase A	Carboxypeptidase E	Metalloendopeptidase
E.C.	3.4.12A.1	3.4.17.10	3.4.24.15
pH Optima	5.5	5.6	7.6
M_r (kDa)	>100	70	
Specificity	Z-Glu-Tyr, Dyn A_{1-6}, Dyn A_{1-12}, Dyn A_{1-13}, Dyn A_{1-10}, Dyn A_{1-11}, Dyn A_{1-8}, Dyn A_{1-9} Met-Enk-Arg-Phe (Phe,Arg), angiotensin-II (Phe release)	Bz-Ala-Arg, Dyn A_{1-7}, Dyn A_{1-13}, Dyn A_{1-6}, Dyn A_{1-9}, Dyn A_{1-10}	tBOC-FAAF-p-AB-, Dyn_{1-8}, Dyn_{1-13}, neoendorphins substance P, LH-RH, bradykinin
Inhibitors	benzylsuccinic acid,[b] 3-phosphonopropionic acids[b] SQ 1603[b]	GEMSA, APMSA, GEMPA, GEMA	cFP-AAF-pAB, cFP-AAY-pAB
	Membrane bound[a] or soluble[b]	Membrane or soluble	Membrane or soluble
		Activated by Co^{2+}	Converts dynorphins to Dyn_{1-5}
Other names	Cathepsin A+	Enkephalin convertase, carboxypeptidase H	Pz-peptidase oligopeptidase-A

Note: Pentapeptide enkephalins are not substrates for these enzymes. Dynorphins are listed in descending rates of conversion by carboxypeptidases A and B with release of single C-terminal residues. 24.15 acts on the dynorphins listed to release Leu-enkephalin (Dyn_{1-5}) directly. The *N*-carboxyalkyl peptides listed selectively inhibit 24.15 and can be compared to those in Table 4 inhibiting 24.11. GEMSA, APMSA, GEMPA, and GEMA are sulfur analogs of arginine and lysine. For structures see text or Refs. 33, 187, 197, 198, 212, 216, 222.

[a] Active at pH 5.5.
[b] Active at pH 7.6.

reduced by pretreatment with 20 mg/kg naloxone.[199,200] At orally administered doses of up to a 1000 mg/day, DPA is available as a dietary supplement for possible treatment of lower back pain in humans. Interpretation of clinical effects of DPA as an inhibitor of pentapeptide enkephalins is disputed since no cleavage of site-4 by tissue enzymes has been observed (Figure 1).[180,201,202]

B. CARBOXYPEPTIDASE B-LIKE (E.C. 3.4.17.10)

Exopeptidases, present in rat pituitary or adrenal chromaffin granules, converting Met-enkephalin-Arg to Met-enkephalin at pH 5 to 6.0 have properties similar to carboxypeptidases B or H (Table 5).[203-209] The enzyme is found in soluble and particulate fractions of tissues and is secreted by cultured astrocytes and neurons.[210] "Enkephalin convertase" or carboxypeptidase E is nonspecific and can be assayed with Dns-Phe-Ala-Arg, Met-enkephalin-Arg, or insulin peptide intermediates, among other substrates. Northern blot analysis of mRNAs shows that it is present in hypothalamus, hippocampus, midbrain, striatum, and cortex, although lower levels are found in brainstem, cerebellum, heart, adrenal gland, and eye of rat. Rat or bovine striata are reported to also contain a carboxypeptidase B-like enzyme activated by a cAMP protein kinase and shown to convert enkephalin precursors.[211]

1. Inhibitors and Pharmacology

Sulfur-containing analog of lysine and arginine amino acids (GEMSA, APMSA, GEMPA, GEMA) potentially inhibit carboxypeptidase B-like enzymes (Table 5).[197,212] GEMSA (guanidinoethylmercaptosuccinic acid) administered at 3 to 6 μg icv or 12 to 50 μg intrathecally in rats elicited naloxone-reversible analgesia based on tail-flick latencies, or potentiated the effects of Met-enkephalin-Arg-Phe or Arg-Gly-Leu when coadministered.[213,214] Labeled inhibitors have been used to map enzyme distribution and they appear to indicate the presence of enzyme in areas rich in opioid receptors.[208,213] Based on changes in mRNAs in adrenal medulla following splanchnic stimulation, there is some evidence that enzyme secretion is correlated with biosynthesis of opioid peptides.[215]

C. METALLOENDOPEPTIDASE (E.C. 3.4.24.15)

This brain metalloendopeptidase, present in soluble and membrane fraction, can be differentiated from 24.11 by its insensitivity to phosphoramidon.[216,217] The enzyme nonspecifically cleaves bradykinin, LH-RH, dynorphin$_{1-8}$, dynorphin B, and substance P, among others. Cleavage of the dynorphins is accompanied by release of Leu-enkephalin as one of the products, suggesting a role as a convertase. More recently, it was shown to resemble PZ-peptidase of bacteria or tissues, and a brain oligopeptidase A degrading neurotensin (Arg^8-Arg) and bradykinin (Phe^5-Ser).[218-221] Carboxyphenylpropyl-Ala-Ala-Phe-*p*-aminobenzoate, a carboxylalkyl peptide administered icv to rats, protected the breakdown of dynorphin A$_{1-8}$ better than captopril.[222]

VI. CONCLUDING REMARKS

The use of enzyme inhibitors as antinociceptive reagents has been disappointing and none have supplanted traditional opiates. In some respects, research in this area reached its zenith in 1984 and very little new can be added to conclusions reached at that time by Hansen and Morgan[36] and Clement-Jones and Bazaar[27] concerning the potential role for inhibitors as pharmacological agents. The earlier assumptions that enkephalins were degraded by highly specific enzymes proved to be incorrect and this fact combined with conceptual difficulties in assigning roles for enkephalins in mediating nociception have hampered progress in preparing inhibitors. Enkephalins were discovered shortly after identification of opiate receptors and, at one time, were considered the native ligands. An intricate network of neural fibers containing enkephalins, endorphins, and dynorphins is described in other chapters in this series, but their functional significance has not been defined. These peptides may function as putative neurotransmitters or neuromodulators. Progress has been made in determining their biosynthetic origin, and there is new evidence that processing at the level of the genome is of considerable functional significance. Receptors, however, have not been cloned and the reasons for the existence of multiple forms is enigmatic. Morphine, the most studied opiate, binds preferentially to μ receptors which also are recognized by β-endorphin rather than smaller enkephalins or dynorphins which preferentially recognize δ or κ receptors.[41,42] There are thousands of reports describing the diverse pharmacological properties of opioid peptides; these have been reviewed on an annual basis by Olsen et al.[28,29]

The number of useful outcomes of studies on opioid peptides and their metabolism auger well for the future. As noted, they were instrumental in identifying new receptor subtypes and confirming the existence of others postulated by studies of Martin et al.[25] in whole animals. Inhibitors are useful as probes to dissect peptidergic pathways in the CNS, explore catalytic mechanisms of enzymes degrading opioid peptides, and provide evidence linking changes in opioid peptides or their metabolism with antinociceptive responses. Thousands of inhibitors have been synthesized, some of which are active orally, or active centrally in potentiating the effects of applied opioid peptides. Most of these have been synthesized by using general principles for targeting metalloenzymes. Earlier clues on the synthesis of such agents were derived by the use of microbial inhibitors, particularly phosphoramidon and bestatin.[22] It is possible that synthetic analogs, as was the case for captopril in treatment of hypertension, will find therapeutic applications. The field has broadened to include effects of inhibitors on catabolism of ANF and other vasoactive peptides suggesting applications in cardiovascular disorders, in addition to pain. Since peptides poorly penetrate membranes, and particularly the CNS, it is to be hoped that computer modeling of agonists or inhibitors will yield nonmetabolizable agents suitable for oral applications along lines discussed by Hruby and his group.[225]

ACKNOWLEDGMENTS

Some of the work reported in this review was supported by ADAMHA DA 04179, the Health Research Council of New York (20-017), and the Kroc Foundation.

REFERENCES

1. **Crout, J. R.,** Enzyme inhibitors as Drugs, in *Enzyme as Targets for Drug Design,* Palfreyman, M. G., McCann, P. P., Lovenberg, W., Temple, J. G., Sjoerdsma, A., Eds., Academic Press, San Diego, 1989, 1.
2. **Schwartz, J.-C.,** Enkephalinase inhibitors as Drugs, in *Design of Enzyme Inhibitors as Drugs,* Sandler, M. and Smith, H. J., Eds., Oxford University Press, New York, 1989, 206.
3. **Reggiani, A., Carenzi, A., Frigeni, V., and Della Bella, D.,** Effect of bestatin and thiorphan on [met^5]enkephalin-arg^6-phe^7-induced analgesia, *Eur. J. Pharmacol.,* 105, 361, 1984.
4. **Fournie-Zaluski, M. C., Coulaud, A., Bouboutou, R., Chaillet, P., Devin, J., Wakesman, G., Costentin, J., and Roques, B. P.,** New bidentates as full inhibitors of enkephalin-degrading enzymes: synthesis and analgesic properties, *J. Med. Chem.,* 28, 1158, 1985.
5. **Jin, W.-Q., Zhou, Z.-F., and Han, J.-S.,** Electroacupuncture and morphine analgesia potentiated by bestatin and thiorphan administered to the nucleus accumbens of the rabbit, *Brain Res.,* 380, 317, 1986.
6. **Morton, C. R., Zhou, Z.-Q., and Duggan, A. W.,** Kelatorphan potentiates the effect of [met^5]enkephalin in the substantia gelatinosa of the cat spinal cord, *Eur. J. Pharmacol.,* 140, 195, 1987.
7. **Gros, C., Giros, B., Schwartz, J.-C., Vlaiculescu, A., Costentin, J., and Lecomte, J.-M.,** Potent inhibition of cerebral aminopeptidases by carbaphethiol, a parenterally active compound, *Neuropeptides,* 12, 111, 1988.
8. **Livingstone, S. J., Sewell, R. D. E., Rooney, K. F., and Smith, H. J.,** Amelioration of naloxone-precipitated opioid withdrawal symptoms by peripheral administration of the enkephalinase inhibitor "acetorphan", *Psychopharmacology,* 94, 540, 1988.
9. **Kayser, V., Fournie-Zaluski, M. C., Guilbaud, G., and Roques, B. P.,** Potent antinociceptive effects of kelatorphan a highly efficient inhibitor of multiple enkephalin-degrading enzymes systemically administered in normal and arthritic rats, *Brain Res.,* 497, 94, 1989.
10. **Llorens-Cortes, C., Gros, C., Schwartz, J. C., Clot, A. M., and Le-Bars, D.,** Changes in levels of the tripeptide tyr-gly-gly as an index of enkephalin release in the spinal cord effects of noxious stimuli and parenterally–active peptidase inhibitors, *Peptides,* 10, 609, 1989.
11. **Nakazawa, T., Ikeda, M., Kaneko, T., Yamatsu, K., Kitagawa, K., and Kiso, Y.,** Bestatin potentiates the antinociception but not the motor dysfunction induced by intracerebrally administered dynorphin-B in mice, *Neuropeptides,* 10, 277, 1989.
12. **Malin, D. H., Leke, J. T., Hamilton, R. F., and Skolnick, M. H.,** Augmented analgesic effects of enkephalinase inhibitors combined with transcranial electrostimulation, *Life Sci.,* 44, 1371, 1989.
13. **Devi, L.,** Consensus sequence for processing of peptide precursors at monobasic sties, *FEBS Lett.,* 280, 189, 1991.
14. **Numa, S.,** Opioid peptide precursors and their genes, in *The Peptides, Analysis, Synthesis, Biology VI,* Udenfriend, S. and Meienhofer, J., Eds., Academic Press, San Diego, 1984, 1.
15. **Udenfriend, S. and Kilpatrick, D. L.,** Proenkephalin and the products of its processing: chemistry and biology, in *The Peptides, Analysis, Synthesis, Biology VI,* Udenfriend, S. and Meienhofer, J., Eds., Academic Press, San Diego, 1984, 26.
16. **Civelli, O., Douglass, J., and Herbert, E.,** Pro-opiomelanocortin: a polyprotein at the interphase of endocrine and nervous systems; in *The Peptides, Analysis, Synthesis, Biology VI,* Udenfriend, S. and Meienhofer, J., Eds., Academic Press, San Diego, 1984, 70.

17. **Goldstein, A.,** Biology and chemistry of the dynorphin peptides, in *The Peptides, Analysis, Synthesis, Biology VI,* Udenfriend, S. and Meienhofer, J., Eds., Academic Press, San Diego, 1984, 95.
18. **Perara, E. and Lingappa, V. R.,** Transport of protein into and across the endoplasmic reticulum membrane, in *Protein Transfer and Organelle Biogenesis,* Das, R. C. and Robbins, P. W., Eds., Academic Press, San Diego, 1988, 3.
19. **Parent, J. B.,** Role of carbohydrate in glycoprotein traffic and secretion, in *Protein Transfer and Organelle Biogenesis,* Das, R. C. and Robbins, P. W., Eds., Academic Press, San Diego, 1988, 51.
20. **Robbins, A. R.,** Endocytosis and compartmentalization of lysosomal enzymes in normal and mutant mammalian cells: mannose-6- phosphtate-independent pathways, in *Protein Transfer and Organelle Biogenesis,* Das, R. C. and Robbins, P. W., Eds., Academic Press, San Diego, 1988, 464.
21. **Walker, J. M., Moises, H. C., Coy, D. H., Bairighi, G., and Akil, H.,** Non-opiate effects of dynorphine and des-tyr-dynorphine, *Science,* 218, 1136, 1982.
22. **Powers, J. C. and Harper, J. W.,** Inhibitors of metalloproteases, *Res. Monogr. Cell Tissue Physiol.,* 12, 219, 1986.
23. **Moore, H.-P. H., Orci, L., and Oster, G. F.,** Biogenesis of secretory vesicles in *Protein Transfer and Organelle Biogenesis,* Das, R. C. and Robbins, P. W., Eds., Academic Press, San Diego, 1988, 521.
24. **Benuck, M., Berg, M. J., and Marks, N.,** Membrane bound enzymes and their role in processing of the dynorphins and the proenkephalin met-enkephalin-arg-gly-leu, *Neurochem. Res.,* 8, 731, 1984.
25. **Martin, W. H., Aedes, C. G., Thompson, J. A., Huppler, R. E., and Gilbert, P. E.,** The effects of morphine and nalorphine-like drugs in the nondependent and morphine-dependent chronic spinal dog, *J. Pharmacol. Exp. Ther.,* 197, 517, 1976.
26. **Lord, J. A. H., Waterfield, A. A., Hughes, I., and Kosterlitz, H. W.,** Endogenous opioid peptides: multiple agonists and receptors, *Nature (London),* 267, 495, 1977.
27. **Clement-Jones, V. and Besser, G. M.,** Opioid peptides in humans and their clinical significance, in *The Peptides, Analysis, Synthesis, Biology VI,* Udenfriend, S. and Meienhofer, J., Eds., Academic Press, San Diego, 1984, 323.
28. **Olson, G. A., Olson, R. D., and Kastin, A. J.,** Endogenous Opiates: 1986, *Peptides,* 8, 1135, 1987.
29. **Olson, G. A., Olson, R. D., and Kastin, A. J.,** Endogenous Opiates: 1989, *Peptides,* 11, 1277, 1990.
30. **Ruda, M. A., Iadorola, M. J., Cohen, L. V., and Young, W. S.,** *In situ* hybridization histochemistry and immunocytochemistry reveal an increase in spinal dynorphin biosynthesis in a rat model of peripheral inflammation and hyperalgesia, *Proc. Natl. Acad. Sci. U.S.A.,* 85, 622, 1988.
31. **Marks, N.,** Peptide hydrolases of the CNS, *Int. Rev. Neurobiol.,* 11, 57, 1968.
32. **Marks, N.,** Biotransformation and degradation of corticotropins, lipotropins, and hypothalamic peptides, *Front. Neuroendocrinol.,* 5, 329, 1978.
33. **Marks, N., Benuck, M., and Berg, M. J.,** Enzymes in the metabolism of opioid peptides: isolation, assay and specificity, *NIDA Res. Monogr.,* 70, 66, 1986.
34. **Gainer, H.,** Peptides and neuronal function, *Adv. Biochem. Psychopharmacol.,* 15, 193, 1976.
35. **Chen, Z. R., Irvine, R. J., Somogyi, A. A., and Bochner, F.,** Mu receptor binding of some commonly used opioids and their metabolites, *Life Sci.,* 48, 2165, 1991.
36. **Hanson, P. E. and Morgan, B. A.,** Structure-activity relationships in enkephalin peptides, in *The Peptides, Analysis, Synthesis, Biology VI,* Udenfriend, S. and Meienhofer, J., Eds., Academic Press, San Diego, 1984, 269.
37. **Fennessy, M. R. and Lee, J. R.,** The assessment of and the problems involved in the experimental evaluation of narcotic analgesics, in *Methods in Narcotics Research, Vol. 5.,* Ehrenpreis, S. and Neidle, A., Eds., Marcel Dekker, New York, 1975, 73.

38. **Jacquet, Y. and Marks, N.,** The C-fragment of b-lipotropin, an endogenous neuroleptic or an antipsychotogen?, *Science,* 194, 632, 1976.
39. **Li, C. H., Barnafi, L., Chretien, M., and Chung, D.,** Isolation and amino-acid sequence of β-LPH from sheep pituitary glands, *Nature (London),* 208, 1093, 1965.
40. **Hughes, J., Smith, T. W., Kosterlitz, H. W., Fothergill, L. A., Morgan, B. A., and Morris, H. R.,** Identification of two related pentapeptides from the brain with potent opiate agonist activity, *Nature (London),* 258, 577, 1975.
41. **Smith, J. R. and Simon, E. J.,** Endorphins, opiate receptors, and their evolving biology, *Pathobiol. Annu.,* 11, 87, 1981.
42. **Simon, E. J.,** Progress in the characterization of the opioid receptor subtypes: peptides as probes. Future directions, *NIDA Res. Monogr.,* 70, 155, 1986.
43. **Civelli, O., Machida, C., Bunzow, J., Albert, P., Hanneman, E., Salon, J., Bidlock, J., and Grandy, D.,** The next frontier in the molecular biology of the opioid system, *Mol. Neurobiol.,* 1, 373, 1987.
44. **Versteg, D. H. G., De Kloet, E. R., and De Wied, D.,** Effects of alpha-endorphin, beta-endorpin, and (desTyr[1])-gamma-endorphin on a alpha1-MPT-induced catecholamine disappearance in discrete regions of the rat brain, *Brain Res.,* 179, 85, 1979.
45. **Verhoeven, W. M. A., Van Ree, J. M., De Wieid, D., and Van Praag, H. M.,** gamma-Type endorphins and schizophrenia, *Arch. Gen. Psychiatry,* 38, 1182, 1981.
46. **Marks, N., Datta, R., and Lajtha, A.,** Partial resolution of brain arylamidases and aminopeptidases, *J. Biol. Chem.,* 243, 2882, 1968.
47. **Dyer, S. H., Slaughter, C. A., Orth, K., Moomaw, C. R., and Hersh, L. B.,** Comparison of the soluble and membrane bound forms of the puromycin sensitive aminopeptidases from rat, *J. Neurochem.,* 54, 547, 1990.
48. **Johnson, G. D. and Hersh, L. B.,** Studies on the subsite specificity of rat brain puromycin-sensitive aminopeptidases, *Arch. Biochem. Biophys.,* 276, 305, 1990.
49. **Berg, M. J. and Marks, N.,** Formation of desTyr-dynorphins 5-17 by a purified cytosolic aminopeptidase of rat brain, *J. Neurosci. Res.,* 11, 313, 1984.
50. **Gros, C., Giros, B., and Schwartz, J.-C.,** Identification of aminopeptidase M as an enkephalin-inactivating enzyme in rat cerebral membranes, *Biochemistry,* 24, 2179, 1985.
51. **Hui, K.-S., Hui, M. P.-P., and Lajtha, A.,** Major rat brain membrane-associated and cytosolic enkephalin-degrading aminopeptidases: comparison studies, *J. Neurosci. Res.,* 20, 231, 1988.
52. **McLellan, S., Dyer, S. H., Rodgriguez, G., and Hersh, L. B.,** Studies on the tissue distribution of the puromycin enkephalin-degrading aminopeptidases, *J. Neurochem.,* 51, 1552, 1988.
53. **Barnes, K., Matsas, R., Hooper, N. M., Turner, A. J., and Kenny, A. J.,** Endopeptidase-24.11 is striosomally ordered in pig brain, and in contrast to aminopeptidase N and peptidyl dipeptidase A (''angiotensin converting enzyme''), is a marker for a set of striatal efferent fibers, *Neuroscience,* 27, 799, 1988.
54. **Hui, K.-S., Hui, M., Lajtha, A., Saito, M., and Saito, M.,** Cellular localization of puromycin sensitive aminopeptidase isozymes, *Neurochem. Res.,* 15, 1447, 1990.
55. **Benuck, M., Berg, M. J., and Marks, N.,** Separate metabolic pathways for leu-enkephalin and met-enkephalin-arg-phe degradation by rat striatal synaptosomal membranes, *Neurochem. Int.,* 4, 389, 1982.
56. **Pert, C. B., Pert, A., Chang, J. K., and Fong, B. T. W.,** (D-Ala^2)-metenkephalinamide; a potent long lasting synthetic pentapeptide analgesic, *Science,* 194, 330, 1976.
57. **Bajusz, S., Ronai, A. Z., Szekely, J. I., Graf, L., Duani-Kovacs, Z., and Berzetei, I.,** A superactive antinociceptive pentapeptide, (D-Met^2, Pro^5)-enkephalinamide, *FEBS Lett.,* 76, 91, 1977.
58. **Marks, N., Benuck, M., Berg, M. J., and Sachs, L.,** Metabolism of enkephalin analogues and surrogates having enhanced pharmacologic activities, *Ann. N.Y. Acad. Sci.,* 398, 308, 1982.

59. **Weinberger, S. B. and Martinez, J. L.,** Characterization of hydrolysis of (leu)enkephalin and [D-ala^2]-(L-leu) enkephalin in rat plasma, *J. Pharmacol. Exp. Ther.*, 247, 129, 1988.
60. **Benter, I. F., Hirsch, E. M., Tuchman, A. J., and Ward, P. E.,** N-terminal degradation of low molecular weight opioid peptides in human cerebrospinal fluid, *Biochem. Pharmacol.*, 40, 465, 1990.
61. **Churchill, L., Bausback, H. H., Gerritsen, M. E., and Ward, P. E.,** Metabolism of opioid peptides by cerebral microvascular aminopeptidase M, *Biochim. Biophys. Acta*, 923, 35, 1987.
62. **Solhonee, B., Gros, C., Pollard, H., and Schwartz, J.-C.,** Major localization of aminopeptidases M in rat brain microvessels, *Neuroscience*, 22, 225, 1987.
63. **Hersh, L. B., Aboukhair, N., and Watson, S.,** Immunohistochemical localization of aminopeptidase M in rat brain and periphery: relationship of enzyme localization and enkephalin metabolism, *Peptides*, 8, 523, 1987.
64. **Olsen, J., Cowell, G. M., Knigshifer, E. M., Danielson, E. M., Miller, J., Laustasen, L., Hansen, O. C., Wellinder, K. G., Engberg, J., and Hunziker, W.,** Complete amino acid sequence of human intestinal aminopeptidase M as deduced from cloned cDNA, *FEBS Lett.*, 238, 307, 1988.
65. **Olsen, J., Sjostrom, H., and Noren, O.,** Cloning of the pig aminopeptidase N gene. Identification of possible regulatory elements and the exon distribution in relation to the membrane spanning region, *FEBS Lett.*, 251, 275, 1989.
66. **Malfroy, B., Kado-Fong, H., Gros, C., Giros, B., Schwartz, J.-C., and Hellmiss, R.,** Molecular cloning and amino acid sequence of rat kidney aminopeptidase M: a member of a superfamily of Zn-metallohydrolases, *Biochem. Biophys. Res. Commun.*, 161, 236, 1989.
67. **Look, A. T., Ashmun, R. A., Shapiro, L. H., and Peiper, S. C.,** Human myeloid plasma membrane glycoprotein CD13 (gp 150) is identical to aminopeptidase M, *J. Clin. Invest.*, 83, 1299, 1989.
68. **Hachisu, M., Hiranuma, T., Shibazaki, Y., Uotani, K., Murata, S., Aoyagi, T., and Umezawa, H.,** Composite effects of actinonin when inhibiting enkephalin degrading enzymes, *Eur. J. Pharmacol.*, 137, 59, 1987.
69. **Nakajima, S., Kaya, K., and Hazato, T.,** Protection of endogenous enkephalin degradation from peptidases in human serum by actinonin, *Neuropeptides*, 13, 201, 1989.
70. **Wagner, G. W. and Dixon, J. C.,** Inhibitors of a rat brain enkephalin aminopeptidase, *J. Neurochem.*, 37, 709, 1981.
71. **Nishizawa, R., Saino, T., Takita, T., Suda, H., Aoyagi, T., and Umezawa, H.,** Synthesis and structure-activity relationships of bestatin analogues, inhibitors of aminopeptidase B, *J. Med. Chem.*, 20, 510, 1977.
72. **Ocain, T. D. and Rich, D. H.,** Synthesis of sulfur-containing analogues of bestatin. Inhibition of aminopeptidases by alpha-thiolbestatin analogues, *J. Med. Chem.*, 31, 2193, 1988.
73. **Gordon, E. M., Godfrey, J. D., Delaney, N. G., Asaad, M. M., Von Langen, D., and Cushman, D. W.,** Design of novel inhibitors of aminopeptidases. Synthesis of peptide-derived diamino thiols and sulfur replacement analogues of bestatin, *J. Med. Chem.*, 31, 2199, 1988.
74. **Harbeson, S. L. and Rich, D. H.,** Inhibition of arginine aminopeptidase by bestatin and arphenamine analogues. Evidence for a new mode of binding to aminopeptidase, *Biochemistry*, 27, 7301, 1988.
75. **Schwartz, J.-C., De La Baume, S., Yi, C. C., Malfroy, B., Chaillet, P., Marcais-Collado, H., and Costentin, J.,** Physiological roles of "enkephalinase" (enkephalin dipeptidyl carboxypeptidase) and of a bestatin sensitive aminopeptidase in the inactivation of enkephalins, in *Degradation of Endogenous Opioids: Its Relevance in Human Pathology and Therapy*, Ehrenpreis, S. and Sicuteri, F., Eds., Raven Press, New York 1983, 51.

76. **Suh, H. H. and Tseng, L. L.,** Intrathecal administration of thiorphan and bestatin enhances the antinociception and release of met-enkephalin induced by beta-endorphin intraventricularly in anesthetized rats, *Neuropeptides,* 16, 91, 1990.
77. **Harding, J. W. and Felix, D.,** The effects of the aminopeptidase inhibitors amastatin and bestatin on angiotensin-evoked neuronal activity in rat brain, *Brain Res.,* 424, 299, 1987.
78. **Abhold, R. H., Sullivan, M. J., Wright, J. W., and Harding, J. W.,** Binding, degradation and pressor activity of angiotensins II and III after aminopeptidase inhibition with amastatin and bestatin, *J. Pharmacol. Exp. Ther.,* 242, 957, 1987.
79. **Quirk, W. S., Harding, J. W., and Wright, J. W.,** Amastatin and bestatin-induced dipsogenicity in the Sprague-Dawley rat, *Brain Res. Bull.,* 19, 145, 1987.
80. **Batt, C. M., Klein, E. W., Harding, J. W. and Wright, J. W.,** Pressor responses to amastatin, bestatin and Plummer's inhibitors are suppressed by pretreatment with the angiotensin receptor antagonist sarthran, *Brain Res. Bull.,* 21, 731, 1988.
81. **Batt, C. M., Jenson, L. L., Hansewother, J. M., Harding, J. W., and Wright, J. W.,** Intracerebroventricular applied peptidase inhibitors increase endogenous angiotensin levels, *Brain Res.,* 529, 126, 1990.
82. **Rabkin, S. W. and Redston, M.,** The effect of inhibitors of endogenous opioid degradation, bacitracin, bestatin, captopril, and D-phenylalanine, on digoxin-induced arrhythmias in guinea pigs, *Can. J. Physiol. Pharmacol.,* 67, 857, 1989.
83. **Takahashi, S., Kato, H., Takahashi, A., Noguchi, T., and Naito, H.,** Mode of action of bestatin and leupeptin to induce the accumulation of acid soluble peptides in rat liver *in vivo* and the properties of the accumulated peptides. The important role of bestatin- and leupeptin-sensitive proteases in the protein degradation pathway *in vivo, Int. J. Biochem.,* 19, 401, 1987.
84. **Botbol, V. and Scornick, O. A.,** Measurement of instant rates of protein degradation in the livers of intact mice by the accumulation of bestatin-induced peptides, *J. Biol. Chem.,* 266, 2151, 1991.
85. **Roques, P. B. and Fournie-Zaluski, C.,** Enkephalin degrading enzyme inhibitors: a physiological way to new analgesics and psychoactive agents, *NIDA Res. Monogr.,* 70, 128, 1986.
86. **Roques, P. B.,** Physiological role of endogenous peptide effectors studied with peptidase inhibitors, *Kidney Int.,* 26(Suppl.), 527, 1988.
87. **Xie, J., Soleilhec, J. M., Renwart, N., Peyrous, J., Roques, B. P., and Fournie-Zaluski, M. C.,** Inhibitors of the enkephalin degrading enzymes. Modulation of activity of hydroxamate containing compounds by modifications of the C-terminal residue, *Int. J. Pept. Protein Res.,* 34, 246, 1989.
88. **Lejczak, B., Kafarski, P., and Zygmunt, J.,** Inhibition of aminopeptidases by aminophosphonates, *Biochemistry,* 28, 3549, 1989.
89. **Yu, Z., Williams, P. S., Patel, A., Smith, H. J., Sewell, R. D., and Gonzalez, J. P.,** 2-Substituted N5-(2′-aminophenyl)-4-thiodydantoic acids as aminopeptidase inhibitors with pro-antinociceptive activity, *Prog. Clin. Biol. Res.,* 328, 481, 1990.
90. **Beattie, R. E., Elmore, D. T., Williams, C. H., and Guthrie, D. J. S.,** The behavior of leucine aminopeptidase towards thionopeptides, *Biochem, J.,* 245, 285, 1987.
91. **Giannousis, P. P. and Bartlett, P. A.,** Phosphorous amino acids analogues as inhibitors of leucine aminopeptidase, *J. Med. Chem.,* 30, 1603, 1987.
92. **McDermott, J. R., Mantle, D., Lauffart, B., Gibson, A. M., and Biggins, J. A.,** Purification and characterization of two soluble Cl^- activated arginyl aminopeptidases from human brain and their endopeptidase action on neuropeptides, *J. Neurochem.,* 50, 176, 1988.
93. **Marks, N. and Berg, M. J.,** Rat brain cathepsin L: characterization and differentiation from cathepsin B utilizing opioid peptides, *Arch. Biochem. Biophys.,* 259, 131, 1987.
94. **Lee, C.-M. and Synder, S. H.,** Dipeptidyl-aminopeptidase II of rat brain, *J. Biol. Chem.,* 257, 12043, 1982.

95. **Inaoka, Y., Tomaoki, H., Takahashi, S., Enokita, R., and Okazaki, T.,** Propioxatins A and B, new enkephalinase B inhibitors. I. Taxonomy, fermentation, isolation and biological properties, *J. Antibiot. (Japan),* 39, 1368, 1986.
96. **Inaoka, Y., Takahashi, S., and Kinoshita, T.,** Propioxatins A and B, new enkephalinase B inhibitors. II. Structural elucidation, *J. Antibiot. (Japan),* 39, 1378, 1986.
97. **Inaoka, Y. and Naruto, S.,** Propioxatins A and B, new enkephalinase B inhibitors. IV. Characterization of the active site of the enzyme using synthetic propioxatin analogues, *J. Biochem. (Japan),* 104, 706, 1988.
98. **Marks, N., Berg, M. J., and Benuck, M.,** Preferential action of rat brain cathepsin B as a peptidyl dipeptidase converting pro-opioid oligopeptides, *Arch. Biochem. Biophys.,* 249, 489, 1986.
99. **Marks, N., Stern, F., Chi, L.-M., and Berg, M. J.,** Diversity of rat brain cysteine proteinase inhibitors: isolation of low-molecular weight cystatins and higher molecular weight T-kininogen-like glycoprotein, *Arch. Biochem. Biophys.,* 267, 448, 1988.
100. **Ghiso, J., Jensson, O., and Frangione, B.,** Amyloid fibrils in hereditary cerebral hemorrhage with amyloidosis of Icelandic type is a variant of γ-trace basic protein (cystatin C), *Proc. Natl. Acad. Sci. U.S.A.,* 83, 2974, 1986.
101. **Nakanishi, S.,** Substance P precursors and kininogen: their structures, gene organizations, and regulation, *Physiol. Rev.,* 67, 1117, 1987.
102. **Machleidt, W., Thiele, U., Laber, B., Asfalg-Machleidt, I., Esterl, A., Wiegand, G., Kos, J., Turk, V., and Bode, W.,** Mechanism of inhibition of papain by chicken egg white cystatin: inhibition constants of N-terminally truncated forms and cyanogen bromide fragments of inhibitor, *FEBS Lett.,* 243, 234, 1989.
103. **Marks, N., Berg, M. J., Makofske, R. C., and Danho, W.,** Synthetic domains of cystatins linked to enkephalins are novel inhibitors of brain catepsins L/B, *Peptides,* 11, 679, 1990.
104. **Kerr, M. A. and Kenny, A. J.,** The purification and specificity of a neutral endopeptidase from rabbit kidney brush border, *Biochem. J.,* 137, 477, 1974.
105. **Matsas, R., Kenny, A. J., and Turner, A. J.,** The metabolism of neuropeptides. The hydrolysis of peptides including enkephalins, tachykinins and their analogues by endopeptidase 24.11, *Biochem. J.,* 223, 433, 1984.
106. **Schwartz, J.-C., Gros, C., Lecomte, J.-M., and Bralet, J.,** Enkephalinase (E.C. 3.4.24.11) Inhibitors: protection of endogenous ANF against inactivation and potential therapeutic applications, *Life Sci.,* 47, 1279, 1990.
107. **Orlowski, M. and Wilk, S.,** Purification and specificity of a membrane bound metalloendopeptidase from bovine pituitaries, *Biochemistry,* 20, 4942, 1981.
108. **Matsumura, Y., Ikegawa, R., Tsukahara, Y., Takaoka, M., and Morimoto, S.,** Conversion of big endothelin-1 to endothelin-1 by two types of metalloproteinases derived from porcine aortic endothelial cells, *FEBS Lett.,* 272, 166, 1990.
109. **McMahon, E. G., Palomo, M. A., Moore, W. M., McDonald, J. F., and Stern, M. K.,** Phosphoramidon blocks the pressor activity of porcine bid endothelin 1 (1-39) *in vivo* and conversion of big endothelin-1 (1-39) to endothelin-1 (1-21) *in vitro, Proc. Natl. Acad. Sci. U.S.A.,* 88, 703, 1991.
110. **Danho, W., Makofske, R. C., Svistok, J., Michalewsky, J., Gabriel, T., Marks, N., Berg, M. J., Baird, L., Geiler, V., Mackie, G., Nelson, D., and Triscari, J.,** Synthesis and biological activity of novel, potent and long-acting analogues of Ac-CCK-7 with high affinity for peripheral (type A) receptors, *Peptide Res.,* 4, 59, 1991.
111. **Vijayaraghavan, J., Scicli, A. G., Cannetero, C. A., Slaughter, C., Moomaw, C., and Hersh, L. B.,** The hydrolysis of endothelins by neutral endopeptidases 24.11, *J. Biol. Chem.,* 265, 14150, 1990.
112. **Skidgel, R. A., Engelbrecht, S., Johnson, A. R., and Erdos, E. G.,** Hydrolysis of substance P and neurotensin by converting enzyme and neutral endopeptidase, *Peptides,* 5, 769, 1984.

113. **Goetzl, E. J., Sreedharan, S. P., Turck, C. W., Bridenbaugh, R., and Malfroy, B.,** Preferential cleavage of amino- and carboxyl-terminal oligopeptides from vasoactive intenstinal polypeptide by human recombinant enkephalinase (neutral endopeptidase, E.C. 3.4.24.11), *Biochem. Biophys. Res. Commun.,* 158, 850, 1989.
114. **Malfroy, B., Kuang, W.-J., Seeburg, P. H., Mason, A. J., and Schofield, P. R.,** Molecular cloning and amino acid sequence of human enkephalinase (neutral endopeptidase), *FEBS Lett.,* 229, 206, 1988.
115. **Devault, A., Lazure, C., Nault, C., LeMoual, H., Seidah, N. G., Chretien, M., Kahn, M., Powell, P., Mallet, J., Beaumont, A., Roques, B. P., Crine, P., and Bolleau, G.,** Amino acid sequence of rabbit kidney neutral endopeptidase 24.11 (enkephalinase) deduced from a complementary DNA, *EMBO J.,* 6, 1317, 1987.
116. **Shipp, M. A., Vijayaraghavan, J., Schmidt, E. V., Masterller, E. L., D'Adamio, L., Hersh, L. B., and Reinherz, E. L.,** Common acute lymphoblastic leukemia antigen (CALLA) is active neutral endopeptidase 24.11 (''enkephalinase''): direct evidence by cDNA transfection analysis, *Proc. Natl. Acad. Sci. U.S.A.,* 86, 297, 1989.
117. **Devault, A., Sales, V., Nault, C., Beaumont, A., Roques, B., Crine, P., and Boileau, G.,** Exploration of the catalytic site of endopeptidase 24.11 by site-directed mutagenesis. Histidine residues 583 and 587 are essential for catalysis, *FEBS Lett.,* 231, 54, 1988.
118. **Beaumont, A., Hernandez, J. F., Chaillet, P., Crine, P., and Roques, B. P.,** Irreversible photolabeling of active site of neutral endopeptidase-24.11 ''enkephalinase'' by azidothiorphan and [^{14}C]-azidothiorphan, *Mol. Toxicol.,* 32, 594, 1987.
119. **Vijayaraghavan, J., Kim, Y. A., Jackson, D., Orlowski, M., and Hersh, L. B.,** Use of site-directed mutagenesis to identify valine-573 in the S'1 binding site of rat neutral endopeptidase 24.11 (enkephalinase), *Biochemistry,* 29, 8052, 1990.
120. **Bateman, R. C., Jackson, D., Slaughter, C. A., Unnithean, S., Chai, Y. G., Moomaw, C., and Hersh, L. B.,** Identification of the active-site arginine in rat neutral endopeptidase 24.11 (enkephalinase) as arginine 102 and analysis of a glutamine 102 mutant, *J. Biol. Chem.,* 264, 6151, 1989.
121. **Bateman, R. C. and Hersh, L. B.,** Evidence for an essential histidine in neutral endopeptidase 24.11, *Biochemistry,* 26, 4237, 1987.
122. **Kenny, A. J.,** Proteinases associated with cells membranes, *Res. Monogr. Cell Tissue Physiol.,* 2, 393, 1977.
123. **Lorkowski, G., Zijderhand-Bleekemolen, J. W., Erdos, E. G., von Figura, K., and Hasilik, A.,** Neutral endopeptidase-24.11 (enkephalinase). Biosynthesis and localization in human fibroblasts, *Biochem. J.,* 248, 345, 1987.
124. **Beaumont, A., Brouet, J. C., and Roques, B. P.,** Neutral endopeptidase 24.11 and angiotensin in converting enzyme like activity in CALLA positive and CALLA negative cells, *Biochem. Biophys. Res. Commun.,* 160, 1323, 1989.
125. **Shipp, M. A., Stefano, R. B., D'ADamio, L., Switzer, S. N., Howard, F. D., Sinisterre, J., Scharrer, B., and Reinharz, R. L.,** Down regulation of enkephalin-mediated responses by CD10/neutral endopeptidase 24.11, *Nature (London),* 347, 391, 1990.
126. **Kenny, A. J., O'Hare, M. J., and Gusterson, B. A.,** Cell-surface peptidases as modulators of growth and differentiation, *Lancet,* 2, 785, 1989.
127. **Arock, M., Gros, C., Devillier, P., Guillosson, J. J., and Schwartz, J.-C.,** Characterization of enkephalinase (E.C. 3.4.24.11) activity on various leukaemic cells expressing the common acute lymphocytic leukemia antigen (CALLA), *FEBS Lett.,* 248, 123, 1989.
128. **Kenny, A. J.,** Regulatory peptide metabolism at cell surfaces: the key role of endopeptidase 24.11, *Biochem. Biochim. Acta,* 45, 1503, 1986.
129. **Wakesman, G., Hamel, E., Fournie-Zaluski, M. C., and Roques, B. P.,** Autoradiographic comparison of the distribution of the neutral endopeptidase ''enkephalinase'' and μ and δ-opioid receptors in rat brain, *Proc. Natl. Acad. Sci. U.S.A.,* 83, 1523, 1986.

130. **Graybiel, A. M. and Chesselet, M. F.,** Compartmental distribution of striatal cell bodies expressing met-enkephalin-like immunoreactivity, *Proc. Natl. Acad. Sci. U.S.A.,* 81, 7980, 1984.
131. **Wilcox, J. N., Pollard, H., Moreau, J., Schwartz, J.-C., and Malfroy, B.,** Localization of enkephalinase mRNA in rat brain by *in situ* hybridization: comparison with immunohistochemical localization of the protein, *Neuropeptides,* 14, 77, 1989.
132. **Kester, W. R. and Matthews, B. J.,** Crystallographic study of the binding of dipeptide inhibitors to thermolysin: Implications for the mechanism of catalysis, *Biochemistry,* 16, 2506, 1977.
133. **Christianson, D. W. and Lipscomb, W. N.,** Binding of a possible transition state analogue to the active site of carboxypeptidase A, *Proc. Natl. Acad. Sci. U.S.A.,* 82, 6840, 1985.
134. **Cushman, D. W., Cheung, H. S., Sabo, E. F., and Ondetti, M. A.,** Design of potent competitive inhibitors of angiotensin-converting enzyme. Carboxyalkylnoyl and mercaptoalkynoyl amino acids, *Biochemistry,* 16, 5484, 1977.
135. **Malfroy, B., Swerts, J. P., Guyon, A., Roques, B. P., and Schwartz, J.-C.,** High affinity enkephalin degrading peptidase in brain is increased after morphine, *Nature (London),* 276, 523, 1978.
136. **Roques, P. P., Fournie-Zaluski, M. C., Soroca, E., Lecomte, J.-M., Malfroy, B., Llorens, C., and Schwartz, J.-C.,** The enkephalinase inhibitor thiorphan shows antinociceptive activity in mice, *Nature (London),* 280, 286, 1980.
137. **Hudgin, R. L., Charleson, S. E., Zimmerman, M., Mumford, R., and Wood, P. L.,** Enkephalinase: selective peptide inhibitors, *Life Sci.,* 29, 2593, 1981.
138. **Chu, T. G. and Orlowski, M.,** Active site directed N-carboxymethyl peptide inhibitors of a soluble metalloendopeptidase from rat brain, *Biochemistry,* 23, 3603, 1984.
139. **Elliott, R. L., Marks, N., Berg, M. J., and Portoghese, P.,** Synthesis and biological evaluation of phosphonamide peptide inhibitors of enkephalinase and angiotensin converting enzyme, *J. Med. Chem.,* 28, 1208, 1985.
140. **Fournie-Zaluski, M. C., Hernandez, J. D., Soeilhac, J. M., Renwart, N., Peyroux, J., Xie, J., and Roques, B. P.,** Enkephalin degrading enzyme inhibitors. Crucial role of the C-terminal residue on the inhibitory potencies of retrohydroxymate dipeptides, *Int. J. Pept. Protein Res.,* 33, 146, 1989.
141. **Danilewicz, J. C., Barcly, P. L., Barnish, I. T., Brown, D., Campbell, S. F., James, K., Samuels, G. M., Terrett, N. K., and Wythes, M. J.,** UK-69,578 a novel inhibitor of E.C. 3.4.24.11 which increases endogenous ANF levels and is natriuretic and diuretic, *Biochem. Biophys. Res. Commun.,* 164, 58, 1989.
142. **Northridge, D. B., Jardine, A. G., Findlay, I. N., Archibald, M., Dilly, S. G., and Dargie, H. J.,** Inhibition of the metabolism of atrial natriuretic factor causes diuresis and natriuresis in chronic heart failure, *Am. J. Hyperten.,* 3, 682, 1990.
143. **al-Rodhan, N., Chipkin, R., and Yaksh, T. L.,** The antinociceptive effects of SCH-32615, a neutral endopeptidase (enkephalinase) inhibitor, microinjected into the periaqueductal, ventral medulla and amygdala, *Brain Res.,* 520, 123, 1990.
144. **Richards, M., Espiner, E., Frampton, C., Ikram, H., Yandle, T., Sopwith, M., and Cussans, N.,** Inhibition of endopeptidase E.C. 24.11 in humans. Renal and endocrine effects, *Hypertension,* 16, 269, 1990.
145. **Zhou, Y., Luo, F., and Han, J.,** Enkephalinase inhibitors acetorphan and SCH 34826 potentiate the analgesic effect of low frequency electroacupuncture in rats, *Chen Tzu Yen Chiu,* 15, 310, 1990.
146. **Cavero, P. G., Margulies, K. B., Winaver, J., Seymour, A. A., Delaney, N. G., and Burnett, J. C.,** Cardiorenal actions of neutral endopeptidase inhibition in experimental congestive heart failure, *Circulation,* 82, 196, 1990.
147. **Sybertz, E. J., Chiu, P. J., Vemulacpalli, S., Watkins, R., and Haslanger, M. F.,** Atrial natriuretic factor-potentiating and antihypertensive activity of SCH 34826. An orally active neutral metalloendopeptidase inhibitor, *Hypertension,* 15, 152, 1990.

148. **Fournie-Zaluski, M. C., Chaillet, P., Bouboutou, R., Coulaud, A., Cherot, P., Wakesman, G., Costentin, J., and Roques, B. P.,** Analgesic effects of kelatorphan, a new highly potent inhibitor of multiple enkephalin degrading enzymes, *Eur. J. Pharmacol.,* 102, 525, 1984.
149. **Lasdun, A. M. and Orlowski, M.,** Inhibition of endopeptidase 24.15 greatly increases the release of luteinizing hormone and follicle stimulating hormone in response to luteinizing hormone-releasing hormone, *J. Pharmacol. Exp. Ther.,* 253, 1265, 1990.
150. **Lecomte, J.-M., Costentin, J., Vlaisculescu, A., Chaillet, P., Marcais-Collado, H., Llorens-Cortes, C., Leboyer, M., and Schwartz, J.-C.,** Pharmacological properties of acetorphan, a parenterally active "enkephalinase" inhibitor, *J. Pharmacol. Exp. Ther.,* 237, 937, 1986.
151. **Fournie-Zaluski, M. C., Lucassoroca, E., Devin, J., and Roques, B. P.,** H-NMR configurational correlation for retro-inverso dipeptides: applications to the determination of the absolute configuration of enkephalinase inhibitors. Relationship between stereochemistry and enzyme recognition, *J. Med. Chem.,* 29, 751, 1986.
152. **Hernandez, J. F., Soleihac, J. M., Roques, B. P., and Fournie-Zaluski, M. C.,** Retro-inverso concept applied to the complete inhibitors of enkephalin-degrading enzymes, *J. Med. Chem.,* 31, 1825, 1988.
153. **Lecomte, J.-M., Baumer, P., Lim, C., Duchier, J., Cournot, A., Dussaule, J.-C., Ardaillou, R., Gros, C., Chaignon, B., Souque, A., and Schwartz, J.-C.,** Stereoselective protection of exogenous and endogenous atrial natriuretic factor by enkephalinase inhibitors in mice and humans, *Eur. J. Pharmacol.,* 179, 65, 1990.
154. **Chipkin, R. E., Latranyi, M. B., and Iorio, L. C.,** Potentiation of stress-induced analgesia (SIA) by thiorphan and its block by naloxone, *Life Sci.,* 31, 1189, 1982.
155. **Haffmans, J., Walsum, M. V., van Amsterdam, J. C., and Dzoljic, M. R.,** Phelorphan, an inhibitor of enzymes in the biodegradation of enkephalin, affected the withdrawal symptoms in chronic morphine-dependent rats, *Neuroscience,* 22, 233, 1987.
156. **Van Amsterdam, J. G. and Llorens-Cortez, C.,** Inhibition of enkephalin degradation by phelorphan: effects on striatal (met-enkephalin) levels and jump latency in mouse hot plate test, *Eur. J. Pharmacol.,* 154, 319, 1988.
157. **Maldonado, R., Daugie, V., Celebert, J., Villette, J. M., Fournie-Zaluski, C., Feger, J., and Roques, B. P.,** Comparison of selective and complete inhibitors of enkephalin-degrading enzymes on morphine withdrawal syndrome, *Eur. J. Pharmacol.,* 165, 199, 1989.
158. **Malin, D. H., Lake, J. R., Hamilton, R. F., and Skolnick, M. H.,** Augmented analgesic effects of enkephalinase inhibitors combined with transcranial electrostimulation, *Life Sci.,* 44, 1371, 1989.
159. **Parsons, C. B. and Herz, A.,** Peripheral opioid receptors mediating antinociception in inflammation. Evidence for activation by enkephalin-like opioid peptides after cold water swim stress, *J. Pharmacol. Exp. Ther.,* 255, 795, 1990.
160. **Knisely, J. S., Beardsly, P. M., Aceto, M. D., Balster, R. L., and Harris, L. S.,** Assessment of the abuse potential of 'acetorphan' an enkephalinase inhibitor, *Drug Alcohol Depend.,* 23, 143, 1989.
161. **deWitte, P., Heidbreder, C., and Roques, B. P.,** Kelatorphan, a potent enkephalinase inhibitor, and opioid receptor agonists DAGA and DTLET, differentially modulate self-stimulating behavior depending on the site of administration, *Neuropharmacology,* 28, 667, 1989.
162. **Shimarura, M., Kawamuki, K., and Hazato, T.,** Angiotensin III: a potent inhibitor of enkephalin-degrading enzymes and an analgesic agent, *J. Neurochem.,* 49, 536, 1987.
163. **Bado, A., Roze, C., Lewin, M. J. M., and Dubraquet, M.,** Endogenous opioid peptides in the control of food intake in cats, *Peptides,* 10, 967, 1989.
164. **Chaussade, Hamm, R., Lecomte, J.-M., Couturier, D., and Guere, J.,** Effects of an enkephalinase inhibitor on esophageal motility in man, *Gastroenterol. Clin. Biol.,* 12, 793, 1988.

165. **Chicau-Chovet, M., Dubrasequet, M., Chariot, J., Tsocas, A., Lecomte, J.-M., and Roze, C.,** Thiorphan and acetorphan inhibit gastric secretion by a central non-opioid mechanism in the rat, *Eur. J. Pharmacol.,* 154, 247, 1988.
166. **Libergem, M., Riviere, P. M. J., and Bueno, L.,** Influence of enkephalinase inhibitors on gastric emptying in mice depends on the nature of the meal, *Life Sci.,* 42, 2047, 1988.
167. **Jivegard, L. Pollard, H., Moreau, J., Schwartz, J.-C., Thune, A., and Svanvik, J.,** Naloxone reversible inhibition of gall-bladder mucosal fluid secretion in experimental cholecystitis in the case by acetorphan' an enkephalinase inhibitor, *Clin. Sci.,* 77, 49, 1989.
168. **Maggi, C. A., Astolfi, M., Santicioli, F., Tramontana, M., Lioncini, G., Geppetti, P., Giachetti, A., and Meli, A.,** The effect of thiorphan on the response of guinea-pig isolated urinary bladder to exogenous and endogenous tachykinins, *J. Urol.,* 144, 1546, 1990.
169. **Michael-Titus, A., Dourmap, N., Caline, H., Costentin, J., and Schwartz, J.-C.,** Role of endogenous enkephalins in locomotion and nociception studied with peptidase inhibitors in two inbred strains of mice C57BL/6J and DBA/2J, *Neuropharmacology,* 28, 117, 1989.
170. **Michael-Titus, Dourmap, N., Costentin, J., and Schwartz, J.-C.,** Role of delta opioid receptors in the effects of inhibitors of enkephalin-degrading peptidases on the horizontal and vertical components of locomotion in mice, *Neuropeptides,* 15, 89, 1990.
171. **Nabeshima, T., Ketoh, A., and Kameyama, T.,** Inhibition of enkephalin degradation attenuated stress-induced motor supression (conditioned suppression of motility), *J. Pharmacol. Exp. Ther.,* 244, 303, 1988.
172. **Ben-Natan, L., Chaillet, P., Lecomte, J.-M., Marcais, H., Uchida, G., and Costentin, J.,** Involvement of endogenous enkephalins in mouse behavioral despair' test, *Eur. J. Pharmacol.,* 97, 301, 1984.
173. **Dourmap, N., Michael-Titus, A., and Costentin, J.,** 'Acetorphan' an enkephalinase inhibitor modulates dopaminergic transmission in rat olfactory tubercle but not in the nucleus accumbens and striatum, *Eur. J. Neurosci.,* 2, 783, 1990.
174. **Gros, C., Souque, A., Schwartz, J.-C., Duchier, J., Cournot, A., Baumer, P., and Lecomte, J.-M.,** Protection of atrial natriuretic factor against degradation: diuretic and natriuretic responses after *in vivo* inhibition of enkephalinase (E.C. 3.4.24.11) by acetorphan, *Proc. Natl. Acad. Sci. U.S.A.,* 86, 7580, 1989.
175. **Schwartz, J.-C., Gros, C., Lecomte, J.-M., and Bralet, J.,** Enkephalinase (EC 3.4.24.11) inhibitors: protection of endogenous ANF against inactivation and potential therapeutic applications, *Life Sci.,* 47, 1279, 1990.
176. **Soffer, R. L.,** Angiotensin converting enzyme, in *Biochemical Regulation of Blood Pressure,* Soffer, R. L., Ed., John Wiley and Sons, New York, 1981, 123.
177. **Strittmatter, S. M., Lo, M. S., Javitch, J. A., and Snyder, S. H.,** Autoradiographic visualization of angiotensin-converting enzyme in rat brain with (^{3}H) captopril: localization to a striatonigral pathway, *Proc. Natl. Acad. Sci. U.S.A.,* 81, 1599, 1984.
178. **Zajac, J. M., Charney, Y., Soleilhic, J. M., Sales, N., and Roques, B. P.,** Enkephalin-degrading enzymes and angiotensin-converting enzyme in human and rat meninges, *FEBS Lett.,* 216, 118, 1987.
179. **Soubrier, F., Alrenc-Gelas, F., Hubert, C., Allegrini, J., John, M., Tregar, G., and Corvol, P.,** Two putative active centers in human angiotensin I-converting enzyme revealed by molecular cloning, *Proc. Natl. Acad. Sci. U.S.A.,* 85, 9386, 1988.
180. **Marks, N., Berg, M. J., Benuck, M., Lo, E.-S., Novachenko, H., and Seyfried, C. A.,** Prodynorphin processing by rat CNS fractions and purified enzymes: formation of dynorphin A 1-8 by sulfhydryl-activated carboxypeptidase and peptidyl dipeptidase, *Neurochem. Int.,* 10, 413, 1987.
181. **Mauborgne, A., Bourgoin, S., Benoliel, U. J., Hirsch, M., Berthier, J. L., Hamon, M., and Cesselin, F.,** Enkephalinase is involved in the degradation of endogenous substance P released from slices of rat substantia nigra, *J. Pharmacol. Exp. Ther.,* 243, 674, 1987.

182. **Skidgel, R. A. and Erdos, E. G.,** The broad substrate specificity of human angiotensin converting enzyme, *Clin. Exp. Hyperten. A,* 9, 243, 1987.
183. **Oblin, A., Danse, M. J., and Zivkovic, B.,** Metalloendopeptidase (E.C. 3.4.24.11) but not angiotensin converting enzyme is involved in the inactivation of substance P by synaptic membranes of the rat substantia nigra, *Life Sci.,* 44, 1467, 1989.
184. **Geppetti, P., Santicioli, P., Rubini, I., Spillantini, M. G., Maggi, C. A., and Sicuteri, F.,** Thiorphan increases capsaicin-evoked release of substance P from slices of dorsal spinal cord of guinea pig, *Neurosci. Lett.,* 14, 69, 1989.
185. **Umeno, E., Nadel, J. A., Huang, H. T., and McDonald, D. M.,** Inhibition of neutral endopeptidase potentiates neurogenic inflammation in the rat trachea, *J. Appl. Physiol.,* 66, 2647, 1989.
186. **Djokic, T. D., Nadel, J. A., Dusser, D. J., Sekizawa, K., Graf, P. D., and Borson, D. B.,** Inhibitors of neutral endopeptidase potentiate electrically and capsaicin-induced noncholinergic contraction in guinea pig bronchi, *J. Pharmacol. Exp. Ther.,* 248, 7, 1989.
187. **Warner, E. A., Krell, R. D., and Buckner, C. K.,** Pharmacologic studies on the differential influence of inhibitors of neutral endopeptidase on nonadrenergic, noncholinergic contractile responses of the guinea pig isolated hilar bronchus to transmural electrical stimulation and exogenously applied tachykinins, *J. Pharmacol. Exp. Ther.,* 254, 824, 1990.
188. **Kondo, M., Tamaoki, J., and Takizawa, T.,** Neutral endopeptidase inhibitor potentiates the tachykinin-induced increase in ciliary beat frequency in rabbit trachea, *Am. Rev. Respir. Dis.,* 142, 403, 1990.
189. **Kazim Turker, R.,** Analgesia from captopril, an inhibitor of angiotensin converting enzyme and enkephalinase, in *Degradation of Endogenous Opioids: Its Relevance in Human Pathology and Therapy,* Ehrenpreis, S. and Sicuteri, F., Eds., Raven Press, New York 1983, 171.
190. **Numata, H., Hiranuma, T., and Oka, T.,** Inactivation of dynorphin-(1-8) in isolated preparations by three peptidases, *Jpn. J. Pharmacol.,* 47, 417, 1988.
191. **Uchiyama, Y., Nakajima, M., Muno, D., Watanabe, T., Ishii, Y., Waguri, S., Sata, N., and Kominami, E.,** Immunocytochemical localization of cathepsins B and H in corticotrophs and melanotrophs or rat pituitary gland, *J. Histochem. Cytochem.,* 38, 633, 1990.
192. **Docherty, K., Carroll, R., and Steiner, D. F.,** Identification of a 31,500 molecular weight islet cell protease as cathepsin B, *Proc. Natl. Acad. Sci. U.S.A.,* 80, 3245, 1983.
193. **Kudo, T., Wei, E. Q., and Inoki, R.,** Is cathepsin B an enkephalin processing enzyme in the tooth pulp? *Prog. Clin. Biol. Res.,* 328, 271, 1990.
194. **Barrett, A. J., Rawlings, N. D., Davies, M. E., Machleidt, W., Salvesen, G., and Turk, V.,** Cysteine proteinase inhibitors of the cystatin superfamily, *Res. Monogr. Cell Tissue Physiol.,* 12, 515, 1986.
195. **Grynbaum, A. and Marks, N.,** Characterization of a rat brain catheptic carboxypeptidase (cathepsin A) inactivating angiotensin-II, *J. Neurochem.,* 26, 313, 1976.
196. **Marks, N., Sachs, L., and Stern, F.,** Conversion of met-enkephalin-arg^6-phe^7 by a purified brain carboxypeptidase (cathepsin A), *Peptides,* 2, 159, 1981.
197. **Ondetti, M. A., Condon, M. E., Reid, J., Sabo, E. F., Cheung, H. S., and Cushman, D. W.,** Design of potent and specific inhibitors of carboxypeptidase A and B, *Biochemistry,* 18, 1427, 1979.
198. **Grobelny, D., Goli, U. B., and Galardy, R. E.,** 3-Phosphonopropionic acids inhibit carboxypeptidase A as multisubstrate analogues or transition-state analogues, *Biochem. J.,* 232, 15, 1985.
199. **Ehrenpreis, S., Balagot, R. C., Greenberg, J., Myles, S., and Ellyin, F.,** Analgesic and other pharmacological properties of D-phenylalanine, in *Degradation of Endogenous Opioids: Its Relevance in Human Pathology and Therapy,* Ehrenpreis, S. and Sicuteri, F., Eds., Raven Press, New York 1983, 207.

200. **Balagot, R. C., Ehrenpreis, S., Greenberg, J., and Hyodo, M.,** D-phenylalanine in human chronic pain, in *Degradation of Endogenous Opioids: Its Relevance in Human Pathology and Therapy,* Ehrenpreis, S. and Sicuteri, F., Eds., Raven Press, New York 1983, 207.
201. **Carenzi, A., Frigeni, V., Reggiani, A., and Della Bella, D.,** On the possibility of influencing enkephalin system: studies *in vitro* and *in vivo*, in *Degradation of Endogenous Opioids: Its Relevance in Human Pathology and Therapy,* Ehrenpreis, S. and Sicuteri, F., Eds.,1983, 107.
202. **Della Bella, D., Carenzi, A., Frigeni, V., and Santini, V.,** Effect of carboxypeptidase inhibition on the *in vivo* and *in vitro* pharmacological properties of morphine and enkephalins, *Neuropharmacology,* 18, 719, 1979.
203. **Fricker, L. D. and Snyder, H.,** Enkephalin convertase purification and characterization of a specific enkephalin synthesizing carboxypeptidase localized in adrenal chromaffin granules, *Proc. Natl. Acad. Sci. U.S.A.,* 79, 3886, 1982.
204. **Hook, V. Y. H., Eidn, L. E., and Brownstein, M. J.,** A carboxypeptidase processing enzyme for enkephalin precusors, *Nature (London),* 295, 341, 1982.
205. **Hook, V. Y. H. and Loh, Y. P.,** Carboxypeptidase-B like converting enzyme like activity in secretory granules of rat pituitary, *Proc. Natl. Acad. Sci. U.S.A.,* 81, 2776, 1984.
206. **Supattapone, S., Fricker, L. D., and Snyder, S. H.,** Purification and characterization of a membrane-bound enkephalin forming enkephalin carboxypeptidase 'enkephalin convertase', *J. Neurochem.,* 42, 1017, 1984.
207. **Davidson, H. W. and Hutton, J. C.,** The insulin-secretory-granule carboxypeptidase H. Purification and demonstration of involvement in proinsulin processing, *Biochem. J.,* 245, 575, 1987.
208. **Fricker, L. D.,** Carboxypeptidase E, *Annu. Rev. Physiol.,* 50, 309, 1988.
209. **Fricker, L. D., Adelman, J. P., Douglass, J., Thompson, R. C., von Strandmann, R. P., and Hutton, J.,** Isolation and sequence analysis of cDNA for rat carboxypeptidase E (E.C. 3.4.17.10), a neuropeptide processing enzyme, *Mol. Endocrinol.,* 3, 666, 1989.
210. **Vilijn, M. H., Das, B., Kessler, J. A., and Fricker, L. P.,** Cultured astrocytes and neurons synthesize and secrete carboxypeptidase E, a neuropeptide-processing enzyme, *J. Neurochem.,* 53, 1487, 1989.
211. **Udea, H., Fukushima, N., and Takagi, H.,** A novel proenkephalin processing carboxypeptidase and its inactivation by cyclic AMP dependent protein kinase, *Biochem. Biophys. Res. Commun.,* 142, 595, 1987.
212. **McKay, T. J. and Plummer, T. H.,** By-product analogues for bovine carboxypeptidase B, *Biochemistry,* 17, 401, 1978.
213. **Silberring, J., Lason, W., Przewlocka, B., and Przewlocka, R.,** Enkephalin convertase in the rat spinal cord, *Neuropeptides,* 8, 367, 1986.
214. **Przewlocka, B., Dziedzicka, M., Silberring, J., and Lason, V.,** Analgesic and convulsant effects of guanidinoethylmercaptosuccinic acid (GEMSA): a potent enkephalin convertase inhibitor, *Neuropeptides,* 8, 354, 1986.
215. **Fricker, L. D., Rigual, R. J., Diliberto, E. J., and Viveros, O. H.,** Reflex splanchnic nerve stimulation increases levels of carboxypeptidase E mRNA and enzymatic activity in the rat adrenal medulla, *J. Neurochem.,* 55, 461, 1990.
216. **Chu, T. G. and Orlowski, H.,** Soluble metallopeptidase from rat brain: action on enkephalin-containing peptides and other bioactive peptides, *Endocrinology,* 116, 1418, 1985.
217. **Acker, G. R., Molineaux, C. J. and Orlowski, M.,** Synaptosomal membrane bound form of endopeptidase 24.15 generates leu-enkephalin from dynorphin 1-8, a- and b-neoendorphin and met-enkephalin from met-enkephalin-arg^{6}-gly^{7}-leu^{8}, *J. Neurochem.,* 48, 284, 1987.
218. **Barrett, A. J.,** A new look at PZ-Peptidase, *Physiol. Chem., Hoppe-Seylers,* 371, (Suppl.), 311, 1990.

219. **Camargo, A. C. M., Oliveira, E. B., Toffoletto, O., Metters, K. M., and Rossier, J.,** Brain endo-oligopeptidase A, a putative enkephalin converting enzyme, *J. Neurochem.,* 48, 1258, 1987.
220. **Toffoletto, O., Metters, K. M., Oliveira, E. B., Carmargo, A. C., and Rossier, J.,** Enkephalin is liberated from metorphamide and dynorphin A 1-8 by endo-oligopeptidase A, but not by metalloendopeptidase 24.15, *Biochem. J.,* 252, 35, 1988.
221. **Toffoletto, O., Camargo, A. C., Oliveira, E. B., Metters, K. M., and Rossier, J.,** Liberation of enkephalins from enkephalin containing peptides by brain endooligopeptidase, *Biochimie,* 70, 47, 1988.
222. **Molineaux, C. J. and Ayala, J. M.,** An inhibitor of endopeptidase-24.15 blocks the degradation of intraventricularly administered dynorphins, *J. Neurochem.,* 55, 611, 1990.
223. **Marks, N., Berg, M. J., and Danho, W.,** Enkephalin analogues as substrates for the assay of brain cysteine proteinase (cathepsin L) and its endogenous inhibitors, *Peptides,* 10, 391, 1989.
224. **Hruby, V. J., Al-Obeidi, F., and Kazmicrski, W.,** Emerging approaches in the molecular design of receptor-selective peptide ligands: conformational, topographical and dynamic considerations, *Biochem. J.,* 268, 249, 1990.
225. **Evans, S. M., Lenz, G. R., and Lesser, R. A.,** Analgesics, *Annu. Rep. Med. Chem.,* 25, 11, 1990.
226. **Chipkin, R. E., Berger, J. G., Billard, W., Iorio, L. C., Chapman, E., and Barnett, A.,** Pharmacology of SCH 34826, an orally active enkephalinase inhibitor analgesic, *J. Pharmacol. Exp. Ther.,* 245, 829, 1988.
227. **Xie, J., Soleilhac, J.-M., Schmidt, C., Peyoux, J., Roques, B. P., and Fournie-Zaluski, M. C.,** New kelatorphan related inhibitors of enkephalin metabolism: improved antinociceptive properties, *J. Med. Chem.,* 32, 1497, 1989.

INDEX

A

D

G

H

I

O

U

V

W

1032101

[redacted]